AF349973

Protection and Management of Species, Habitats, Ecosystems and Landscapes

Protection and Management of Species, Habitats, Ecosystems and Landscapes

Editors

Panayotis Dimopoulos
Ioannis P. Kokkoris

MDPI • Basel • Beijing • Wuhan • Barcelona • Belgrade • Manchester • Tokyo • Cluj • Tianjin

Editors

Panayotis Dimopoulos
Laboratory of Botany
Department of Biology
University of Patras
Greece

Ioannis P. Kokkoris
Laboratory of Botany
Department of Biology
University of Patras
Greece

Editorial Office
MDPI
St. Alban-Anlage 66
4052 Basel, Switzerland

This is a reprint of articles from the Special Issue published online in the open access journal *Forests* (ISSN 1999-4907) (available at: https://www.mdpi.com/journal/forests/special_issues/forest_species_habitats).

For citation purposes, cite each article independently as indicated on the article page online and as indicated below:

LastName, A.A.; LastName, B.B.; LastName, C.C. Article Title. *Journal Name* **Year**, *Volume Number*, Page Range.

ISBN 978-3-0365-0176-5 (Hbk)
ISBN 978-3-0365-0177-2 (PDF)

Cover image courtesy of Charitakis Papaioannou.

Contents

About the Editors

Panayotis Dimopoulos is a Professor in Botany and Ecology at the Department of Biology of the University of Patras in Greece. His research interests focus on biodiversity and community analysis, monitoring, conservation status assessment and mapping of habitat and ecosystem types, as well as on the mapping and assessment of the ecosystem's condition and their ecosystem services. In his academic career, he has been actively involved as an expert and supervisor of field vegetation research and as scientific coordinator of various mapping and monitoring projects for habitat types and species related to the implementation of the Dir/ 92/43/EEC in Greece. He has participated in 35 national and 15 European research projects, 40 for which he has acted as principal investigator and scientific coordinator. He has served as Vice-Chair (2010–2013) and Chair (2020-2023) of the National Committee for the Natura 2000 Protected Areas in Greece. Since 1990, he has published 127 articles in peer-reviewed international journals. Since 2014, he has been the national representative in EU MAES Working Group of the DG ENV; until 2015, he was coordinating a scientific working group in the Ministry of the Environment on conservation management of the protected areas of Greece. Currently, he is coordinating the MAES implementation in Greece, in the frame of the approved LIFE-IP 4 Greece and other projects related to biodiversity and the management of protected areas in Greece.

Ioannis P. Kokkoris is a senior researcher at the University of Patras, Greece (Department of Biology and Department of Economics). He holds a PhD and MSc in Conservation Biology, an MSc in Applied Economics and Big Data Analysis, and a Diploma in Forestry and Natural Environment. His research interests are within the spectrum of conservation ecology, ecosystem and ecosystem services mapping and assessment. In this perspective, he is participating and contributing, as a Greek Member, to the MAES, ESMERALDA (H2020) and MAIA (H2020) activities and workshops. He is responsible for the MAES implementation action of the national LIFE-IP 4 Natura project. He is also involved in habitats conservation projects and ecosystem service mapping assessments combing classical methodology with various remote sensing and GIS techniques. He has worked in several national and European projects on conservation, biomonitoring and ecosystem services mapping and assessment.

Preface to "Protection and Management of Species, Habitats, Ecosystems and Landscapes"

Earth has entered a new geological epoch, the Anthropocene, characterized by a human-induced temperature increase, thus placing immense impacts upon natural, atmospheric and hydrological processes, and possibly leading to a substantial global biodiversity decrease and ecosystem function degradation. The environmental preservation and sustainable use of natural resources are global goals that must be achieved in the coming decades. The capitalization of the international partnerships to promote the biodiversity strategy and the European green deal is creating a decisive political framework to tackle the challenges ahead. Understanding the natural environment and socio-ecological interactions is the main objective of strategic planning, management practice and policy decisions in the sustainable development era. However, the ever-growing earth population and the subsequent overexploitation of natural resources (species, habitats, ecosystems, and landscapes) to provide space for human activities is leading to a bottleneck for sustainable environmental management and human well-being. Simultaneously, climate change impact and future climate projections provoke urgent action when it comes to conservation and restoration practices. This leads to targeting the mitigation of pressures and threats, in order to protect and appropriately manage species, habitats, ecosystems and landscapes, alongside the maintenance of provided ecosystem services. By this, it is evident that multidisciplinary scientific, as well as applied approaches are needed to understand the rapidly changing environmental, cultural, social and economic conditions and tackle contemporary environmental and social challenges. This publication tries to contribute in this "battle" and present modern aspects on handling sensitive environmental resources, while simultaneously pinpointing the importance of a sustainable human–nature interaction, which guarantees environmental health as well as societal well-being.

Panayotis Dimopoulos, Ioannis P. Kokkoris

Editors

Editorial

Protection and Management of Species, Habitats, Ecosystems and Landscapes: Current Trends and Global Needs

Panayotis Dimopoulos * and Ioannis P. Kokkoris

Department of Biology, Laboratory of Botany, University of Patras, 26504 Patras, Greece; ipkokkoris@upatras.gr
* Correspondence: pdimopoulos@upatras.gr; Tel.: +30-2610-996777

Received: 8 November 2020; Accepted: 23 November 2020; Published: 25 November 2020

Abstract: Human well-being and the prerequisite sustainable environmental management are currently at stake, reaching a bottleneck when trying to cope with (i) the ever-growing world population, (ii) the constantly increasing need for natural resources (and the subsequent overexploitation of species, habitats, ecosystems, and landscapes) and (iii) the documented and on-going impacts of climate change. In developed societies, the concern about environmental protection is set high in the public dialogue, as well as to management and policy agendas. The recently constituted Intergovernmental Science—Policy Platform on Biodiversity and Ecosystem Services (IPBES) urges transformative changes for technological, economic, and social factors aiming to tackle both direct and indirect drivers of biodiversity loss. By this, the role of conservation and management practices for the environment is characterized as a crucial and top issue and should deal with (a) promoting best practices from the local to the global level, (b) identifying spatial and temporal knowledge gaps, (c) multidisciplinary aspects for sustainable management practices, (d) identifying and interpreting the role of stakeholders and socio-economic parameters in the decision-making process, and (e) methods and practices to integrate the concept of ecosystem services into natural capital assessment and accounting, conservation and management strategies. Modern literature highlights that land-use change and prioritization, restoration of natural areas, cultural landscape identification and maintenance, should be considered to the top of the scientific and policy agenda, as well as to the epicenter of novel awareness-raising strategies for the environment in the near future.

Keywords: conservation management; conservation status assessment; ecosystem condition assessment; ecosystem services; natural capital; spatial patterns; species and habitats monitoring; sustainable practices; temporal patterns

---

During the last two decades our understanding about how ecosystems and the relevant biodiversity attributes interact with human society has been scientifically increased, highlighting their importance to the quality of life and human well-being [1]. Important initiatives triggered science—policy interactions worldwide, based on the results of multidisciplinary approaches and case studies, e.g., [2,3]. The importance of this interaction and co-existence of biodiversity, ecosystems in good condition and socio-economic prosperity is recently studied by the Global Assessment of Biodiversity and Ecosystem Services, under the Intergovernmental Science—Policy Platform on Biodiversity and Ecosystem Services (IPBES) initiative [4]; the relevant summary of the assessment was approved in 2019, by more than 130 Governments, which constitute the Members of IPBES [1]. The results of this extensive review and these analyses support the findings of the Millennium Ecosystem Assessment, presented 15 years ago (in 2005) [5], encouraging relevant, ongoing, national and local projects, and urges for new efforts on pilot studies and implementation actions. Already, many countries are on this track, implementing projects regarding capacity building, methodological standardization and drafting typologies (e.g., [6–11]), while 'umbrella' projects provide guidance to standardize and mainstream the efforts in a wider scale (e.g., the ESMERALDA [12] and MAIA [13] Horizon 2020 projects in the European Union). Even more multidisciplinary studies are published during the last decade [14], indicating an increased trend; numerous future-projection assessments are also provided from global (e.g., [15–17]) to local level (e.g., [18,19]), based mainly on climatic and/or management scenarios, aiming to support decision making and policy needs for a sustainable management drafting.

Simultaneously, the United Nations Statistical Commission endorsed the System of Environmental-Economic Accounting—Experimental Ecosystem Accounting (SEEA EEA) as the basis for commencing testing and further development of natural capital accounting [20]. Further work has been conducted aiming to raise awareness about, propose and demonstrate a set of concepts that are needed to understand ecosystems as assets in natural capital accounting [21]. To fulfill this promising task (even at its initial stages), (a) aspects of ecosystem extent, ecosystem condition and ecosystem services (capacity, supply and potential supply) at different scales (i.e., local, regional and national) should be considered and (b) baseline assessments are needed starting with relevant data gathering, modelling, assessments and time series development. This need calls for robust scientific information on various aspects of ecosystems' characteristics, including area cover, biodiversity and functioning, as well as of human activities and interactions ranging from industrial activity to intellectual use and aesthetic value.

It is evident that this complex scientific environment and the multi-disciplinarity needed for this holistic approach presupposes robust spatially assigned scientific evidence from natural and environmental sciences, as well as from humanitarian sciences and the socio-economic filed. Moreover, examples from case-studies implemented in different places around the globe should be considered for knowledge transfer and replication to meet local needs in similar conditions. The purpose of this Special Issue (SI) is to contribute to this effort by presenting case examples from different regions of the world dealing with biodiversity, habitat dynamics, management scenarios, human-nature interaction and policy initiatives on biodiversity and ecosystem services. We believe that the papers published in this SI provide scientific evidence on various aspects of nature conservation and management aiming to support win–win opportunities for human activities within the environment, which are considered as increasingly scarce in a 'full' global, ecological–economic system [22].

Author Contributions: Both authors contributed to the writing of this editorial. Both authors have read and agreed to the published version of the manuscript.

Funding: This research received no external funding.

Conflicts of Interest: The authors declare no conflict of interest.

References

1. IPBES. *Global Assessment Summary for Policymakers*; Díaz, S., Settele, J., Brondízio, E.S., Ngo, H.T., Guèze, M., Agard, J., Arneth, A., Balvanera, P., Brauman, K.A., Butchart, S.H.M., Eds.; IPBES Secretariat: Bonn, Germany, 2019.
2. TEEB. *The Economics of Ecosystems and Biodiversity: Mainstreaming the Economics of Nature: A synthesis of the Approach, Conclusions and Recommendations of TEEB*; TEEB: Madhya Pradesh, India, 2010; ISBN 9783981341034.
3. Maes, J.; Teller, A.; Erhard, M.; Liquete, C.; Braat, L.; Berry, P.; Egoh, B.; Puydarrieus, P.; Fiorina, C.; Santos, F.; et al. *Mapping and Assessment of Ecosystem and their Services. An Analytical Framework for Ecosystem Assessments under Action 5 of the EU Biodiversity Strategy to 2020*; Publications office of the European Union: Luxemburg, 2013; ISBN 9789279293696.
4. IPBS. *Global Assessment Report on Biodiversity and Ecosystem Services of the Intergovernmental Science-Policy Platform on Biodiversity and Ecosystem Services*; Brondizio, E.S.J., Settele, S.D., Ngo, H.T., Eds.; IPBES Secretariat: Bonn, Germany, 2019.
5. Millenium Ecosystem Assessment. *Ecosystems & Human Well-Being*; Island Press: Washington, DC, USA, 2005.
6. Haines-Young, R.; Paterson, J.; Potschin, M.; Wilson, A.; Kass, G. The UK NEA Scenarios: Development of Storylines and Analysis of Outcomes. In *The UK National Ecosystem Assessment Technical Report*; UK National Ecosystem Assessment, United Nations Environment Programme/World Conservation Monitoring Centre: Cambridge, UK, 2011; pp. 1195–1264.

7. Spanish National Ecosystem Assessment. *Ecosystems and Biodiversity for Human Wellbeing: Synthesis of Key Findings*; Ministerio de Agricultura, Alimentación y Medio Ambien: Madrid, Spain, 2014.

8. Van Reeth, W. *Ecosystem Service Indicators in Flanders: Are we Measuring What we Want to Manage?* Instituut voor Natuuren Bosonderzoek: Brussel, Belgium, 2014.

9. Albert, C.; Bonn, A.; Burkhard, B.; Daube, S.; Dietrich, K.; Engels, B.; Frommer, J.; Götzl, M.; Grêt-Regamey, A.; Job-Hoben, B.; et al. Towards a national set of ecosystem service indicators: Insights from Germany. *Ecol. Indic.* **2016**, *61*, 38–48. [CrossRef]

10. Grunewald, K.; Syrbe, R.U.; Walz, U.; Richter, B.; Meinel, G.; Herold, H.; Marzelli, S. Germany's ecosystem services—state of the indicator development for a nationwide assessment and monitoring. *One Ecosyst.* **2017**, *2*, e14021. [CrossRef]

11. Kokkoris, I.P.; Mallinis, G.; Bekri, E.S.; Vlami, V.; Zogaris, S.; Chrysafis, I.; Mitsopoulos, I.; Dimopoulos, P. National set of MAES indicators in Greece: Ecosystem services and management implications. *Forests* **2020**, *11*, 595. [CrossRef]

12. ESMERALDA (Enhancing Ecosystem Services Mapping for Policy And Decision Making). Available online: http://www.esmeralda-project.eu/ (accessed on 12 April 2020).

13. Mapping and Assessment For Integrated Ecosystem Accounting (MAIA). Available online: https://maiaportal.eu/ (accessed on 7 November 2020).

14. Costanza, R.; de Groot, R.; Braat, L.; Kubiszewski, I.; Fioramonti, L.; Sutton, P.; Farber, S.; Grasso, M. Twenty years of ecosystem services: How far have we come and how far do we still need to go? *Ecosyst. Serv.* **2017**, *28*, 1–16. [CrossRef]

15. Schröter, D.; Cramer, W.; Leemans, R.; Prentice, I.C.; Araújo, M.B.; Arnell, N.W.; Bondeau, A.; Bugmann, H.; Carter, T.R.; Gracia, C.A.; et al. Ecology: Ecosystem service supply and vulnerability to global change in Europe. *Science* **2005**, *310*, 1333–1337. [CrossRef] [PubMed]

16. Di Marco, M.; Harwood, T.D.; Hoskins, A.J.; Ware, C.; Hill, S.L.L.; Ferrier, S. Projecting impacts of global climate and land-use scenarios on plant biodiversity using compositional-turnover modelling. *Glob. Chang. Biol.* **2019**, *25*, 2763–2778. [CrossRef] [PubMed]

17. Nunez, S.; Arets, E.; Alkemade, R.; Verwer, C.; Leemans, R. Assessing the impacts of climate change on biodiversity: Is below 2 °C enough? *Clim. Change* **2019**, *154*, 351–365. [CrossRef]

18. Dai, E.; Wu, Z.; Ge, Q.; Xi, W.; Wang, X. Predicting the responses of forest distribution and aboveground biomass to climate change under RCP scenarios in southern China. *Glob. Chang. Biol.* **2016**, *22*, 3642–3661. [CrossRef] [PubMed]

19. Kokkoris, I.P.; Bekri, E.S.; Skuras, D.; Vlami, V.; Zogaris, S.; Maroulis, G.; Dimopoulos, D.; Dimopoulos, P. Integrating MAES implementation into protected area management under climate change: A fine-scale application in Greece. *Sci. Total Environ.* **2019**, *695*, 133530. [CrossRef] [PubMed]

20. United Nations; European Commission; Food and Agricultural Organization of the United Nations; Organization for Economic Co-Operation and Development; World Bank. *System of Environmental-Economic Accounting 2012: Experimental Ecosystem Accounting*; United Nations: New York, NY, USA; European Commission: Brussels, Belgium; Food and Agricultural Organization of the United Nations: Rome, Italy; Organization for Economic Co-Operation and Development: Paris, France; World Bank: Washington, DC, USA, 2014; ISBN 9789210559263.

21. Hein, L.; Bagstad, K.; Edens, B.; Obst, C.; De Jong, R.; Lesschen, J.P. Defining ecosystem assets for natural capital accounting. *PLoS ONE* **2016**, *11*, e0164460. [CrossRef] [PubMed]

22. Costanza, R.; Daly, H.E. Natural Capital and Sustainable Development. *Conserv. Biol.* **2007**, *6*, 37–46. [CrossRef]

Publisher's Note: MDPI stays neutral with regard to jurisdictional claims in published maps and institutional affiliations.

 *forests*

Article

Naturalness Assessment of Forest Management Scenarios in *Abies balsamea–Betula papyrifera* Forests

Sylvie Côté [1,*], Louis Bélanger [1], Robert Beauregard [1], Évelyne Thiffault [1] and Manuele Margni [2]

[1] Department of Wood and Forestry Sciences, Université Laval, Quebec City, QC G1V 0A6, Canada; louis.belanger.1@ulaval.ca (L.B.); Robert.Beauregard@sbf.ulaval.ca (R.B.); Evelyne.Thiffault@sbf.ulaval.ca (É.T.)

[2] CIRAIG, Polytechnique Montréal, Department of Mathematical and Industrial Engineering, Montreal, QC H3C 3A7, Canada; manuele.margni@polymtl.ca

* Correspondence: sylvie.cote.14@ulaval.ca; Tel.: +1-418-424-0422

Received: 21 April 2020; Accepted: 21 May 2020; Published: 25 May 2020

Abstract: *Research Highlights:* This research provides an application of a model assessing the naturalness of the forest ecosystem to demonstrate its capacity to assess either the deterioration or the rehabilitation of the ecosystem through different forest management scenarios. *Background and Objectives:* The model allows the assessment of the quality of ecosystems at the landscape level based on the condition of the forest and the proportion of different forest management practices to precisely characterize a given strategy. The present work aims to: (1) verify the capacity of the Naturalness Assessment Model to perform bi-directional assessments, allowing not only the evaluation of the deterioration of naturalness characteristics, but also its improvement related to enhanced ecological management or restoration strategies; (2) identify forest management strategies prone to improving ecosystem quality; (3) analyze the model's capacity to summarize the effect of different practices along a single alteration gradient. *Materials and Methods:* The Naturalness Assessment Model was adapted to the *Abies balsamea–Betula papyrifera* forest of Quebec (Canada), and a naturalness assessment of two sectors with different historical management strategies was performed. Fictive forest management scenarios were evaluated using different mixes of forestry practices. The sensitivity of the reference data set used for the naturalness assessment has been evaluated by comparing the results using data from old management plans with those based on Quebec's reference state registry. *Results:* The model makes it possible to identify forest management strategies capable of improving ecosystem quality compared to the current situation. The model's most sensitive variables are regeneration process, dead wood, closed forest and cover type. *Conclusions:* In the *Abies balsamea–Betula papyrifera* forest, scenarios with enhanced protection and inclusion of irregular shelterwood cuttings could play an important role in improving ecosystem quality. Conversely, scenarios with short rotation (50 years) could lead to further degradation of the ecosystem quality.

Keywords: naturalness; forest management intensity; land use intensity; quality of ecosystems; boreal forest

---

1. Introduction

Green building conception relies on quantitative tools to evaluate the environmental impact of different building materials. For wood products, the environmental impacts include the effects of forest management strategies and practices on ecosystem quality.

A conceptual model for naturalness assessment in boreal forests has been recently proposed for the assessment of the impact of wood harvesting on the quality of ecosystems [1]. This model has been developed from the perspective of being used in life cycle assessment (LCA). LCA, by default, uses biodiversity damage [2,3] to evaluate the quality of ecosystem. However, biodiversity based on species count alone does not reflect the multidimensional character of biodiversity, which includes multiple levels of organization: genetic, species, populations, community and ecosystem [3], and might lead to inappropriate conclusions [4,5]. The model generally used in LCA to establish the relationship between land use and biodiversity (i.e., the species–area relationship (SAR)) presents many limitations, and is not appropriate when the habitat modification does not result in species losses [5]. On the other hand, studies investigating the effects of land use on biodiversity often contrast intensive vs extensive uses [6]. For forestry, such a simplistic approach does not allow for the consideration of the full diversity of practices, each implying a different pressure on the environment [7–9]. To overcome these issues, we developed an alternative approach, based on the naturalness concept [10], which focuses on habitat characteristics, and allows the evaluation of various forest management practices along a single bi-directional alteration gradient [1], i.e., forest ecosystem degradation or restoration, related to given forest management strategies. Generally, models proposed up to now in LCA do not account for a possible improvement of habitat condition related to enhanced ecological management strategies and restoration efforts [2,5], despite the fact that these are seen as crucial actions to enhance ecosystem functioning and halt the decline of biodiversity [11,12].

Many authors have proposed the use of the concepts of naturalness and hemeroby in impact evaluation of land use (such as forestry) on the quality of ecosystems in LCA [13–17]. Naturalness is defined as *"the similarity of a current ecosystem state to its natural state"* [10], whereas hemeroby expresses *"distance to nature"* in landscape ecology [15]. The use of these concepts can provide a management guide that overcomes the challenge of data gaps in biodiversity [1]. However, the use of subjective hemeroby or naturalness classes has been criticized [3]. The model developed here for boreal forest provides a single numerical index, a suitable approach for use in LCA [1].

In order to evaluate forest management scenarios, the assessment should go beyond the gradual transformation related to the progressive implementation of the scenario through time [1], and be placed in the context of its continuous application over the whole productive area.

The aim of this study was to test the bi-directional capacity of the recently proposed model for the assessment of the impact of wood harvesting on the quality of ecosystems [1], to evaluate the performance of distinct enhanced ecological management strategies, including restoration efforts, at the landscape level. For this purpose, we used two adjacent territories located in the boreal eastern *Abies balsamea–Betula papyrifera* ecological bioclimatic domain of Quebec (Canada) with different histories of forest management. The specific objectives of the study were to:

1. Determine the naturalness of different mix of forest management practices to evaluate the bi-directional capacity of the model to assess both ecosystem degradation and restoration;
2. Identify forest management strategies prone to improving ecosystem quality based on a naturalness evaluation;
3. Analyze the model's capacity to summarize the effect of different practices along a single alteration gradient.

2. Materials and Methods

The impact on ecosystem quality of forest management scenarios involving a mix of different proportions of forestry practices is evaluated using the Naturalness Assessment Model initially developed for the *Picea mariana*–feathermoss ecological domain of Quebec [1]. This model uses indicators of condition and pressure to calculate a unique index of naturalness resulting from the combination of management strategies including conservation, and different silvicultural treatments (e.g., careful logging, plantation of indigenous species and partial cutting) (see Côté et al., 2019 for the full description of the naturalness assessment method). For the purpose of this study, the model was adapted to the context of the boreal eastern *Abies balsamea–Betula papyrifera* ecological bioclimatic domain of Quebec (see Appendix A for details related to model's adaptation).

2.1. Test Area

The territory of the Montmorency Experimental Forest, located north of Quebec City and included in the *Abies balsamea–Betula papyrifera* domain, was used as a test area (see Appendix A for localization and historical information). This experimental research station is divided in two sectors, designated as FM-A and FM-B, according their different histories. FM-A has been subject to continuous small-scale commercial harvest since the mid-sixties, while FM-B has been subject to a second wave of large-scale commercial harvest between 1985 and 2008, before being incorporated into the Montmorency Experimental Forest.

2.2. Naturalness Assessment

For the *Abies balsamea–Betula papyrifera* domain, the five naturalness characteristics of the model were evaluated using the same indicators and variables used for the *Picea mariana*–feathermoss domain of the Quebec's boreal forest [1] (Table 1), except for composition where merchant volume proportion of *Picea* spp. was used as a surrogate to obviate the lack of composition data for late successional species (see Appendix A for details). The evaluation is realized in two steps: partial naturalness index for condition indicators (condition_pni: pni in lower case) and naturalness degradation potentials (NDP) are first evaluated. To do so, we use respectively curves, relating measures (percentage of area or volume) to condition_pni, and tables relating percentage of forest area by practices to NDP factors, shown in Appendix A. Then, the partial naturalness for each naturalness characteristic (characteristic_PNI: PNI in capital letters) is calculated using corresponding formula as per Table 1. The final result corresponds to the naturalness index (NI) obtained from the arithmetic mean of the five characteristic_PNI. To ease results interpretation the continuous gradients (partial or global naturalness indexes) can be split in classes of 0.2 (0.0–0.2: very altered; 0.2–0.4: altered; 0.4–0.6: semi-natural; 0.6–0.8: near-natural; 0.8–1: natural).

Table 1. Partial naturalness index equations for each naturalness characteristic (characteristic_PNI) (source: [1]).

Naturalness Characteristic	Characteristic_PNI Equation
Landscape context	$Context\_PNI = CF\_pni \times (1 - (ANT\_NDP + Wm\_NDP + W\_CC\_NDP))$
Forest Composition	$Compo\_PNI = ((CT\_pni + LS\_pni)/2) \times (1 - (exo\_NDP + CS\_NDP))$
Structure	$Struc\_PNI = ((OF\_pni + IR\_pni)/2) \times (1 - HS\_NDP)$
Dead wood	$DW\_PNI = 1 - DW\_NDP$
Regeneration process	$RP\_PNI = 1 - RP\_NDP$

PNI: partial naturalness index for naturalness characteristics; pni: partial naturalness index for condition indicators; NDP: naturalness degradation potential; CF: closed forests; ANT: anthropization; Wm: modified wetlands; W_CC: humid area in clearcut; CT: cover type; LS: late successional species; exo: exotic species; CS: companion species; OF: old forests; IR: irregular stands; HS: horizontal structure; DW: dead wood; RP: regeneration process.

The model's adaptation to a new region requires to ensure the capture of the main ecological issues recognized for the territory under investigation, and to identify appropriate indicators, as well as variables and data sources for the evaluation. We tested two types of data sources for reference data: local historical studies [18–20] or Quebec's reference state registry [21] (Table A1). The curves used for partial naturalness index (condition_pni) evaluation were calibrated according to the reference data set used (Figures A2 and A3 based on reference data from studies and registry respectively), and the factors used for naturalness degradation potential (NDP) evaluation adapted (Tables A2–A5).

2.3. Description of Scenarios

Scenarios correspond to different mixes of practices applied in the context of sustainable timber production. To figure out the result at the landscape level, in the current study, each practice representing a scenario component is applied on a constant basis over a given proportion of the productive area. However, this exercise is highly theoretical, considering that, in reality, each scenario is adjusted through time to maximize the wood production, rather than having each component being applied on a constant basis. Furthermore, the prevalence of spruce budworm epidemics, which affect balsam fir (*Abies balsamea*) landscapes every 30 years [22], is not taken into account here.

Total area is broken down as follows: total area = water area + terrestrial area; terrestrial area = non-forested area + forested area; forested area = protection area + productive area. Protection is applied as a reduction percentage over the total forested area, and practices related to wood procurement are applied over a given proportion of the productive area. The evaluation must cover the whole landscape; therefore, scenario components must encompass 100% of the productive forested area. The forested area excluded from the productive area corresponding to protection will have a natural evolution, for which historical reference data has been used, except for late successional species. The hypothesis for spruce content (LS) in protected areas was reduced to 4%, based on results from secondary forests [23–25], given that protected areas are generally located in previously exploited areas, where spruce seed-trees have been harvested.

The practices considered for scenarios in the productive area were: careful clearcut logging (CL), which corresponds to the cut with regeneration and soil protection required by law for clearcut operations in Quebec [26], forest plantation (PL) and irregular shelterwood cutting (ISC). As a result of the abundance of natural pre-established balsam fir regeneration in these secondary balsam fir forests [20,23,27], careful logging (CL) is the main regeneration method applied in these forests. Plantations (PL) are generally concentrated on rich sites, where natural resinous regeneration is scarce. The irregular shelterwood silvicultural system (ISC) is compatible with forests types driven by partial stand mortality and gap dynamics such as balsam fir forests, and provides a way to maintain old-growth forest attributes [28]. Two levels of protection were considered: initial protection (ip) and enhanced protection (ep). The initial protection represents 24.4% of the forested area for FM-A and 13.3% for FM-B, and corresponds to current protected areas, along with other areas excluded from forest management, due to the various regulations and constraints (e.g., riparian strips, steep slopes etc.). The enhanced protection (ep) corresponds to the initial protection, plus protected areas projects proposed for the Montmorency Forest [29], and represents 31.7% of the forested area for FM-A and 32.2% for FM-B.

For careful clearcut logging (CL), two different rotation lengths were tested: 50 years (CL50) and 70 years (CL70), the first one corresponding to the business-as-usual in commercial balsam fir forests, and the second one to a practice that favors establishment of natural regeneration [23] and carbon sequestration [30]. Plantation with a rotation length of 60 years (PL60) has been used, based on the rotation used in the last sustainable yield calculation. The irregular shelterwood cutting (ISC) corresponds to a sequence of partial cuts organized in space and time to insure the permanency of the forest cover [28]; this practice was simulated as harvests of 33% of the volume every 30 years. Data used to evaluate practices effects come from secondary forests.

Scenario elaboration began with the evaluation of each of the three components with two variants of CL, over the entire productive area separately, in order to evaluate the maximal theoretical effect of each practice. These scenarios were first evaluated with the current level of protection and then with the enhanced protection level. For the scenarios involving practices mixes, the first assessment considered only the enhanced protection level. Each mix involving CL has been evaluated two times: one with CL50 and one with CL70. Proportion of CL tested varied between 90% and 50% of the productive area (by multiple of 10), between 0 and 40% for PL and between 0 and 50% for ISC, and a test considering 50% PL and 50% ISC has also been included (Table 2).

Table 2. Description of management scenarios evaluated.

| | Productive Area Proportions by Scenario Component | | |
Scenario#	ISC	PL60	CL
1	0	1	0
2	0	1	0
3	0	0	1
4	0	0	1
5	0	0.1	0.9
6	0	0.2	0.8
7	0	0.3	0.7
8	0	0.4	0.6
9	0.1	0.1	0.8
10	0.2	0.1	0.7
11	0.1	0.2	0.7
12	0.1	0.3	0.6
13	0.2	0.2	0.6
14	0.3	0.1	0.6
15	0.1	0	0.9
16	0.2	0	0.8
17	0.3	0	0.7
18	0.4	0	0.6
19	0.5	0	0.5
20	0.5	0.5	0
21	1	0	0
22	1	0	0

Scenario #: scenario number; ISC: irregular shelterwood cutting; PL60: plantation with 60 years revolution; CL: careful logging.

2.4. Hypotheses

The hypotheses used for each scenario component are presented in Table 3.

Table 3. Hypotheses used for each scenario component.

Scenario Component	Cover Type (CT: % Prod Area of Coniferous Cover Type)	Late Successional Species (LS: % Merchantable Volume in *Picea* spp.)	Closed Forests (CF: % Productive Area of Forests > 40 Years Old)	Old Forests (OF: % Productive Area of Forests > 80 Years Old)	Irregular Stands (IR: % Productive Area of Irregular Stands)
CL50 [1]		1	20	0	0
FM-A	77.49				
FM-B	79.79				
CL70		3.5	42.85	0	0
FM-A	81.54				
FM-B	83.84				
PL60		50	33.33	0	0
FM-A	92.38				
FM-B	94.64				
ISC		15	90	90	90
FM-A	78.02				
FM-B	80.41				
Protection					
FM-A	79.3	4	79.9	23.7	17.8
FM-B	85.7	4	76.19	57.9	40

[1] CL50: careful logging in 50 years old stands; CL70: careful logging in 70 years old stands; PL60: plantation with 60 years rotation; ISC: irregular shelterwood cutting; prod_area: productive area.

Age structure related to each scenario under sustainable production has been used to evaluate closed and old forests. For example, for a rotation of 50 years, 40% of the productive area will be in the 10 years old class, 40% in the 30 years old class and 20% in the 50 years old class, therefore, the corresponding proportion of closed forests will be 20% and 0% for old forests (>60 years old). For ISC, 10% of the area are permanent skid trails. Despite the fact that these trails will be part of the future stand, the proportion of CF, OF and IR were set to a maximum of 90% used as a security factor. The proportion of coniferous cover type for PL60 and ISC is based on eco-forest map data for the corresponding origin code in each territory. For CL50 and CL70, the proportion of coniferous cover type is based on data (number of stems) gathered in FM-A [31]. The resulting proportion has been raised of 2.3% in FM-B, based on the differences observed in inventory data between the two territories for diverse practices, to be coherent with the coniferous aggressiveness in that territory. The proportion of spruce (*Picea* spp.) for each practice was estimated using different studies in the Montmorency Forest [23,32,33]. Clearcuts on wetlands and anthropization levels were set to current values, and kept constant in all scenarios. For pni's evaluation, the proportion of the productive area was adjusted to represent the proportion of forest area for CT, LS, OF and IR, and the proportion of terrestrial area for CF.

2.5. Sensitivity Analysis

To identify the most sensitive variables of the current naturalness assessment, a sensitivity analysis was performed by varying pni for condition indicators and NDP by ±10%. The same analysis was also performed for each scenario component, using the 100% scenario with enhanced protection, in order to identify the most sensitive variables related to each component.

As naturalness condition indicators are assessed against historical values considered to be representative of the pre-industrial condition, the sensitivity to the choice of reference data set has been evaluated by comparing results obtained when using two different reference data sets (Table A1). The first assessment was performed using values drawn from different studies and old management plans around the territory of Montmorency Forest [18–20]. A second assessment was performed on scenarios initially assessed using values from the reference state Quebec's registry [21], resulting from simulations of the vegetation dynamics considering only natural disturbance regimes. The scenario ranking resulting from the use of the two reference data sets was statistically compared using the Kendall rank correlation coefficient. The statistical test was performed with "R" 3.6.1 [34].

Multiple assessments of the same set of scenarios was then performed to analyze the effect of different combinations of management parameters, such as the proportion of protected areas, plantation rotation lengths and the proportion of *Picea* spp. in plantations.

3. Results

Detailed results of each scenarios evaluation, including those performed for sensitivity analysis, are presented as Supplementary Material where the different assessments (ass) are as follows:

ass_1: initial scenarios assessment with enhanced level of protection (ep);
ass_2: scenarios with enhanced level of protection using registry's reference data set;
ass_3: scenarios with the initial level of protection (initial protection: ip);
ass_4: scenarios with PL50 and enhanced level of protection;
ass_5: scenarios with PL50 and initial level of protection;
ass_6: scenarios with PL50 and enhanced level of protection and 90% of spruce at maturity in plantations;
ass_7: scenarios with enhanced protection and alternate set of NDP factors.

3.1. Naturalness Index for the Current State of the Forest of the Experimental Forest

The naturalness assessment results for the current state of the forest in FM-A and FM-B are presented in Figure 1, showing intermediary results for condition indicators (pni), results for each naturalness characteristics (PNI) and the resulting naturalness index (NI), which corresponds to the arithmetic mean of the five Characteristic_PNI. Results show a naturalness index (NI) of 0.5294 for FM-A and 0.4691 for FM-B. Both territories scored in the semi-natural class (NI: 0.4-0.6), despite their different histories. However, FM-B showed a lower naturalness compared to FM-A, mainly related to a poor structural diversity associated with an important deficit of old forests combined with a lack of irregular stands (Figure 1). The structure is altered in FM-A (PNI_Struc= 0.3332) and very altered in FM-B (PNI_Struc= 0.1931). The alteration of structure in FM-A results from the very low ratio of old forests compared with the historical values; whereas, in FM-B, it is both the low ratio of old forests and the lack of irregular stands.

The landscape context of FM-B is also characterized by an important deficit of closed forests where it reaches the altered level. According to the map information, in FM-A less than 10% of the harvested area was in 50 years old stands, whereas in FM-B this proportion reaches 50%. Precommercial thinning covered 9.2% of the forest area in FM-A compared to 23.4% in FM-B, plantation 6.8% in FM-A and 8.8% in FM-B, and partial cuttings, excluding commercial thinning, 5.5% in FM-A and 1.4% in FM-B.

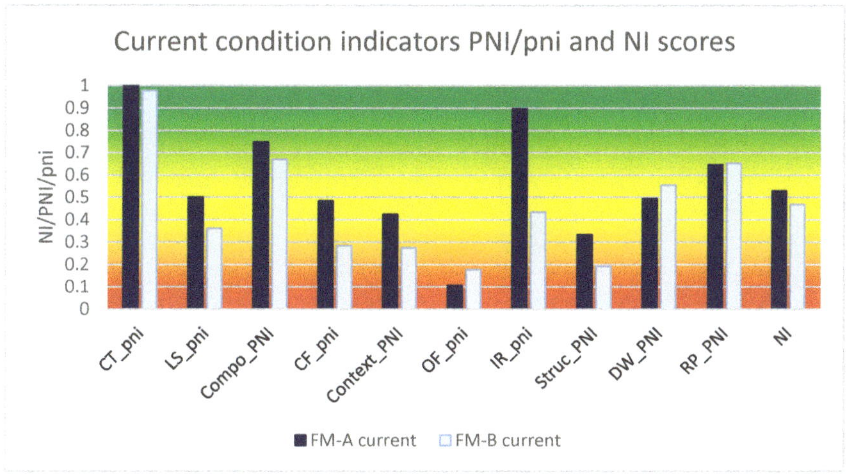

Figure 1. Results for condition indicators and naturalness for FM-A and FM-B. PNI: partial naturalness index for naturalness characteristics; pni: partial naturalness index for condition indicators (intermediary results); CT: cover type; LS: *Picea* spp. corresponding to late successional species; Compo: composition; CF: closed forests; Context: landscape context; OF: old forests; IR: irregular stands; Struc: structure; DW: dead wood; RP: regeneration process; NI: naturalness index.

3.2. Naturalness of Different Forest Management Scenarios

Scenarios results of the initial assessment are presented in Figure 2 (detailed results in Supplementary Material, file results, page ass_1). Compared with results based on the current state of the forest, some scenarios lead to a lower naturalness index and others to a higher level, indicating the model's capacity to detect not only alteration of the ecosystem quality, but also its improvement.

Scenarios featuring the application of only one practice on 100% (of the productive area) were tested to evaluate their respective extreme theoretical effects on the naturalness assessment resulting from the model. Each of these were evaluated twice: once with the current level of protection (initial protection: ip), and once with the inclusion of the protection projects (enhanced protection: ep). These 100% theoretical scenarios are shown as benchmarks in every test performed to see the maximal theoretical effect related with each scenario component, considering the two levels of protection. Figure 2 shows that enhanced protection impact differs among the scenario components. The positive impact related to protection enhancement is more important when applied concurrently with PL60 than with CL50, CL70 and finally ISC.

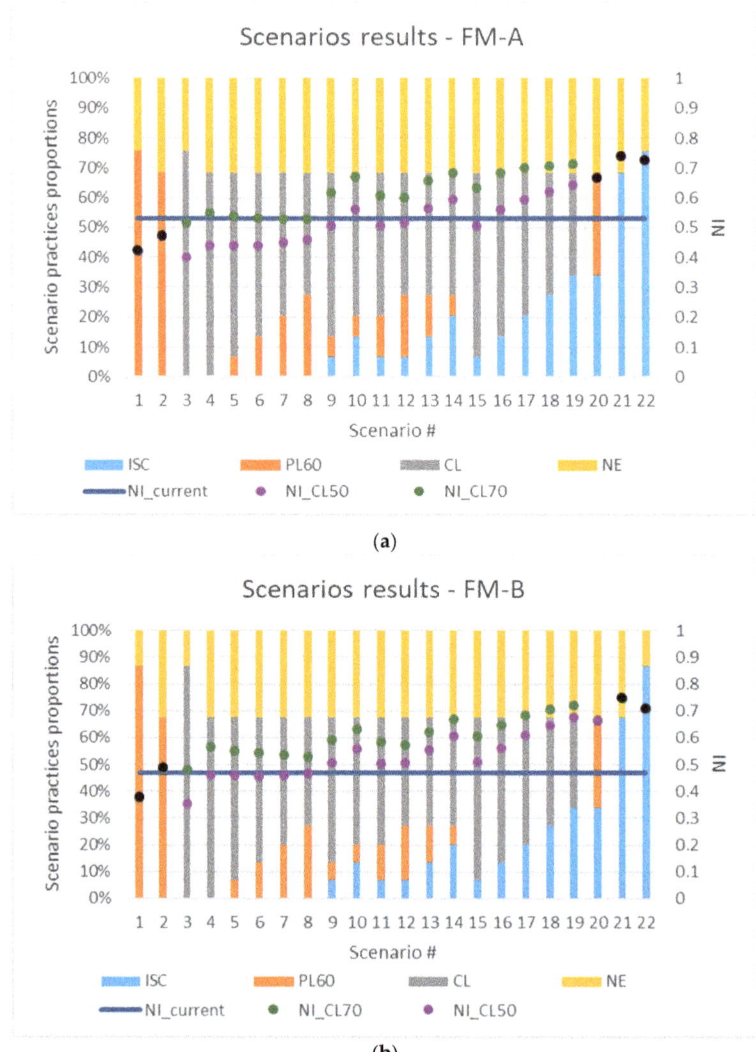

Figure 2. Naturalness assessment of forest management scenarios: (**a**) FM-A; (**b**) FM-B. For each scenario, practices mixes are illustrated using the stacked histogram: ISC: Irregular shelterwood cutting; PL60: plantation with 60 years revolution; CL: careful logging; NE: natural evolution (protection); Resulting naturalness index (NI) is indicated with the dots: NI_CL70 (green): naturalness index for CL applied in 70 years old stands; NI_CL50 (purple): naturalness index for CL applied in 50 years old stands; naturalness index for practices other than CL (black); the horizontal solid line shows the current naturalness index (2018) with the initial level of protection.

The PL60 scenario (Sce#1) produces a degradation of the naturalness in both territories. With the inclusion of protection projects (Sce#2), the 100% PL60 scenario still scores lower than current results for FM-A, but slightly better for FM-B. The 100% CL50 scenario is worse than the 100% PL60 scenario, because of the shorter rotation period and a lower amount of spruce. The 100% ISC scenario (Sce#21 and Sce#22) would have the potential to improve naturalness at the quasi-natural level. The quasi-natural level could also be reached in FM-A, with CL70 combined with at least 30% of ISC and enhanced protection (Figure 2a), and in FM-B with CL70 and enhanced protection with at least 40% ISC (Figure 2b). The 100% CL70 scenario has a NI near to the current level, for the same level of protection. Protection projects represent a greater proportion in FM-B, so the resulting naturalness improvement related to the inclusion of these projects is more important.

The current naturalness in FM-A shows an impact level analog to the 100% CL70 scenario with initial protection, or CL70 with enhanced protection and some plantation. The current level in FM-B presents an impact closer to the 100% CL70 scenario with initial protection, or CL50 with enhanced protection scenarios with some plantation. Considering the intervention mixes, ISC has the potential to compensate at least for a part of the degradation related to forest rejuvenation resulting from careful logging or plantation.

The ISC's potential to improve the overall naturalness is more important when the proportion of ISC is smaller than the historical proportion of irregular stands (FM-A: 17.9%; FM-B: 40%). Above this level, naturalness improvement resulting from a higher level of ISC is less important, as seen in FM-A's CL70 results, where the difference of NI between 40% and 50% of ISC is smaller than the difference between 10% and 20% of ISC. Results suggest that ISC can compensate to some extent for the alteration caused by CL50.

With the hypotheses used, plantations under 60 years rotation, with 50% of the volume in spruce at maturity, could produce an improvement of the naturalness when combined with CL50, mainly related to rotation length.

3.3. Sensitivity Analysis

3.3.1. Test of a Variation of 10% of the Parameter Values

The results of the sensitivity analysis performed on current pni and NDP values for the two territories are provided in Figure 3. Due to the use of non-linear models, a uniform variation of input parameters (10%) can have a non-linear effect on the results, depending on the curve slope around the parameter value [1].

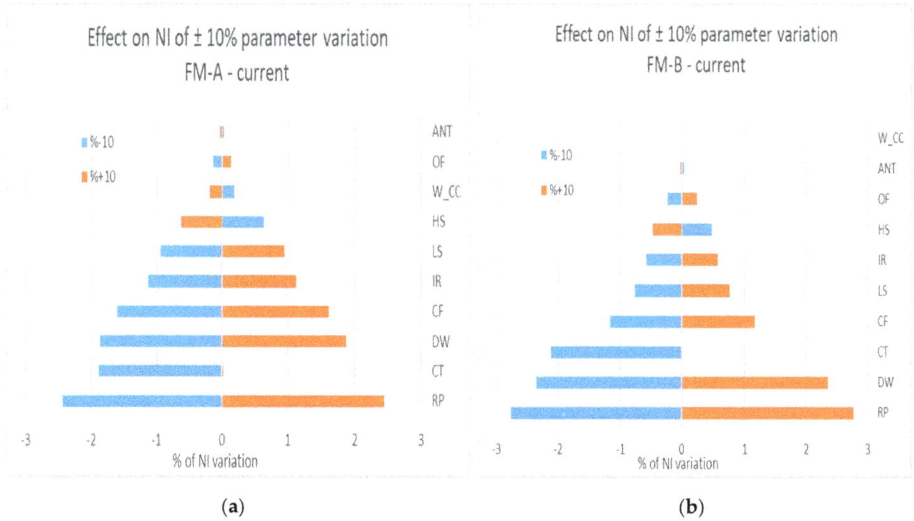

(a) (b)

Figure 3. Results of the sensitivity analysis testing a variation of 10% of the parameter value on the current assessment: CF: closed forests; ANT: anthropization; W_CC: humid area in clearcut; CT: cover type; LS: late successional species; OF: old forests; IR: irregular stands; HS: horizontal structure; DW: dead wood; RP: regeneration process. (a) FM-A, (b) FM-B.

The most sensitive variables for the naturalness assessment of the current state of the forest are the regeneration process (RP), the cover type (CT), the dead wood (DW) and the closed forest (CF) for both territories; however, dead wood is more sensitive than cover type in FM-B.

The results of the sensitivity analysis performed on the scenarios in which a single component is applied over 100% of the productive area are shown in the Supplementary Material (page sensitivity analysis). As seen from these figures, the most sensitive variables for scenarios involving cuttings (CL, ISC) correspond, in varying order, to the regeneration process, the dead wood, the cover type and the closed forests, whereas the plantation influences mainly closed forest, cover type, late successional companion species and the regeneration process.

3.3.2. Test of an Alternative Reference Data Set

Using the registry's reference data set [21], the naturalness of the two territories are closer (FM-A: NIr = 0.4683; FM-B: NIr = 0.4349) (Figure 4) and are both lying in the semi-natural range. The main disparities between the two data sets values are for OF and IR (Table A1) especially for FM-A, leading to a much lower PNI_Struc with the registry's values. The difference related to coniferous cover type between the two data set is not reflected in the model, because of the use of topped curve for CT above the historical value.

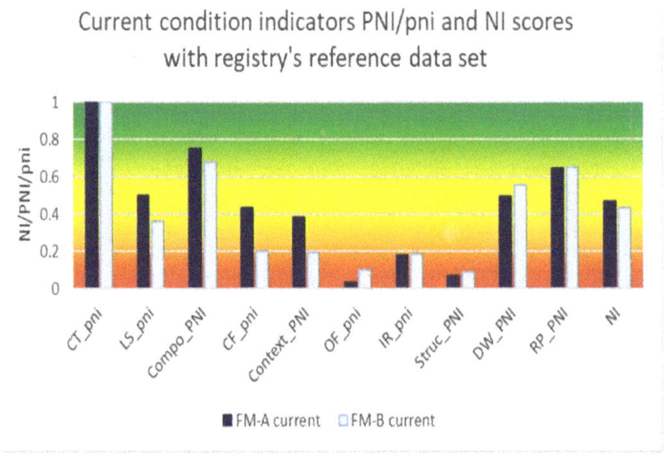

Figure 4. Results for condition indicators and naturalness index (NI) for FM-A and FM-B using registry's reference state data set [21].

Scenario results using registry's data as alternative reference data set are presented in Supplementary Material (page ass_2 and Figure S2). The use of the registry's data set places the current assessment for FM-A to a level below CL70 scenarios with some PL60, with initial protection, as well as enhanced protection. We observe the same trends seen with the use of studies as the reference data set: improvement related to use of some ISC, and a slight degradation related to the use of PL60 combined with CL70, but no deterioration (even a slight improvement) with an additional use of PL60 when combined with CL50. However, the reference data set used might affect the identification of the potentially improving or deteriorating scenarios when compared with the current result.

The scenarios were sorted in descending order of NI results using local studies [18–20] as the reference data set versus the registry's reference data [21] (Table 4). The ranking of scenarios is not the same, depending on the reference data set used, but the four best-scoring scenarios are the same (the 100% ISC with both level of protection, and at least 40% ISC combined with CL70 with enhanced protection), as well as the two worst-scoring scenarios (CL50 and PL60 with initial protection).

The remaining 34 scenarios score in the semi-natural class ($0.4 \leq NI < 0.6$). Among these, CL70 with at least 20% ISC and enhanced protection score in the upper part of the class, while CL50 with PL60 and enhanced protection are in the lower part. Among scenario components, the addition of ISC has the biggest positive impact on naturalness, while the addition of PL60 has a negative impact, but is less important when compared to the importance of the positive impact of the ISC.

Table 4. Scenario ranking resulting from the use of the two reference data sets for FM-A and FM-B, sorted in descending order of their naturalness index.

Scenario Sequential nb	Scenario Description [1]	Rank FM-A Studies	Rank FM-A Registry	Rank FM-B Studies	Rank FM-B Registry
22	100ISC_ep	1	1	1	1
23	100ISC_ip	2	2	3	2
20	50CL70_50ISC_ep	3	3	2	3
19	60CL70_40ISC_ep	4	4	4	4
18	70CL70_30ISC_ep	5	7	5	7
15	60CL70_10PL60_30ISC_ep	6	8	7	8
17	80CL70_20ISC_ep	7	9	10	10
11	70CL70_10PL60_20ISC_ep	8	11	11	11
21	50PL60_50ISC_ep	9	5	8	5
14	60CL70_20PL60_20ISC_ep	10	12	12	12
40	50CL50_50ISC_ep	11	6	6	6
16	90CL70_10ISC_ep	12	13	14	13
39	60CL50_40ISC_ep	13	10	9	9
10	80CL70_10PL60_10ISC_ep	14	14	16	16
12	70CL70_20PL60_10ISC_ep	15	15	17	17
13	60CL70_30PL60_10ISC_ep	16	16	18	19
38	70CL50_30ISC_ep	17	19	13	14
35	60CL50_10PL60_30ISC_ep	18	17	15	15
34	60CL50_20PL60_20ISC_ep	19	25	22	25
37	80CL50_20ISC_ep	20	26	20	22
31	70CL50_10PL60_20ISC_ep	21	27	21	24
5	100CL70_ep	22	18	19	18
6	90CL70_10PL60_ep	23	20	23	20
7	80CL70_20PL60_ep	24	21	24	21
8	70CL70_30PL60_ep	25	22	25	23
1	Current_ip	26	28	33	33
9	60CL70_40PL60_ep	27	23	26	26
4	100CL70_ip	28	24	32	31
33	60CL50_30PL60_10ISC_ep	29	29	28	28
32	70CL50_20PL60_10ISC_ep	30	31	30	30
36	90CL50_10PL60_ep	31	32	27	27
30	80CL50_10PL60_10ISC_ep	32	33	29	29
3	100PL60_ep	33	30	31	32
29	60CL50_40PL60_ep	34	34	34	34
28	70CL50_30PL60_ep	35	35	36	36
27	80CL50_20PL60_ep	36	36	38	38
25	100CL50_ep	37	37	35	35
26	90CL50_10PL60_ep	38	38	37	37
2	100PL60_ip	39	39	39	39
24	100CL50_ip	40	40	40	40

[1] ISC: Irregular shelterwood cutting; PL60: plantation with 60 years revolution; CL50: careful logging applied in 50 years old stands; CL70: careful logging applied in 70 years old stands; ip: initial protection; ep: enhanced protection; Current: current naturalness (2018). The digits preceding component code correspond to the percentage of productive area of the component application (ex:100PL60_ep: plantation with 60 years revolution applied over 100% of the productive area with enhanced level of protection).

The use of registry reference data set induces a negative bias compared with the use of the studies reference data set (FM-A: −0.0591; FM-B: −0.0522), as the naturalness index is generally lower when registry's data are used. This bias is not constant as a result of the use of nonlinear relationships in the model. Nevertheless, the general trends related to the positive effects of ISC and the relatively limited negative effects of mechanically released plantations, including 50% of spruce at maturity, are observed in the ranking arising from both data sets. The ordinal association between the two rankings, as measured with Kendall rank correlation computed with the base cor function in R, showed a coefficient of 90.5% for FM-A and 95.9% in FM-B. Therefore, the model's aptitude at classifying different scenarios along a single alteration gradient is robust, in regard to the reference data set used. However, assessing improvement or degradation against the current value might be affected by the reference data set used.

3.3.3. Naturalness of the Forest Management Scenarios Tested Using Various Hypothesis

Scenarios with Initial Protection

In order to see the impact related to the protection level, scenario evaluation using studies as reference data set was calculated using the current level of protection (ip) (Supplementary Material: page ass_3 and Figure S3). With 24% of the forested area in protection in FM-A, none of the scenarios, except 100% ISC (sce#22), would reach the quasi-natural level, and the scenarios considering CL50 with some PL60 would score around the limit between the altered and the semi-natural levels. The current alteration level is similar to 100% CL70, and scenarios combining with some PL60 are below to the current level even when combined with CL70. In FM-B, with 13% of the forested area in protection, the scenarios considering CL50 with some PL60 would score in the altered level, except when combined with at least 20% of ISC or, at least 10% ISC, if considering CL50 without plantation. The current alteration level in FM-B is similar to 100% CL70, or CL70 with less than 20% PL60, or CL70 with up to 30% PL60, if combined with 10% ISC, with initial protection.

Scenarios with Enhanced Protection and Plantation with 50 Years Rotation

In order to see the impact related to plantation rotation length, an evaluation using studies as the reference data set was calculated using a rotation length of 50 years for plantations (PL50). This test was performed by adjusting the age structure for the planted proportion of the scenario, but keeping all others factors constant (Supplementary Material: page ass_4 and Figure S4).

The 100% PL50 with initial protection scenario would lead to a naturalness index lying in the altered class for both territories (Sce #1: FM-A: NI = 0.3871; FM-B: NI = 0.3308). Adding new protection projects (reaching a total of more than 30% of the forested area) makes it possible to attain the semi-natural class even with 100% PL50 in the productive area (Sce#2: FM-A: NI = 0.4400; FM-B: NI = 0.4511). The use of PL50 instead of PL60 reduces slightly the improvement related to the increasing use of PL, when combined with CL50. We can still observe no further degradation and even a small improvement for Scenario 8 with CL50, compared with Scenario 7 with CL50, related to the increase of spruce species at the landscape level. However, more spruce in the plantations could induce a degradation, as these species were subdominant at the landscape level.

Scenarios with Initial Protection and Plantation with 50 Years Rotation

In order to see the combined impact related to the protection level and plantation rotation, scenario evaluation using studies as reference data set was calculated using the current level of protection and plantation with 50 years rotation (PL50) (Supplementary Material: page ass_5 and Figure S5). Compared with the test considering initial protection only, the use of PL50 instead of PL60 has a limited effect, but pushes more scenarios closer to the altered class in FM-B. With the initial protection and the use of PL50, some ISC is necessary to reach a level above the current naturalness.

Plantation with Enhanced Protection and 50 Years Rotation and 90% of Merchantable Volume in Spruce at Maturity

A test was performed using a proportion of spruce of 90% (instead of 50%) of the merchantable volume at maturity for PL50, in order to analyze the effect of more monospecific plantations with a short rotation (Supplementary Material: page ass_6 and Figure S6). The 100% PL50 with 90% of spruce at maturity scenarios, with initial protection, would lead to a naturalness index lying in the altered class in both territories. The addition of protection projects (reaching more than 30% of the forested area) makes it possible to attain the semi-natural class in FM-B, even with 100% PL50 in the productive area, that would produce 90% of the volume in spruce (Sce#2: FM-A: NI = 0.3998; FM-B: NI = 0.4242). An increasing use of PL50 producing 90% of the volume in spruce, combined with CL50, could improve the naturalness to some extent, until the total spruce proportion of the scenario reaches its historical level. However, above this level, an increase of spruce proportion could lead to a noticeable deterioration, as shown with the 100% PL50 with 90% of spruce (Sce #1 and 2).

Scenarios with Enhanced Protection and a Variant of NDP Factors

The regeneration process and dead wood were among the most sensitive variables for the assessment of the current forest and for the evaluation of scenario components related to harvest (CL, ISC). In order to see the influence of NDP factors, an alternate set of factors, inducing more difference between CL70 and CL50 and ISC, was tested by lowering the three NDP factors (i.e., HS, DW and RP) for CL50 of 0.1, and the NDP factor for partial cutting of 0.1 for HS and RP. Inducing a more important degradation of RP, DW and HS for CL50, and of HS and RP for partial cuttings, results in a slightly higher NI for the assessment of the current forest (FM-A: 0.5550; FM-B: 0.4756). The difference is marginally more important for FM-A, as this territory has less CL50 than FM-B. As expected, this enhances the performance of CL70 scenarios compared to CL50 scenarios, allowing more scenarios to reach the quasi-natural level. On the other hand, the distinction between scenarios that improve naturalness vs. those that degrade it remains the same (Supplementary Material: page ass_7 and Figure S7).

4. Discussion

Scenario evaluation suggests that some combinations of practices could produce an improvement of the naturalness index, compared to the state of the current forest, confirming the model's capacity of performing bi-directional assessment to satisfy the need of assessing restoration efforts.

4.1. Naturalness Assessment of FM-A And FM-B's Current Forest

Current results in FM-A and FM-B lead to a naturalness index lying between 0.4 and 0.6, corresponding to the semi-natural class. These results consider the current level of protection (ip: FM-A: 23.8%; FM-B: 12.7% of the terrestrial area or 24.4% and 13.3% of the forested area). The difference induced by the distinct forest management strategies applied over the last 50 years is a priori small; it is, however, not surprising, considering that the forest management strategies in both territories were prioritizing the liquidation of old forests using principally single aged management. The lower naturalness observed in FM-B is mainly related to the low level of closed forests (only 27% of the forest area is over 40 years old in FM-B), and a lack of structural diversity (the low level of old forests in FM-B being worsened by the scarcity of irregular stands). However, the contrast between the two territories is limited by the use of the same hypothesis for IR and LS. As the forest rejuvenation was applied more uniformly and over a short period in FM-B, compared with FM-A, a better characterization of the landscape context, taking into account the size of the stands by age-class, could be necessary to include that issue and improve the comparison.

4.2. Naturalness Evolution through Time

Results from the Naturalness Assessment Model can be applied to the conceptual time frame of the evolution of land quality with land use intervention generally applied in LCA [35] (Figure 5), considering that the reference state (Qref) corresponds to the natural state (NI = 1). The reference time considered in this study refers to the pre-industrial state, characterized by the spruce budworm epidemics prevailing every 30–40 years in the balsam fir forests [36]. The first cutting cycle (from t_0 to t_1) corresponds to the transformation phase causing the initial decrease in ecosystem quality, which is progressive in forestry, as shown with the assessment performed in the three Forest Management Units in the *Picea mariana*–feathermoss domain [1]. For a given management strategy, the subsequent rotations (from t_1 to t_2) cause further degradation, but this is relatively less important than the impact resulting from the initial transformation [1]. For FM-A and FM-B, as no data was available for t_1, the same level of naturalness after the first cutting cycle has been assumed. FM-B shows a current naturalness index lower than FM-A, which can be related to the forest management regime applied up to now. Evolution during the sustainable management phase depends upon the forest management scenario applied, and can be further degraded or improved, depending on the forest management strategy, as schematized. A full rotation is necessary to reach the naturalness level evaluated for a given scenario (prior to that, the evaluation would include a part of the previous forest management strategy). Therefore, in order to compare different scenarios, the scenario evaluation must be performed as if the scenario would have been applied over the entire productive area, covering a whole cutting cycle. The conceptual framework for the evolution of ecosystem quality [35] considers an hypothetical relaxation of the land use (at t_2), and a progressive return to a level corresponding to the potential natural vegetation (Q$_{PNV}$). The level of quality after relax (Q$_{PNV}$) depends not only on natural vegetation dynamics, but could also be affected by the permanent impact related to practices applied in the past (for example, the introduction of exotic species, disappeared species, forest drainage, etc.). When the land has been used for an extended period of time (t_2>>>t_1), the natural reference could not be appropriate, the PNV represents an alternative [37]. Our evaluation uses historic pre-industrial data as reference, data from secondary forest to quantify the impact of the silvicultural treatments, and to adjust data for protection to reflect their localization in previously harvested sectors by lowering the proportion of *Picea* spp. However, scenario assessments do not consider future changes resulting from other causes, such as pollution, natural disturbance regime modification related to climate change or other modification of pressures.

4.3. Scenario Comparison

The model is used to express the impact level on the environment along a single alteration gradient. Though the choice of the reference data set used to parametrize the curves for condition_pni evaluation might influence the fine ranking of the forestry management practices, in the studied case, the scenarios that could improve naturalness were identified with a good level of confidence.

Among the practices considered, with the hypotheses used, the enhancement of protection combined with the use of ISC provided the best naturalness index scores. Conversely, practices involving short rotation (50 years) lead to the lowest scores. The scenarios are generally leading to the semi-natural range. Reaching of the quasi-natural level is associated with the use of irregular shelterwood cutting (ISC); a proportion close to 100% would be necessary with the initial level of protection. With an enhanced level of protection, the quasi-natural level could be attained when combining CL70 without plantations with at least 30% ISC in FM-A, and at least 40% ISC in FM-B.

On the opposite side of the alteration gradient, with the current level of protection of 13.3% of the forested area in FM-B, the scenario considering clearcuts with a rotation of 50 years (CL50), with or without plantation, would produce an altered level, indicating a potential loss of species, according to the model [1]. In FM-A, with a current level of protection of 24.4% of the forested area, the CL50 scenario with up to 10% PL60 would produce an altered level; scenarios with CL50 combined with more PL60 would also score close to the altered level. The results show the important role of protection in mitigating forest management effects, from the perspective of limiting the negative effects on biodiversity.

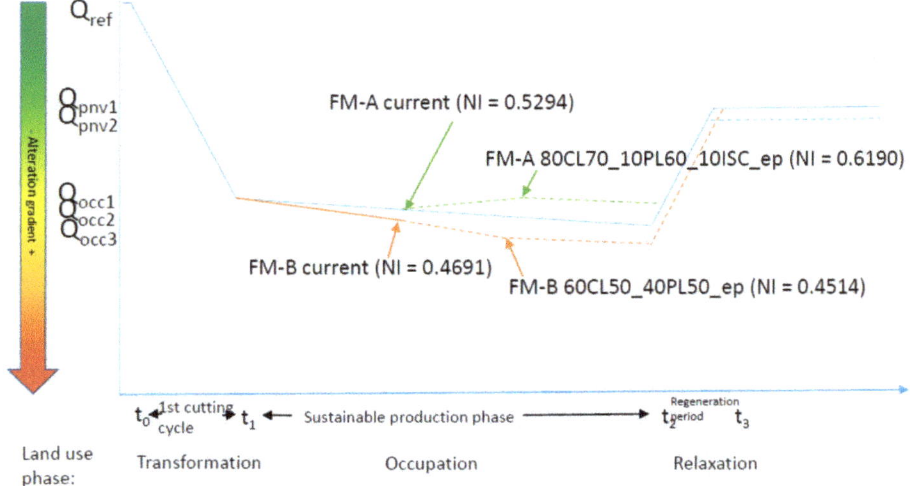

Figure 5. Ecosystem quality evolution thru time. Q_{ref}: Quality at the reference state (natural); Q_{PNV}: Quality of the Potential Natural Vegetation that have been regenerated after land use; Q_{PNV1}: with low permanent impacts; Q_{PNV2}: with higher permanent impact; Q_{occ}: Quality during occupation phase; Q_{occ1}: for the scenario with the lower impact; Q_{occ2}: for the mid-impact scenario; Q_{occ3}: for the lower impact scenario; FM-A 80CL70_10PL60_10ISC_ep: scenario for FM-A figuring 80% of the productive area in clearcut careful logging on a rotation of 70 years, plus 10% in plantation on a 60 years rotation and 10% subject to irregular shelterwood cutting, with the enhanced level of protection; FM-B 60CL50_40PL50_ep: scenario for FM-B figuring 60% of the productive area in clearcut careful logging on a rotation of 50 years, and 40% in plantation on a 50 years rotation, with the enhanced level of protection; NI: Naturalness index.

According to our results, most forest regimes involving a combination of practices (excluding the use of exotic species) applied on a sustainable basis, combined with a significant level of protection (around 30%), would lead to the semi-natural class. However, in FM-B, with a lower naturalness index (NI = 0.4713) compared with FM-A (NI = 0.5294), the threatened boreal caribou has declining populations, though the decline causes are manifold, and result from interacting factors [38,39]. However, in FM-B, where precommercial thinnings were performed over 23.4% of the forested area, the habitat quality of the snowshoe hare, a keystone species in this boreal forest, has been affected; however, effects on populations dynamic depend on multiple factors and are subject to time lag [40].

Our results underline the need for the development of more comprehensive biodiversity indicators that can capture other effects than species loss. The impact on biodiversity of different forest regimes including an important proportion of protection (30%) is related to the impact on species assemblage and populations, rather than to species loss, with the latter being restricted to very specialized and sensitive species. In this respect, a study based on time series showed that changes in assemblage composition has been widespread over the last 40 years, but there was no impact on the number of species [41]. In the *Abies balsamea–Betula papyrifera* forest of low and high altitude, the assemblage of bird species evolves with the vegetation succession [36]. In FM-B, old growth and senescent forests, which are waning, harbor specialized communities of invascular species [37]. In fact, biodiversity issues recognized for the area are mainly related to species associated with old growth and senescent forests or some of their particular features (e.g., dead wood) [38]. This situation raises the issue of the appropriate biodiversity indicators in LCA, for land uses have a low impact where biodiversity erosion is slow, and do not necessarily result in species loss, in comparison with other land uses causing a drastic change of habitat and observable loss of species richness. The model could thus be used to refine the ecosystem quality assessment in a more comprehensive way than the use of the five naturalness classes.

A test on the scenarios confirms the model's capacity of performing bi-directional assessment to satisfy the need of assessing restoration efforts. However, the effects on biodiversity of enhanced forest management strategies and restoration depend not only on the restored presence of essential habitat features, but also the duration of prior intensive management application [12]. This poses another challenge associated with the use of biodiversity as an indicator of ecosystem quality, as proposed for LCA [9].

Despite over 20 years of research on how to include man-made impacts on biodiversity in LCA, no comprehensive biodiversity impact assessment has been performed so far [42]. In LCA, pressures on biodiversity resulting from land use can be represented as a midpoint impact category, whereas biodiversity in general corresponds to an endpoint category related to ecosystem health [42]. The naturalness assessment model evaluated in this research work could be used as a midpoint level indicator to evaluate the pressure of forestry management practices on ecosystem quality. However, linking such a naturalness index with the endpoint biodiversity indicator, representing the potential loss of species as currently used in LCA, is reductive, as it fails to capture other factors influencing ecosystem health, such as species assemblages and populations. Ideally, other earlier warning biodiversity indicators, such as populations of sensitive species, should be taken into account.

4.4. Recommandations for Model's Improvement

As highlighted by the present exercise, here are some recommendations for further use of the model for naturalness assessment:

Historical data should be accurate to avoid bias;
Historical data should include an estimation of the natural variability to improve the setting of the pni's evaluation curves;
Historical data acquisition must match the method used for current evaluation. In the case of methodology improvements—as seen with internal structure evaluation subject to recent technical evolution, or with the new method developed in Quebec's forest inventory for species group identification by 10% of basal area, which would have been appropriate to evaluate the current importance of the white spruce in the stands—the historical data should be accordingly reassessed if possible;
Improve NDP factors evaluation as these are used to evaluate model's most sensitive variables;
Improve the hypotheses used for projection of scenario components on a sustainable basis, considering that the model is applied over the whole landscape (not only on the most important ecological types), as some marginal types in terms of area, such as wetlands, could be important for biodiversity;
Improve the hypotheses used for natural evolution considering that some condition indicators will probably not recover their initial status as a result of past management practices (e.g., some protected area are created in areas which have been subject to harvest in the past) and climate change;
Explore the inclusion of variables characterizing landscape configuration and connectivity [43] in the landscape context.

Tests performed on *Picea* spp. proportion in plantations points to a limit of the model, when using a general indicator, such as the proportion of merchantable volume in spruce, instead of an indicator that would capture the subdominant status of the characteristic late successional companion species.

There is a certain level of uncertainty related to climate change and its potential effects on the natural disturbance regime, which we did not take into account in this study.

Naturalness evaluation should be extended to reach a scale more suitable for LCA. This could be done either by assessing other management units within the same bioclimatic domain [44] (which could also provide a better evaluation of the variability), or be performed at the scale of the bioclimatic sub-domain, depending on the scale of the available data.

Additional research efforts should be dedicated on expanding the assessment on ecosystem alteration levels beyond the sole forestry land use.

5. Conclusions

This study has applied a model assessing the impact on ecosystem quality of different forestry management practices through a naturalness index over two regions in Québec. The most sensitive variables for the current naturalness assessment correspond to the regeneration process, the cover type, the dead wood and the closed forest for both territories. The results show the capacity of the naturalness assessment model to perform bi-directional evaluation, assessing not only deterioration, but also improvement of ecosystem quality related to different enhanced ecological management strategies and restoration efforts. In the *Abies balsamea–Betula papyrifera* forest, scenarios that include irregular shelterwood combined with enhanced protection could play an important role in improving ecosystem quality, whereas scenarios applying short rotation (50 years) could lead to further deterioration. Provided that an enhanced protection level is assured, most management scenarios among those tested would produce a semi-natural environment. The model allows adequate forest management scenarios ranking within the different naturalness classes, leading to a finer characterization of the impact of forestry on ecosystem quality. However, the accuracy of historical reference data is important for a fine characterization of the impact, compared with the assessment of the current state of the forest. As most of the results lie in the semi-natural class with an enhanced level of protection, the effects on biodiversity could be mainly related to impacts on species assemblages and populations of species, but not necessarily leading to species losses. This still needs to be better studied.

6. Acronyms

ANT	Anthropization
ANT_NDP	Naturalness degradation potential from anthropization
CF	Close forests
CF_pni	partial naturalness for close forests
CL	careful clearcut logging
Compo	composition
Compo_PNI	Partial naturalness for composition
Context	landscape context
Context_PNI	Partial naturalness for landscape context
CS	Companion species
CS_NDP	Naturalness degradation potential related to companion species
CT	cover type
CT_pni	partial naturalness index for cover type
DW	dead wood
DW_NDP	Naturalness degradation potential related to dead wood
DW_PNI	Partial naturalness index for dead wood
ep	enhanced protection
exo	exotic species
exo_NDP	Naturalness degradation potential from exotic species
for_area	forest or forested area
HS	horizontal structure
HS_NDP	Naturalness degradation potential related to horizontal structure
ip	initial level of protection
IR	irregular stands

IR_pni	Partial naturalness index for irregular stands
ISC	irregular shelterwood cutting
LCA	Life cycle analysis
LS	Late successional characteristic species (i.e., *Picea* spp.)
LS_pni	Partial naturalness index for late sucessionnal characteristic species
NDP	Naturalness degradation potential
NE	natural evolution
NI	Naturalness Index
OF	Old forests
OF_pni	Partial naturalness index for old forests
PL	Plantation
PNI	Partial naturalness index for a given characteristic of naturalness (characteristic_PNI)
Pni	Partial naturalness index for a given condition indicator (condition_pni)
Prod_area	productive area
RP	regeneration process
RP_NDP	Naturalness degradation potential related to regeneration process
Struc	Structure
Struc_PNI	Partial naturalness index for structure
W_CC	Clearcuts on wetlands
W_CC_NDP	Naturalness degradation potential related to clearcuts on wetlands
Wm	Modified wetlands
Wm_NDP	Naturalness degradation potential related to modified wetlands

Supplementary Materials: The following is available online at http://www.mdpi.com/1999-4907/11/5/601/s1, Results: Scenario_nb_description: scenario numbering and description; ass_1 and Figure S1: Naturalness assessment for scenarios with enhanced protection (ep) using studies for reference data set; ass_2 and Figure S2: Naturalness assessment for scenarios with enhanced protection (ep) using registry as reference data set; ass_3 and Figure S3: Naturalness assessment for scenarios with initial protection (ip) using studies for reference data set; ass_4 and Figure S4: Naturalness assessment for scenarios with enhanced protection (ep) and PL50 using studies for reference data set; ass_5 and Figure S5: Naturalness assessment for scenarios with initial protection (ip) and PL50 using studies for reference data set; ass_6 and Figure S6: Naturalness assessment for scenarios with enhanced protection (ep) and PL50 with 90% of spruce at maturity using studies for reference data set; ass_7 and Figure S7: Naturalness assessment for scenarios with enhanced protection (ep) and alternate set of NDP factors using studies for reference data set; sensitivity_analysis: results of sensitivity analysis for scenario involving 100% of each practice tested.

Author Contributions: Conceptualization, S.C. and L.B.; methodology, S.C. and L.B.; validation, S.C.; formal analysis, S.C.; investigation, S.C.; resources, R.B.; data curation, S.C. and É.T. for dead wood.; writing—original draft preparation, S.C.; writing—review and editing, S.C., L.B., M.M., R.B. and É.T.; visualization, S.C.; supervision, R.B., L.B. and M.M.; project administration, R.B. and M.M.; funding acquisition, M.M. and E.T. All authors have read and agreed to the published version of the manuscript.

Funding: This research was funded by the Natural Sciences and Engineering Research Council of Canada (NSERC) through a grant to Manuele Margni, grant number CRD-462197-13, with the collaboration of Cecobois, Canadian Wood Council, Desjardins, GIGA, Hydro-Québec and Pomerleau. The first author received scholarships from: the Discovery Grant to E. Thiffault from NSERC (RGPIN-2018-05755); as well as from the Jean-Claude-and-Lisette-Mercier Fund.

Acknowledgments: The authors would like to thank Julie Bouliane from Montmorency Forest for providing data and information, Talagbé Gabin Akpo from Laval University, Mathematics and statistics department consulting service for his help with the data analysis, Stefano Biondo and his team form Laval University GeoStat Center for their help with maps data handling and Naïm Perrault for the localization map.

Conflicts of Interest: The authors declare no conflict of interest. The funders had no role in the design of the study; in the collection, analyses, or interpretation of data; in the writing of the manuscript, or in the decision to publish the results.

Appendix A. Model Adaptation to *Abies balsamea–Betula papyrifera* Domain

Appendix A.1. Test Area Description and Localization

The territory of the Montmorency Experimental Forest, located north of Quebec City and included in the *Abies balsamea–Betula papyrifera* domain, was used as a test area (Figure A1). Most of the current territory of the Experimental Forest was part of a forest concession allocated to the Anglo Canadian Pulp and Paper company in 1926. The territory has been subject to a first commercial harvest in the 1930s and 1940s. At the time of its creation in 1963, the experimental forest covered a total area of 66 km^2, now designated as FM-A. In 2014, an adjacent territory of 348 km^2, formally part of a public forest management unit with a forest company, has been added to the Experimental Research station and is now designated as FM-B. Therefore, between the 1960s and the beginning of 2010s, the two territories were subject to two different management scenarios. In FM-A, small-scale commercial harvest was continuous since the mid-1960s, while in FM-B, the territory has been subject to a second wave of harvest between 1985 and 2008. At the time of its creation, FM-A was poorly stocked as a result of the young age of the previously harvested stands. Therefore, the initial forest management plan for FM-A focused on the liquidation of the older stands as the harvest priority. Partial cutting was also performed on around 5.5% of the productive area in FM-A and 1.4% in FM-B. Plantations cover 6.8% and 8.8% of the productive area of FM-A and FM-B, respectively. Precommercial thinning has been performed over 9.2% of the productive area for FM-A compared to 23.4% for FM-B. At the time of its inclusion to the experimental station, most of FM-B had been recently harvested; parts of the remaining old forests were saved for protection.

The territory is subject to regular outbreaks of spruce budworm. Past management activities raised the following ecological issues in both territories: decrease of old forests, especially in FM-A, and overabundance of young forests, particularly in FM-B. Prior to the first harvest, stands were characterized by the presence of several old and large white spruce trees. The decrease of spruce in these stands resulting from the first commercial harvest has been described in an experimental design installed on the territory in the 1950s [45]. The remeasurement of some of these plots in the mid 1980s confirmed the scarcity of spruce among the pre-established regeneration of the second-growth forest; spruce regeneration was found only where adult individuals were present, related to residual small trees left at the time of harvest [23]. It also indicated a high risk of broadleaf invasion after clearcutting on the most fertile sites, as a result of the low abundance of coniferous seedlings combined with their small size [23,27].

Appendix A.2. Model Adaptation to Balsam Fir-White Birch Domain

For the *Abies balsamea–Betula papyrifera* domain, the five naturalness characteristics of the model were evaluated using the same indicators and variables used for the *Picea mariana*–feathermoss domain of the Quebec's boreal forest [1], except for composition (Table 1). In this case, the decreasing late successional species (LS) are the spruce species. In *Abies balsamea–Betula papyrifera* domain, black spruce (*Picea mariana*) can be dominant on humid stations, but white spruce is a companion species, and has not been quantified at the stand level in the historical data. The percentage of merchantable volume of spruce species for late successional characteristic species was used to assess its presence. However, this measure does not allow controlling for the companion status of the white spruce. As no other companion trees species presents signs of reduction and exotic species were never used in Montmorency Forest, the corresponding values for naturalness degradation potential from exotic species (exo_NDP) and naturalness degradation potential related to companion species diminution (CS_NDP) are set to 0. We kept these variables in the Forest Composition formula (Table 1), despite their null effect for the time being, making provision for future assessment.

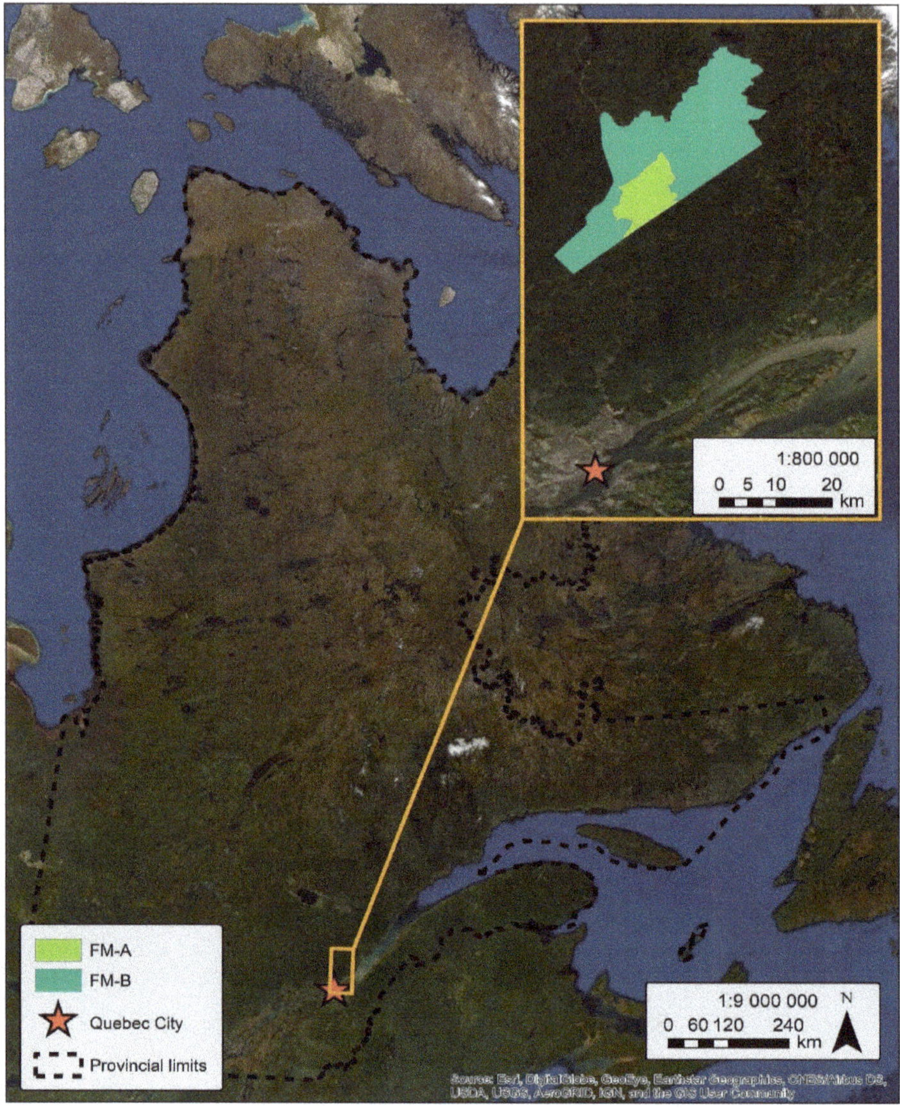

Figure A1. Montmorency Forest localization.

Appendix A.3. Determining Condition Indicator Curves

Assessing naturalness requires definition of a reference state of the condition indicators, i.e., the natural undisturbed habitat [10]. For Quebec's context, the model application requires pre-industrial data, which can be found in local studies or in the Quebec's reference state registry [21]. We tested both sources of reference data in order to analyze the sensitivity to the historical data set. Model adaptation involves resetting the curves used for partial naturalness index of condition indicators, corresponding to variables related to "pni" in lower cases in the Table 1 equations, using historical values considered valid for the territory. For the original evaluation, reference data used for naturalness assessment were drawn from studies performed in the vicinity of Montmorency Forest [18–20] (lines –s in Table A1).

However, in cases where historical information is not available, studies based on forest dynamic simulation, such as Quebec's reference state registry [21], can be used as an alternate source of historical data. The registry provides reference values for cover types, irregular stands, closed and old forests by homogenous vegetation regions. For the evaluation based on the registry, values for our two sectors were obtained by weighing by the area proportion in each vegetation unit and were used to reset the condition_pni curves accordingly. Compared with the reference data set from studies, the data set from the registry shows, for both territories, less of the coniferous cover type (below current coverage), but a more important coverage of closed forests, old forests and irregular stands (lines –r in Table A1). The spruce proportion was not evaluated in the registry data set; therefore, the same proportion used for the initial assessment was applied for the test using registry's data.

Data used for current naturalness evaluation were taken from the 2018 version of the Quebec eco-forest map [46], except for *Picea* spp. volume proportion [33].

Table A1. Historical values from studies (s) in Forest Montmorency vicinity or reference state registry (r) used as reference data, and current values from 2018 eco-forest map.

Territory	Cover Type (CT: % Forest Area of Coniferous Cover Type)	Late Successional Species (LS: % Merchantable Volume in *Picea* spp.)	Closed Forests (CF: % Terrestrial Area of Forests > 40 Years Old)	Old Forests (OF: % Forest Area of Forests > 80 Years Old)	Irregular Stands (IR: % Forest Area of Irregular Stands)
FM-A—historical-s	79.3 [1]	32 [2]	79.9 [1]	23.7 [1]	17.8 [1]
FM-A—historical-r	63.2	32 [2]	85.5	71.0	51.1
FM-A—current	79.1 [3]	16 [4]	39.1 [3]	4.9 [3]	14.2 [3]
FM-B—historical-s	85.7 [5]	39 [2]	76.19 [1]	57.9 [6]	40 [7]
FM-B—historical-r	77.8	39 [2]	90.5	81.1	64.4
FM-B—current	81.9 [3]	16	27.4 [3]	16.1 [3]	18.3 [3]

FM-A: Montmorency Forest sector a; FM-B: Montmorency Forest sector b; [1] [18]; [2] [20]; [3] Eco-forest map; [4] [33]; [5] Anglo's data in [18]; [6] [19]; [7] Donnacona's data in [18].

Curves used for the evaluation of partial naturalness index of condition indicators evaluation using local studies for reference data are presented in Figure A2 for FM-A and Figure A3 for FM-B. The curves were topped to one for irregular stands (IR), old forests (OF) and coniferous cover type (CT) for the following reasons. Descending curve past the historical value has not been used for IR, because of the evolving evaluation of stands with irregular structure related to improvements of technological identification capacities. For OF, the natural variability is important, as a result of spruce budworm epidemics, and the proportion used as the historical value is based on aerial photography taken around 20 years after an epidemic [18]; the ratio used as the historical value is therefore conservative and values over that ratio are considered natural. The cover type in the Forest Montmorency area was mainly coniferous, and the presence of deciduous cover was related to previous fires and showed an important variability [18]. As the proportion of coniferous cover observed in the ancient forest could reach 97.8% [18], the curve used for pni_CT evaluation was also topped to one. This could theoretically hinder a diagnosis of broadleaf species decrease; however, as no such issue has been identified for that area and no hypothesis used for scenario evaluation goes beyond this proportion, an adjustment for the curve end was not considered necessary. Nevertheless, this situation indicates that it could be appropriate to set both ends of the natural class, considering the natural variability range instead of using a proportion of a unique value. For late successional species (LS), we used the percentage of merchantable volume of spruce species, and descending the curve past the historical value was applied (Figures A2b and A3b), as too much spruce in the landscape would be outside of the natural range of variability.

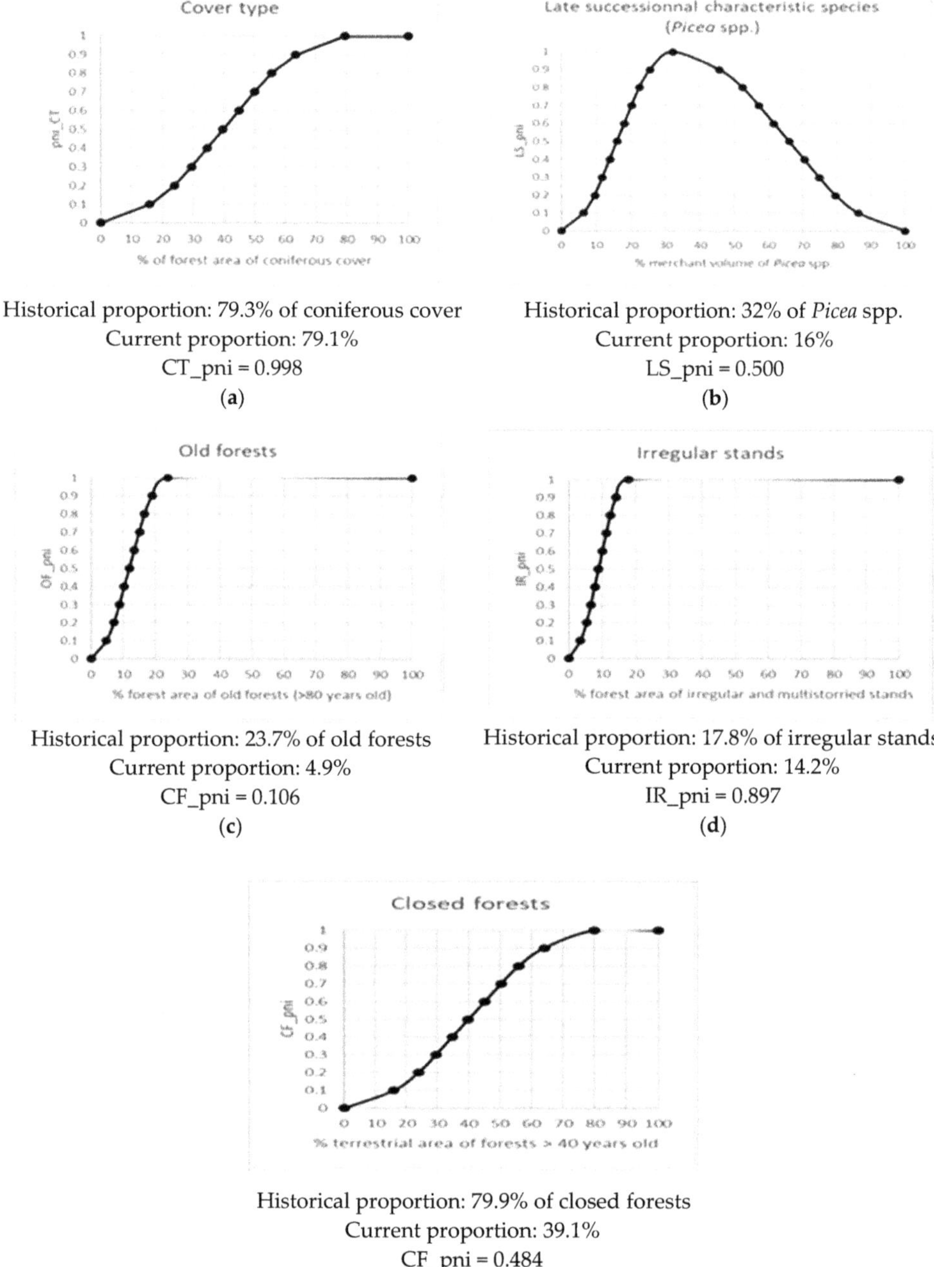

Historical proportion: 79.3% of coniferous cover
Current proportion: 79.1%
CT_pni = 0.998
(**a**)

Historical proportion: 32% of *Picea* spp.
Current proportion: 16%
LS_pni = 0.500
(**b**)

Historical proportion: 23.7% of old forests
Current proportion: 4.9%
CF_pni = 0.106
(**c**)

Historical proportion: 17.8% of irregular stands
Current proportion: 14.2%
IR_pni = 0.897
(**d**)

Historical proportion: 79.9% of closed forests
Current proportion: 39.1%
CF_pni = 0.484
(**e**)

Figure A2. Curves determining the potential naturalness index to evaluate the condition indicators (condition_pni) for FM-A using local studies for reference data: (**a**) coniferous cover type; (**b**) late successional characteristic species; (**c**) old forests; (**d**) irregular stands; (**e**) closed forests.

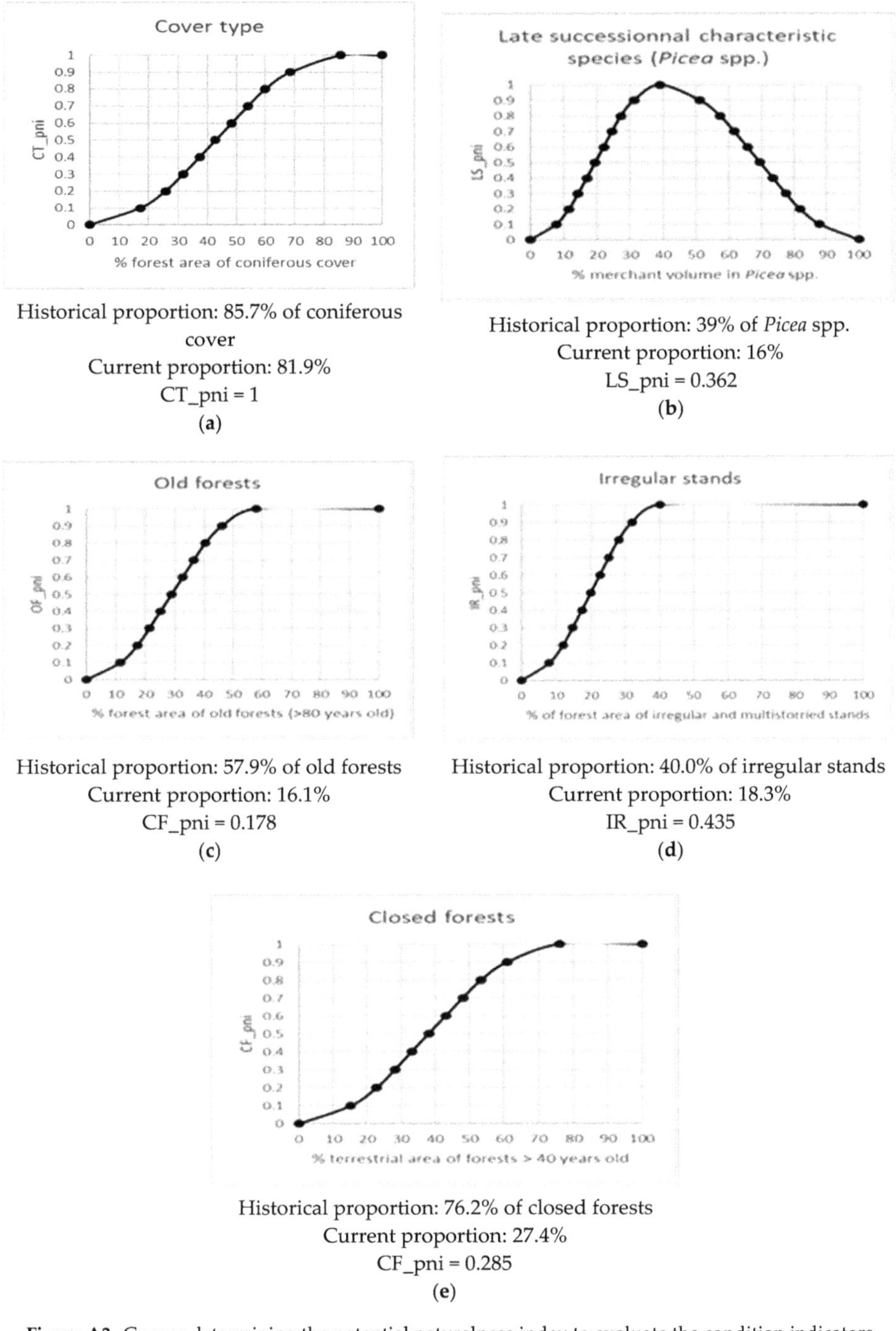

Historical proportion: 85.7% of coniferous cover
Current proportion: 81.9%
CT_pni = 1

(**a**)

Historical proportion: 39% of *Picea* spp.
Current proportion: 16%
LS_pni = 0.362

(**b**)

Historical proportion: 57.9% of old forests
Current proportion: 16.1%
CF_pni = 0.178

(**c**)

Historical proportion: 40.0% of irregular stands
Current proportion: 18.3%
IR_pni = 0.435

(**d**)

Historical proportion: 76.2% of closed forests
Current proportion: 27.4%
CF_pni = 0.285

(**e**)

Figure A3. Curves determining the potential naturalness index to evaluate the condition indicators (condition_pni) for FM-B using local studies for reference data: (**a**) coniferous cover type; (**b**) late successional characteristic species; (**c**) old forests; (**d**) irregular stands; (**e**) closed forests.

Appendix A.4. Determining the Naturalness Degradation Potentials

The evaluation of naturalness degradation potentials (NDP) was adapted by resetting the NDP factors related to practices. NDP calculation for horizontal structure, dead wood and regeneration process are presented in Tables A2–A4, respectively, for each territory. For careful logging effects, a distinction has been made between a CL performed in a 50-year-old stand (CL50) and a 70-year-old stand (CL70). Stands resulting from logging performed in 50-year-old stands were identified using an overlay of the eco-forest map and the SIFORT1 map, a tessellation of provincial forest inventory maps for the first measure, considering clear-cuts or careful logging performed between 1981 and 2002, located in sectors having a development stage designed as "young" at the time of the first forest inventory.

The others NDP factors evaluated based on current values and kept constant for all scenarios are showed in Table A5.

Table A2. Naturalness degradation potential for horizontal structure (HS_NDP) by silvicultural treatment in Montmorency Forest.

Territory:		FM-A		FM-B	
Practice	NDP Factors	% Forest_Area	NDPx	% Forest_Area	NDPx
Plantation – thinning	1	2.50%	0.0250	5.58%	0.0558
Plantation	0.9	4.32%	0.0388	3.21%	0.0289
Thinning (natural), strip cutting	0.8	1.45%	0.0116	0.17%	0.0014
Precom. thinning (natural), release	0.75	9.24%	0.0693	23.40%	0.1755
CL50	0.5	7.01%	0.0351	13.28%	0.0664
CL70	0.3	48.34%	0.1451	13.17%	0.0395
Partial cutting	0.2	5.48%	0.0110	1.38%	0.0028
Undisturbed or natural disturbances	0	21.64%	0.0000	39.80%	0.0000
Current HS_NDP			**0.3359**		**0.3702**

Note: NDP_factors: naturalness degradation potential factors related to practices; %for_area: percentage of forested area; NDPx: portion of the naturalness degradation potential for the xth practice; CL50: careful logging (CL) and clearcut of 50-year-old stands; CL70: careful logging (CL) and clearcut of 70-year-old stands.

Table A3. Naturalness degradation potential for dead wood (DW_NDP) by silvicultural treatment in Montmorency Forest.

Territory:		FM-A		FM-B	
Practice	NDP Factors	% Forest_Area	NDPx	% Forest_Area	NDPx
Biomass harvesting	1	0.00%	0.0000	0	0.0000
Plantation + thinnings	0.95	2.50%	0.0237	5.58%	0.0530
Plantation – no thinnings	0.85	5.76%	0.0490	3.39%	0.0288
Partial cutting and precom. thinnings	0.8	14.72%	0.1178	24.78%	0.1983
CL50	0.7	7.01%	0.0491	13.28%	0.0930
CL70	0.55	48.35%	0.2660	13.17%	0.0725
Undisturbed or natural disturbances	0	21.64%	0.0000	39.80%	0.0000
DW_NDP			**0.5056**		**0.4454**
Current *DW_PNI*			**0.4944**		**0.5546**

Note: NDP_factors: naturalness degradation potential factors related to practices; %for_area: percentage of forested area; NDPx: portion of the naturalness degradation potential for the xth practice; CL50: careful logging (CL) and clearcut of 50 years old stands; CL70: careful logging (CL) and clearcut of 70-year-old stands.

Table A4. Naturalness degradation potential for regeneration process (RP_NDP) by silvicultural treatment in Montmorency Forest.

Territory:		FM-A		FM-B	
Practice	NDP Factors	% Forest_Area	NDPx	% Forest_Area	NDPx
Exotic plantations, afforestation	1	0.00%	0.0000	0.00%	0.0000
Plantation	0.9	6.82%	0.0613	8.79%	0.0791
Seeding	0.7	2.78%	0.0195	0.00%	0.0000
Precommercial thinning	0.65	9.24%	0.0601	23.40%	0.1521
In-fill planting	0.6	2.48%	0.0149	0.00%	0.0000
CL50	0.5	4.53%	0.0226	13.28%	0.0664
Commercial thinning (natural)	0.4	1.45%	0.0058	0.17%	0.0007
CL70	0.35	45.58%	0.1595	13.17%	0.0461
Partial cut	0.2	5.48%	0.0110	1.38%	0.0028
Undisturbed or natural disturbances	0	21.64%	0.0000	39.80%	0.0000
RP_NDP			**0.3547**		**0.3472**
Current *RP_PNI*			**0.6453**		**0.6528**

Note: NDP_factors: naturalness degradation potential factors related to practices; %for_area: percentage of forested area; NDPx: portion of the naturalness degradation potential for the xth practice; CL50: careful logging (CL) and clearcut of 50-year-old stands; CL70: careful logging (CL) and clearcut of 70-year-old stands.

Table A5. Naturalness degradation potential for companion species (CS_NDP), exotic species (exo_NDP), wetlands with clear cuts (W_CC_NDP) and anthropization (ANT_NDP) in Montmorency Forest.

Territory:	FM-A		FM-B	
Item	% Area [1]	NDPx	% Area [1]	NDPx
Companion species	0.00%	0.0000	0.00%	0.0000
Exotic species	0.00%	0.0000	0.00%	0.0000
Wetlands with clear cuts	42.37%	0.2118	2.66%	0.0133
Anthropization	1.72%	0.0172	2.68%	0.0268

[1] % of forested area for companion species and exotic species; % of terrestrial area for wetlands with clear cuts and anthropization.

References

1. Côté, S.; Bélanger, L.; Beauregard, R.; Thiffault, É.; Margni, M. A Conceptual Model for Forest Naturalness Assessment and Application in Quebec's Boreal Forest. *Forests* **2019**, *10*, 325. [CrossRef]
2. Life Cycle Initiative. *Global Guidance for Life Cycle Impact Assessment Indicators*; UNEP DTIE: Paris, France, 2016; Volume 1, p. 159.
3. Curran, M.; De Souza, D.M.; Anton, A.; Teixeira, R.F.M.; Michelsen, O.; Vidal-Legaz, B.; Sala, S.; Mil, I.; Canals, L. How Well Does LCA Model Land Use Impacts on Biodiversity?—A Comparison with Approaches from Ecology and Conservation. *Environ. Sci. Technol.* **2016**, *50*, 2782–2795. [CrossRef] [PubMed]
4. Gaudreault, C.; Wigley, T.B.; Margni, M.; Verschuyl, J.; Vice, K.; Titus, B. Addressing Biodiversity Impacts of Land Use in Life Cycle Assessment of Forest Biomass Harvesting. *Wiley Interdiscip. Rev. Energy Environ.* **2016**. [CrossRef]
5. Souza, D.M.; Teixeira, R.F.; Ostermann, O.P. Assessing biodiversity loss due to land use with Life Cycle Assessment: Are we there yet? *Glob. Chang. Biol.* **2015**, *21*, 32–47. [CrossRef]
6. Gossner, M.M.; Schall, P.; Ammer, C.; Ammer, U.; Engel, K.; Schubert, H.; Simon, U.; Utschick, H.; Weisser, W.W. Forest management intensity measures as alternative to stand properties for quantifying effects on biodiversity. *Ecosphere* **2014**, *5*, 1–111. [CrossRef]
7. Gabel, V.M.; Meier, M.S.; Kopke, U.; Stolze, M. The challenges of including impacts on biodiversity in agricultural life cycle assessments. *J. Environ. Manag.* **2016**, *181*, 249–260. [CrossRef] [PubMed]
8. Chaudhary, A.; Brooks, T.M. Land Use Intensity-specific Global Characterization Factors to Assess Product Biodiversity Footprints. *Environ. Sci. Technol.* **2018**, *52*, 5094–5104. [CrossRef]

9. Jolliet, O.; Antón, A.; Boulay, A.-M.; Cherubini, F.; Fantke, P.; Levasseur, A.; McKone, T.E.; Michelsen, O.; Milà i Canals, L.; Motoshita, M.; et al. Global guidance on environmental life cycle impact assessment indicators: Impacts of climate change, fine particulate matter formation, water consumption and land use. *Int. J. Life Cycle Assess.* **2018**. [CrossRef]

10. Winter, S. Forest naturalness assessment as a component of biodiversity monitoring and conservation management. *Forestry* **2012**, *85*, 293–304. [CrossRef]

11. Hekkala, A.-M. Restoration of the Naturalness of Boreal Forests. Ph.D. Thesis, University of Oulu, Oulu, Finland, 2015.

12. Kouki, J.; Hyvärinen, E.; Lappalainen, H.; Martikainen, P.; Similä, M. Landscape context affects the success of habitat restoration: Large-scale colonization patterns of saproxylic and fire-associated species in boreal forests. *Divers. Distrib.* **2012**, *18*, 348–355. [CrossRef]

13. Brentrup, F.; Küsters, J.; Lammel, J.; Kuhlmann, H. Life Cycle Impact Assessment of Land Use Based on the Hemeroby Concept. *Int. J. Life Cycle Assess.* **2002**, *7*, 339–348. [CrossRef]

14. Michelsen, O. Assessment of land use impact on biodiversity. *Int. J. Life Cycle Assess.* **2008**, *13*, 22–31. [CrossRef]

15. Fehrenbach, H.; Grahl, B.; Giegrich, J.; Busch, M. Hemeroby as an impact category indicator for the integration of land use into life cycle (impact) assessment. *Int. J. Life Cycle Assess.* **2015**, *20*, 1511–1527. [CrossRef]

16. Rossi, V.; Lehesvirta, T.; Schenker, U.; Lundquist, L.; Koski, O.; Taylor, R.; Humbert, S. Capturing the potential biodiversity effects of forestry practices in life cycle assessment. *Int. J. Life Cycle Assess.* **2017**. [CrossRef]

17. Farmery, A.K.; Jennings, S.; Gardner, C.; Watson, R.A.; Green, B.S. Naturalness as a basis for incorporating marine biodiversity into life cycle assessment of seafood. *Int. J. Life Cycle Assess.* **2017**, *22*, 1571–1587. [CrossRef]

18. Leblanc, M.; Bélanger, L. *La Sapinière Vierge de la Forêt Montmorency et de sa Région: Une Forêt Boréale Distincte*; Gouvernement du Québec: Québec, QC, Cannada, 2000; p. 91.

19. Boucher, Y.; Grondin, P. Impact of logging and natural stand-replacing disturbances on high-elevation boreal landscape dynamics (1950–2005) in eastern Canada. *For. Ecol. Manag.* **2012**, *263*, 229–239. [CrossRef]

20. Bouliane, J.; Bélanger, L.; Sansregret, H.; Pineault, P. *Plan D'aménagement Forestier Intégré Tactique 2014–2019*; Forêt Montmorency. Faculté de foresterie, de géographie et de géomatique, Université Lava; Presses de l'Université Laval: Quebec, QC, Canada, 2014; p. 200.

21. Boucher, Y.; Bouchard, M.; Grondin, P.; Tardif, P. *Le Registre des États de Référence: Intégration des Connaissances sur la Structure, la Composition et la Dynamique des Paysages Forestiers Naturels du Québec Méridional*; Mémoire de recherche forestière No 161; Gouvernement du Québec, Ministère des Ressources Naturelles et de la Faune, Direction de la Recherche Forestière: Québec, QC, Canada, 2011; p. 21.

22. MFFP. La Tordeuse des Bourgeons de L'épinette. Available online: https://mffp.gouv.qc.ca/forets/fimaq/insectes/fimaq-insectes-insectes-tordeuse.jsp (accessed on 24 October 2019).

23. Côté, S.; Bélanger, L. Variations de la régénération préétablie dans les sapinières boréales en fonction de leurs caractéristiques écologiques. *J. Can. De La Rech. For.* **1991**, *21*, 1779–1795. [CrossRef]

24. Thiffault, E. *Data from Montmorency Forest*; Université Laval: Québec, QC, Canada, 2019.

25. Grondin, P.; Cimon, A. *Les Enjeux de Biodiversité Relatifs à la Composition Forestière*; Ministère des Ressources Naturelles, de la Faune et des Parcs: Québec, QC, Canada, 2003; p. 200.

26. Harvey, B.; Brais, S. Effects of mechanized careful logging on natural regeneration and vegetation competition in the southeastern Canadian boreal forest. *Can. J. For. Res.* **2002**, *32*, 653–666. [CrossRef]

27. Déry, S.; Bélanger, L.; Marchand, S.; Côté, S. Succession après épidémie de la tordeuse des bourgeons de l'épinette (*Choristoneura fumiferana*) dans des sapinières boréales pluviales de seconde venue. *Can. J. For. Res.* **2000**, *30*, 801–816. [CrossRef]

28. Raymond, P.; Bédard, S.; Roy, V.; Larouche, C.; Tremblay, S. The Irregular Shelterwood System: Review, Classification, and Potential Application to Forests Affected by Partial Disturbances. *J. For.* **2009**, *107*, 405–413. [CrossRef]

29. Bouchard, C.; Frédéric, C.; Langlais, D.; Leguerrier, J.; Lessard, F.; Venne, F. *Réserve aquatique de la Forêt Montmorency*; Rapport remis au Comité scientifique et d'aménagement de la Forêt Montmorency; Université Laval: Québec, QC, Canada, 2017.

30. Paradis, L.; Thiffault, E.; Achim, A. Comparison of carbon balance and climate change mitigation potential of forest management strategies in the boreal forest of Quebec (Canada). *For. Int. J. For. Res.* **2019**, *92*, 264–277. [CrossRef]

31. Senez-Gagnon, F.; Thiffault, E.; Paré, D.; Achim, A.; Bergeron, Y. Dynamics of detrital carbon pools following harvesting of a humid eastern Canadian balsam fir boreal forest. *For. Ecol. Manag.* **2018**, *430*, 33–42. [CrossRef]

32. Raymond, P.; Ruel, J.-C.; Pineau, M. Effet d'une coupe d'ensemencement et du milieu de germination sur la régénération des sapinières boréales riches de seconde venue du Québec. *For. Chron.* **2000**, *76*, 643–652. [CrossRef]

33. Urli, M.; Thiffault, N.; Barrette, M.; Bélanger, L.; Leduc, A.; Chalifour, D. Key ecosystem attributes and productivity of boreal stands 20 years after the onset of silviculture scenarios of increasing intensity. *For. Ecol. Manag.* **2017**, *389*, 404–416. [CrossRef]

34. R Core Team. *R 3.6.1: A Language and Environment for Statistical Computing*; R Foundation for Statistical Computing: Vienna, Austria. Available online: https://www.r-project.org/ (accessed on 24 August 2019).

35. Milà i Canals, L.; Bauer, C.; Depestele, J.; Dubreuil, A.; Freiermuth Knuchel, R.; Gaillard, G.; Michelsen, O.; Müller-Wenk, R.; Rydgren, B. Key Elements in a Framework for Land Use Impact Assessment Within LCA (11 pp). *Int. J. Life Cycle Assess.* **2007**, *12*, 5–15. [CrossRef]

36. Morin, H.; Laprise, D.; Simard, A.-A.; Amouch, S. Régime des épidémies de la tordeuse des bourgeons de l'Épinette dans l'Est de l'Amérique du Nord. In *Aménagement Écosystémique en Forêt Boréale*; Gauthier, S., Ed.; Presses de l'Université du Québec: Québec, QC, Canada, 2008; pp. 167–192.

37. Milài Canals, L.; Michelsen, O.; Teixeira, R.F.; Souza, D.M.; Curran, M.; Anton, A. Building consensus for assessing land use impacts on biodiversity in LCA. In Proceedings of the 9th International Conference on Life Cycle Assessment in the Agri-food Sector (LCA Food 2014), San Francisco, CA, USA, 8–10 October 2014; ACLCA: Vashon, WA, USA; pp. 823–834.

38. MFFP. Espèce Faunique Menacée—Caribou des Bois. Available online: https://www3.mffp.gouv.qc.ca/faune/especes/menacees/fiche.asp?noEsp=53 (accessed on 16 September 2019).

39. Leblond, M.; Dussault, C.; Ouellet, J.-P.; St-Laurent, M.-H. Caribou avoiding wolves face increased predation by bears—Caught between Scylla and Charybdis. *J. Appl. Ecol.* **2016**, *53*, 1078–1087. [CrossRef]

40. Kawaguchi, T.; Desrochers, A. A time-lagged effect of conspecific density on habitat selection by snowshoe hare. *PLoS ONE* **2018**, *13*, e0190643. [CrossRef]

41. Dornelas, M.; Gotelli, N.J.; McGill, B.; Shimadzu, H.; Moyes, F.; Sievers, C.; Magurran, A.E. Assemblage time series reveal biodiversity change but not systematic loss. *Science (N. Y.)* **2014**, *344*, 296. [CrossRef]

42. Winter, L.; Lehmann, A.; Finogenova, N.; Finkbeiner, M. Including biodiversity in life cycle assessment—State of the art, gaps and research needs. *Environ. Impact Assess. Rev.* **2017**, *67*, 88–100. [CrossRef]

43. Kuipers, K.J.J.; May, R.F.; Graae, B.J.; Verones, F. Reviewing the potential for including habitat fragmentation to improve life cycle impact assessments for land use impacts on biodiversity. *Int. J. Life Cycle Assess.* **2019**. [CrossRef]

44. MFFP. Système Hiérarchique de Classification Écologique du Territoire. Available online: https://mffp.gouv.qc.ca/forets/inventaire/inventaire-systeme.jsp (accessed on 24 May 2020).

45. Hatcher, R.J. *Croissance du Sapin Baumier Après une Coupe Rase Dans le Québec*; Sylvicoles, D.d.r., Ed.; Ministère du Nord Canadien et des Ressources Nationales du Canada: Chicoutimi, QC, Canada, 1960; p. 24.

46. MFFP. QUEBEC'S ECOFOREST Map with Disturbances. Available online: https://www.donneesquebec.ca/recherche/fr/dataset/carte-ecoforestiere-avec-perturbations (accessed on 24 August 2018).

Article

Biodiversity of the Cocoa Agroforests of the Bengamisa-Yangambi Forest Landscape in the Democratic Republic of the Congo (DRC)

Germain Batsi [1,*], Denis Jean Sonwa [2], Lisette Mangaza [3], Jérôme Ebuy [1] and Jean-Marie Kahindo [4]

[1] Faculty of Renewable Natural Resources Management, The University of Kisangani, Avenue Kitima 5110400, PO Box 2012 Kisangani, Democratic Republic of Congo; jebuyalip5@gmail.com
[2] Center for International Forestry Research (CIFOR), Yaoundé PO Box 2008 Messa, Cameroon; d.sonwa@cgiar.org
[3] Faculty of Science, University of Goma, Avenue de la Paix 6110103, PO Box 204 Goma, Democratic Republic of Congo; lisemangaza@gmail.com
[4] Faculty of Science, University of Kisangani, Avenue Kitima 5110400, PO Box 2012 Kisangani, Democratic Republic of Congo; jkahindo2@yahoo.fr
* Correspondence: germainbatsi@gmail.com

Received: 9 August 2020; Accepted: 25 September 2020; Published: 15 October 2020

Abstract: Cocoa agroforestry has evolved into an accepted natural resource conservation strategy in the tropics. It is regularly proposed as one of the main uses for REDD+ projects (Reducing Emissions from Deforestation and forest Degradation and the role of conservation, sustainable management of forests, and enhancement of forest carbon stocks in developing countries) in the Democratic Republic of the Congo. However, few studies have characterized the cocoa agroforestry systems in this country. Hence, this research proposes to determine the impact of distance from Kisangani (the unique city in the landscape) and land-use intensity on the floristic composition of cocoa agroforests in Bengamisa-Yangambi forest landscape in the Congo Basin. The results revealed that species diversity and density of plants associated with cocoa are influenced by the distance from Kisangani (the main city in the landscape and province). Farmers maintain/introduce trees that play one or more of several roles. They may host caterpillars, provide food, medicine, or timber, or deliver other functions such as providing shade to the cocoa tree. Farmers maintain plants with edible products (mainly oil palms) in their agroforests more than other plants. Thus, these agroforests play key roles in conserving the floristic diversity of degraded areas. As cocoa agroforestry has greater potential for production, biodiversity conservation, and environmental protection, it should be used to slow down or even stop deforestation and forest degradation.

Keywords: floristic diversity; cocoa agroforests; Bengamisa-Yangambi; landscape; Democratic Republic of the Congo

---

1. Introduction

Land-use activities, such as clearing tropical forests, practicing subsistence agriculture, and intensifying farmland production, are the most important drivers of biodiversity loss and the associated ecosystem services on the local- and landscape-scale [1,2]. Although the rate of tropical forest loss is alarming, some agricultural systems offer a glimmer of hope. Systems using shade species offer greater potential for production, biodiversity conservation, and long-term environmental protection [3]. Thus, agroforestry is proposed as one of the strategies for conserving natural resources in the tropics [4]. The agroforestry practice provides a potentially valuable conservation tool that can be useful for reducing land-use pressure and enhancing income from rural livelihoods in tropical countries [5].

Several examples across the tropics have shown that agroforests represent a substantial proportion of biodiversity of forest reserves [3,6,7].

Cocoa is one of the most important crops in agroforests [8]. Cocoa farming has played an important role in the conservation of lowland tropical forest landscapes in Latin America, Africa and Asia over the past centuries and continues to do so today [9]. It helps to reduce land-use pressure through the availability of useful tree species and other non-timber forest products (NTFPs) and to improve rural livelihoods [10]. Shade tree systems provide habitat diversity for plant and animal species that do not strictly depend on natural forest. They also connect otherwise disjunctive fragments of the remaining forest patches in the landscape [11]. Cocoa agroforests systems with a mixture of diverse tree species provide more functions than other land-uses in forest landscapes. These functions include maintenance of carbon stocks, biodiversity conservation, and locally relevant ecosystem services, such as protection of the soil [12] and better water management [13]. The introduction of sustainable shade tree management can make cocoa agroforestry an important agent of reforestation [14]. Shade tree management has positive effects on pest outbreaks that may hold the key to breaking cocoa production cycles and helping conserve valuable tropical biodiversity in agroforestry systems [15,16].

Despite the benefits of agroforestry practices, cocoa agroforestry in natural habitats is an important driver of forest degradation and deforestation [14,17]. Land-use planning is needed to reduce further deforestation for the expansion of cocoa land. Such planning can determine areas to preserve under forest cover for ecological reasons and also areas where cocoa might be planted [18].

Introduced in Africa more than a century ago, cocoa production is a major contributor to the economies of many African countries [19]. The continent supplies more than two-thirds of the world's cocoa, the majority being produced by Côte d'Ivoire and Ghana [20]. Full-sun systems are found mostly in the Lower Guinean forest systems of Liberia, Ghana, Côte d'Ivoire and Nigeria; the more complex systems are in the Congo Basin countries, mainly in Cameroon and the Democratic Republic of the Congo (DRC) [21]. As cocoa production increases in DRC, the sector can learn a lot from its West African neighbours about mistakes to avoid and priorities to emphasize, including the importance of sustainable and climate-smart practices and good governance [22]. In DRC, cocoa agroforests are frequently proposed in projects for REDD+ (reducing emissions from deforestation and forest degradation and the role of conservation, sustainable management of forests, and enhancement of forest carbon stocks in developing countries). The Wildlife Conservation Society, for example, has adopted the idea of growing cocoa as a tool in forest conservation. In the Mambasa region, 1250 hectares (ha) of forest-cover cocoa has planted within the context of the national REDD+ program [23].

In DRC, conflicts and political instability have deeply affected the agricultural sector. For example, there are insufficient data on initiatives in the cocoa sector [20]. Cocoa is generally cultivated by small farmers, most often alongside other crops [22]. However, the quantities produced are small in comparison to West Africa [24]. On the other hand, in Tshopo province, cocoa has been promoted since the colonial era when it was planted under controlled forest cover from which cocoa pest species had been eliminated [23]. However, cocoa cultivation has never been fully developed. Today, most farmers spontaneously plant cocoa in the region by sourcing from former plantations in response to market signals and rumors of market development in the east of DRC [25]. With adequate assistance, cocoa can be produced sustainably without clearing new forest land and can help reduce household poverty [21]. Further efforts are needed to rehabilitate existing cocoa farms to develop sustainable cocoa agroforestry [25].

In cocoa agroforests, species richness and vegetation structure are key components of structural complexity and form the basis of biodiversity [26]. Therefore, good knowledge is needed of the plants associated with cocoa trees in cocoa agroforests in DRC. However, no study on the contributions of cocoa agroforests to the conservation of floristic diversity has focused on DRC.

The choice of Tshopo province, more specifically the Bengamisa-Yangambi landscape, is based on its inclusion of the Yangambi Biosphere Reserve. The United Nations Educational, Scientific and Cultural Organization (UNESCO) declared Yangambi a biosphere reserve in 1977 [27]. Thus, to preserve the important biodiversity of the Yangambi reserve, agroforestry systems in the Bengamisa-Yangambi forest landscape should be studied to understand their influence on the conservation of floristic diversity.

This study paid special attention to how distance from Kisangani (The main city in the landscape) and thus the related disturbance and land-use intensity have affected the floristic composition of cocoa agroforests in the forest landscape. The study is based on the hypothesis that native forest cover, disturbance, land-use intensity, and market access influenced floristic composition of cocoa agroforests. We thus recorded floristic composition of cocoa agroforest in four zones of 15 km each, defined along the main road, from Kisangani city to Yangambi forest reserve landscape, via Bengamisa village.

2. Materials and Methods

2.1. Study Area

This study was conducted between June and July 2018 in 16 villages and recorded floristic composition of cocoa agroforest in 4 zones of 15 km each, defined along the main road, from Kisangani city to Yangambi forest reserve landscape, via Bengamisa village (Figure 1). This area is located between the city of Kisangani (N 00°31′; E 25°11′) [28] and the Yangambi Biosphere Reserve (N 00°48′; E 24°29′) [29] in Banalia territory, Tshopo province, DRC. The Tshopo province, with an area of 200,240 km^2, is in the northeast of DRC and includes seven territories (Bafwasende, Banalia, Basoko, Isangi, Opala, Ubundu and Yahuma) [28,30]. Kisangani, founded in 1883, is the capital of the Tshopo province and is the main city of this province. Its population is estimated at about 1 million (around 20% of the 5,032,472 inhabitants of the province) [28,31]. It is among the five main cities of DRC. Kisangani have an international airport, is the end of the water navigation road on the Congo river and it is crossed by 4 main roads leading to different directions in the country. It has a road connection with the other east part of the DRC and bordering countries such as Uganda. Several smallholders' industries and shops are thus found in the city. In the Bengamisa-Yangambi forest landscape, it is the main city with his urbanisation impacting the forest landscape between the city and the Yangambi Biosphere reserve.

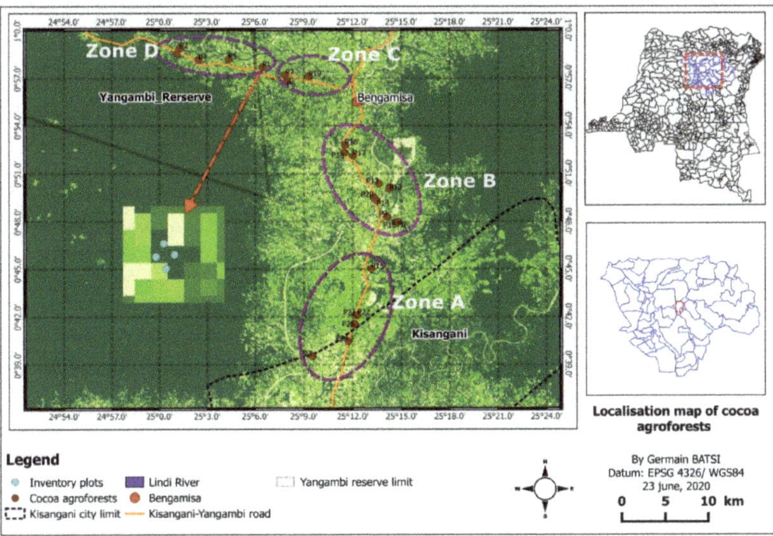

Figure 1. Localization of cocoa agroforests and inventory plots in the Bengamisa-Yangambi-landscape.

The region is still predominantly covered with moist forest and has an average population density of 9.8 people per square kilometer (km^2). The region has a hot and humid climate without a marked dry season classified as Af in Köppen's typology [28,30]. It receives an annual precipitation of 1839.5 ± 205.7 millimeters (mm) with an average sub-dry season length of 3.3 ± 1.3 months with monthly precipitation lower than 100 mm, during December–February. Temperatures are high and constant throughout the year with a minimum of 24.2 ± 0.4 °C in July and a maximum of 25.5 ± 0.6 °C in March [29].

The main economic activity in the region is agriculture, which is practiced in a shifting cultivation system [28,31,32]. Cassava is the main staple crop. Furthermore, exploitation of NTFPs, such as bush meat, caterpillars, and wild edible plants, as well as commercial and artisanal logging, artisanal mining, and petty trade provide sources of income to rural households. Six major road axes cut through the forest in a radial pattern starting from Kisangani, along which nearly all settlements are located [31].

In this region, which has excellent agricultural and climatic conditions, cocoa cultivation has been encouraged since colonial times. In 1979, the African Development Bank (AfDB) financed the development of 1750 ha of small peasant and commercial plantations in Bengamisa called CABEN (Cacaoyères de Bengambisa) to increase Congolese cocoa production. CABEN did not achieve its objectives and most of these plantations were abandoned [23].

2.2. Methods

2.2.1. Study Design

To better understand the role of native forest cover, disturbance and land-use intensity on floristic diversity of cocoa agroforests in the region, 4 Blocs/Zones of 15 km each were defined along the main road, including Zone A (18–33 km from Kisangani city), Zone B (34–49 km from Kisangani city), Zone C (50–65 km from Kisangani city), and Zone D (66–81 km from Kisangani city). These zones are arranged from the highly degraded area nearest Kisangani city (Zone A) to the less degraded area in the forest zone (Zone D) (Figure 1). The Bengamisa-Yangambi landscape is still covered by a vast rainforest [33]. The lowest proportion of this forest is found around Kisangani where we have the nucleus of population pressure [34]. The agricultural system is characterized by slash-and-burn agriculture, resulting from a shifting patchwork of cultivated fields, fallows, secondary forests, and remnants of primary forests [33]. Cash crops such as oil palm and cocoa have been cultivated by small farmers since several decades [23,35]. The forest of this landscape is gradually being converted to agricultural land, roads or modified by timber and charcoal exploitation [27,34]. As with other forests stands in DRC [36], the Bengamisa-Yangambi landscape is thus under the agricultural extension, fuelwood collection, timber exploitation, urbanisation, and demography increase pressures.

2.2.2. Collection of Floristic Data

The criteria used to select agroforestry plantations are the net area of the plantation (at least 0.5 hectares) [37]. Four plots of 25 × 25 m (i.e., 625 m^2 for each plot), corresponding to an area of 2500 m^2 were established within each cocoa agroforest to record plant diversity. In small plantations, the plots were installed successively, spaced 3–5 m apart. In large plantations, the plots were arranged on a diagonal to represent the diversity of the cocoa farm. The diameter was measured at 30 cm from the ground on each cocoa tree and at 1.30 m on the cocoa-associated plants as noted by [38]. A total of 25 cocoa agroforests were surveyed in the Bengamisa-Yangambi forest landscape.

In each cocoa agroforest, and within each plot (25 × 25 m), the number of cocoa trees were counted. Each associated plant with a diameter at breast height (DBH) above/equal to 2.5 centimeter (cm) was recorded. The scientific name of the plant was provided using the Catalogue—Flora of Plants of Kisangani and Tshopo Districts [39]. The main use of each plant is noted. In this study, "Cocoa associated species" refer to all plants (spontaneous or planted) except cocoa trees present in cocoa agroforests. Thus, for their classification, we used the same approach (nature and main use of plants) of previous studies on biodiversity in cocoa agroforests [40–42]. All cocoa-associated plant species inventoried were later sorted according to the list of suitable/useful and unsuitable species to cocoa trees established by the National Institute of Agronomic Studies and Research (INERA). Usefulness/suitability criteria used by INERA include production increase, providing light and cocoa-friendly shade, and protection against pests and diseases. The non-suitability of the plant was generally due to pest criteria, including cocoa tree pest housing, competition with cocoa trees, intense shade, and same diseases with cocoa trees. Such criteria were already been used by many cocoa research/extension services in west and central Africa [7].

2.2.3. Data Analysis

Preliminary analysis of cocoa agroforests help to determine three following models of cocoa agroforests: Cocoa agroforests in which companion plants (associated with cocoa trees) were composed primarily of forest/native species (residual species from the previous natural forest or from regeneration of these species or regeneration of species from the adjacent forest), named in this study as Model F; cocoa agroforests in which companion plants were split equally between forest/native species and oil palms (named in this study as Model FP); lastly, cocoa agroforests in which companion plants were composed primarily of oil palms (named in this study as Model P).

We measured the diversity of cocoa agroforests in the Bengamisa-Yangambi forest landscape by evaluating base on the following: (i) species diversity (species richness, Shannon–Wiener index, Piélou's evenness index, Simpson's index, rarefaction curve); (ii) relative abundance; (iii) structure (density and basal area); and (iv) linkage between density and biodiversity. Each of these parameters is generally used in the characterization of cocoa agroforest in other countries [15,40,41,43–47] and for the characterization of forest stands [29,48,49].

The diversity and structure (basal area and density of associated plants) were calculated for each of the 25 cocoa agroforests. Diversity was expressed using (a) species richness, (b) Shannon–Wiener index, (c) Piélou's evenness index, and (d) Simpson's index [50]. Specific richness (S) is represented by the total or average number of species counted in the cocoa agroforest (obtained by counting the number of species). The Shannon–Wiener index provides an expression of diversity by considering the number of species and the abundance of individuals within each of these species. It is calculated by the following formula: $H' = -\sum_{i=1}^{S} pi \log pi$, where: pi = proportional abundance or percentage of species importance, calculated as follows: $pi = ni/N$; ni is the number of individuals of a species in the plot; N is the total number of individuals of all species in the plot; and S is the total number of species in the plot. The Shannon–Wiener index is often accompanied by the Piélou's evenness index, which is expressed by the following formula: $' = H'/H'max$, where: $H'max = \log S$ and S is the total number of species. This index is a measure of the distribution of individuals within species, independent of species richness. Its value varies from 0 (dominance of a single species) to 1 (equitable distribution of individuals within the species). These two indexes remain dependent on sample size and habitat type. Simpson's index measures the probability that two randomly selected individuals belong to the same species. It is determined by the following formula: $L = \sum \{[ni(ni-1)]/[N(N-1)]\}$, where ni is the number of individuals in the species i and N is total number of individuals. As sometimes different diversity indices do not all lead to the same conclusion [50], the rarefaction curves were associated with these indices to make conclusions more robust.

Relative abundance was calculated (for each zone; for each cocoa agroforest models; and for the suitability of associated plants for the agronomy of cocoa (Suitable species and Unsuitable species)) according to the following formula:

$$A(\%) = 100\left(\frac{Number\ of species\ stems}{All\ stems\ of\ the\ plot}\right) \tag{1}$$

Structure of each cocoa agroforest was evaluated based on the density (number of trees per ha) and the basal area (sums of areas of sections of all trees measured at 1.3 m) of cocoa trees and associated plants. The basal area was calculated by the following formula: $ST = \pi(Dbh)^2/4$, where ST is the basal area expressed in m^2 per ha and *DBH* is expressed in meters. The density calculation concerned first the whole tree population of cocoa agroforest and then the cocoa-associated plants.

Relation between density and species diversity: To establish a possible link between the stem density of species used as associated plant for cocoa trees and species richness, we used the Pearson correlation. This allowed us to correlate density and species richness by zone.

We also used statistical analysis to compare zones (Zones A, B, C, and D) and agroforest models (Model F, Model FP, and Model P). Therefore, descriptive (mean and standard deviation) and inferential analyses (one-way ANOVA, Kruskal test, Wilcoxon–Mann–Whitney test, and simple linear regression) were used. We verified normality using the Shapiro test and verified the homoscedasticity of variance using the Bartlett test. When the data distribution was nonparametric, Kruskal test, and Wilcoxon–Mann–Whitney test were used as appropriate. Otherwise, we used One-way ANOVA and Pearson's correlation as appropriate. All these analyses were performed by the R software version 4.0.2 [51] under its R studio interface. The acceptable error for the statistical analyses was 5%. The graphs were produced using packages ggplot2 [52] (simple linear regression) and BioDiversity Professional version 2.0 software (The Scottish Association for Marine Science, Oban, United Kingdom) (rarefaction curves). Diversity indices were performed using the Biodiversity R Package [53] and the Vegan Package [54] in the R software [51].

Findings are presented in the following way: (a) species richness and diversity; (b) abundance of plants associated with cocoa; (c) suitable and unsuitable plant species for cocoa agronomy; (d) main uses of plants associated with cocoa; (e) structure of cocoa agroforest; and (f) relation between density and species richness of each cocoa agroforest. These results are later discussed in the context of DRC, and in parallel with previous research across other forest landscapes of the tropics.

3. Results

3.1. Species Richness and Diversity Index

A total of 6558 stems, including 996 stems of cocoa-associated plants and 5562 cocoa trees. They are distributed in 90 species including *Theobroma cacao*, and 78 genera and 38 families were inventoried throughout the survey area. Each cocoa agroforest contains on average 13 species associated with cocoa trees (Table 1). Species richness and diversity indexes (Shannon-Wiener and Simpson) and Piélou's equitability (the distribution of stems within species) increase with distance from Kisangani city. Indeed, all these indices are low in Zone A (near Kisangani, between 18 and 33 km from Kisangani city) and increase as one moves away from it. On the other hand, they are all high in Zone D (forest zone, between 66 and 81 km from Kisangani city), except for specific richness, which is higher in Zone C (situated between 50 and 65 km from Kisangani city). The rarefaction curve (Figure 2) shows a low number of species in the area surrounding the city (Zone A). The species richness and indexes are generally low in Model P (cocoa agroforests in which companion plants are dominated by oil palms) compared to Model F (agroforests in which companion plants (associated with cocoa trees) are dominated by forest species) and Model FP (cocoa agroforests in which companion plants are split equally between forest species and oil palms) (Table 2). Looking at the entire landscape, Model FP contains the most species followed by Model F (Figure 3).

Table 1. Average species richness and diversity (±standard deviation) of cocoa-associated plants per agroforest in four zones of the Bengamisa-Yangambi landscape.

Indices/Index	Zone A (*n* = 6 Agroforests)	Zone B (*n* = 11 Agroforests)	Zone C (*n* = 3 Agroforests)	Zone D (*n* = 5 Agroforests)	Whole Region (*n* = 25 Agroforests)	*p*-Value
Species Richness	5.83 (±2.48)	12.55 (±8.71)	20.67 (±14.29)	16 (±6.63)	12.6 (±8.9)	0.0725
Shannon-Wiener index	0.99 (±0.5)	1.65 (±1.2)	2.14 (±1.51)	2.3 (±0.54)	1.68 (±1.05)	0.1836
Piélou's equitability	0.56 (±0.19)	0.6 (±0.37)	0.69 (±0.38)	0.85 (±0.09)	0.65 (±0.3)	0.2828
Simpson's index	0.45 (±0.22)	0.58 (±0.39)	0.68 (±0.45)	0.83 (±0.12)	0.61 (±0.33)	0.2142

Legend: Zone A (18–33 km from Kisangani city), Zone B (34–49 km from Kisangani city), Zone C (50–65 km from Kisangani city), Zone D (66–81 km from Kisangani city) and *n* = number of cocoa agroforests.

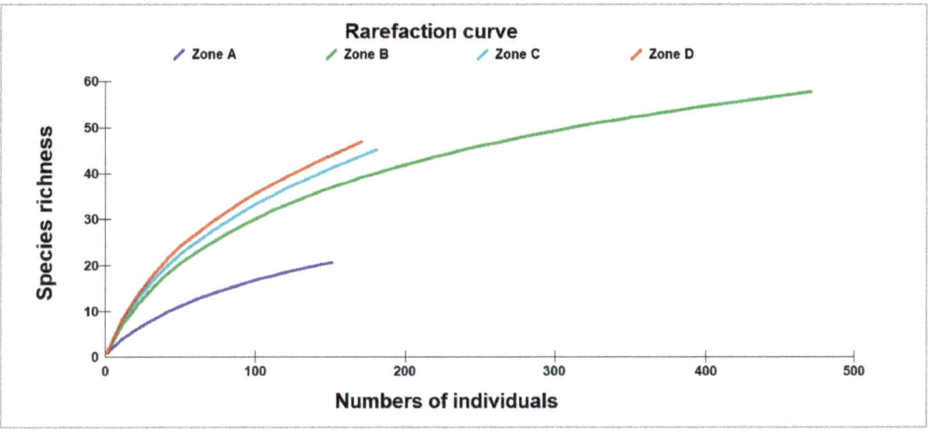

Figure 2. Species rarefaction curve considering the number of individuals sampled and the species richness by cocoa agroforests in four zones of the Bengamisa-Yangambi landscape. Legend: Zone A (18–33 km from Kisangani city), Zone B (34–49 km from Kisangani city), Zone C (50–65 km from Kisangani city) and Zone D (66–81 km from Kisangani city).

Table 2. Average species richness and diversity (±standard deviation) of cocoa-associated plants in three models of cocoa agroforests in Yangambi-Bengamisa landscape.

Indices	Model F (*n* = 10 Agroforests)	Model FP (*n* = 5 Agroforests)	Model P (*n* = 10 Agroforests)	Whole region (*n* = 25 Agroforests)	*p*-Value
Species Richness	16.4 (±6.19) b	17.4 (±10.92) b	6.4 (±6.96) a	12.6 (±8.9)	0.011
Shannon-Wiener index	2.33 (±0.63) b	2.29 (±0.78) b	0.73 (±0.79) a	1.68 (±1.05)	0.00011
Piélou's equitability	0.85 (±0.08) b	0.83 (±0.11) b	0.36 (±0.25) a	0.65 (±0.3)	0.002
Simpson's index	0.83 (±0.14) b	0.81 (±0.16) b	0.29 (±0.27) a	0.61 (±0.33)	8.50×10^{-6}

Legend: Model F (agroforests in which companion plants [associated with cocoa trees] are dominated by forest species); Model FP (cocoa agroforests in which companion plants [associated with cocoa trees] are split equally between forest species and oil palms); and Model P (cocoa agroforests in which companion plants [associated with cocoa trees] are dominated by oil palms). Models not sharing a common letter (a and b) in a row are significantly different at *p* = 0.05.

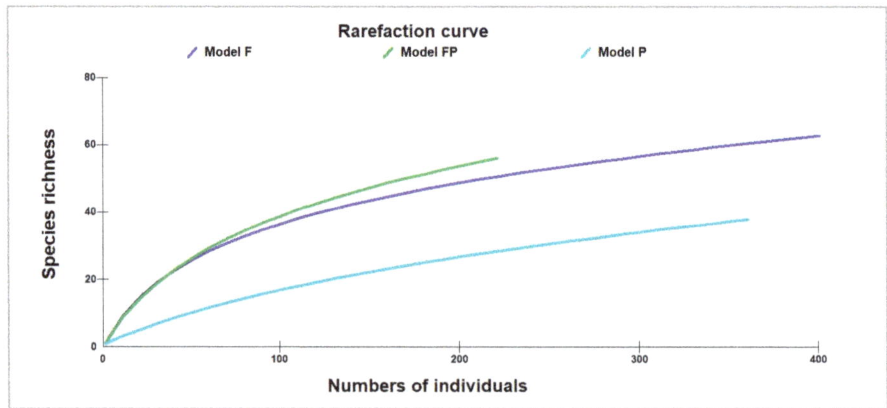

Figure 3. Species rarefaction curve considering the number of individuals and specific richness in three models of cocoa agroforests in the Bengamisa-Yangambi landscape. Legend: Model F (agroforests in which companion plants [associated with cocoa trees] are dominated by forest species); Model FP (cocoa agroforests in which companion plants [associated with cocoa trees] are split equally between forest species and oil palms); and Model P (cocoa agroforests in which companion plants [associated with cocoa trees] are dominated by oil palms).

3.2. Abundance of Plants Associated with Cocoa

The five most abundant species represent around 60% of the plants associated with cocoa. These five most abundant species represented 84%, 59%, 53%, and 47% of plants associated with cocoa in Zones A, B, C, and D, respectively. *Elaeis guineensis* JACQ. (366 plants representing 36.75% of total plants associated to cocoa) is the most abundant species in the entire study area, followed by *Musanga cecropioides* R. BR. (45 plants or 4.52%). Considering each zone, *Elaeis guineensis* (110 plants or 84.62% in Zone A; 179 plants or 63.93 in Zone B; and 51 plants or 51% in Zone C) was most abundant of all species except in the forest area (Zone D), where it was surpassed by *Musanga cecropioides* (35 plants or 42.17%). Species planted by farmers such as *Elaeis guineensis* (oil palm), *Persea americana* MILLER (avocado) and *Dacryodes edulis* (D. DON) H.J. LAM. (African pear) dominate in cocoa agroforests around Kisangani city (Zone A) in contrast to other areas where residual forest species dominate. However, the five most abundant species in the cocoa agroforests of four zones constitute more than half (59.54%) of the cocoa-associated plant individuals in the Bengamisa-Yangambi landscape (Table 3).

Table 3. The five most abundant species by cocoa agroforests in four zones of Bengamisa-Yangambi landscape.

Species	Local Names	Main Uses	Zone A (*n* = 6 Agroforests)	Zone B (*n*= 11 Agroforests)	Zone C (*n* = 3 Agroforests)	Zone D (*n* = 5 Agroforests)	Whole Region (25 Agroforests)
Elaeis guineensis	Adjagale	Edible	110	179	51	26	366
Musanga cecropioides	Kombo	Timber	-	-	10	35	45
Pycnanthus angolensis	Gbotugbu	Timber	-	35	-	-	35
Ficus exasperata	Kasage	Medicinal	3	24	-	-	27
Pseudospondias microcarpa	Bume	Medicinal	-	25	-	-	25
Maesopsis eminii	Ngana	Medicinal	-	-	12	6	18
Petersianthus macrocarpus	Angbeche	Caterpillar	-	17	-	-	17
Carapa procera	Mbindo	Medicinal	-	-	14	-	14

Table 3. *Cont.*

Species	Local Names	Main Uses	Zone A (*n* = 6 Agroforests)	Zone B (*n*= 11 Agroforests)	Zone C (*n* = 3 Agroforests)	Zone D (*n* = 5 Agroforests)	Whole Region (25 Agroforests)
Macaranga monandra	Abou chumbuge	Timber	-	-	13	-	13
Tetrorchidium didymostemon	Aboligi	Timber	-	-	-	9	9
Persea americana	Savoka	Edible	8	-	-	-	8
Bridelia atroviridis	Bubu	Caterpillar	-	-	-	7	7
Dacryodes edulis	Angboka	Edible	5	-	-	-	5
Senna siamea	Ngbangaolaya	Medicinal	4	-	-	-	4
Total of top five species			130	280	100	83	593
Total of all species in study area			155	476	187	178	996
Percentage of top five species			83.87	58.82	53.48	46.63	59.54

Legend: Zone A (18–33 km from Kisangani city), Zone B (34–49 km from Kisangani city), Zone C (50–65 km from Kisangani city), Zone D (66–81 km from Kisangani city) and *n* (number of cocoa agroforests in each zone).

3.3. Suitable and Unsuitable Species for Cocoa

3.3.1. Suitable Species to Cocoa Agronomy

The suitable species for cocoa production (Table 4) represent 27.31% of cocoa-associated plants recorded in the Bengamisa-Yangambi landscape's cocoa agroforests. They belong to 19 species of the 90 species recorded (They thus represent 21.59% of species in the whole region). *Musanga cecropioides* (with 57 plants representing 21% of the total plants associated with cocoa), followed by *Pycnanthus angolensis* (WELW.) EXELL (with 38 plants representing 14% of the total plants associated with cocoa), are the most abundant species for the entire studied area. Considering each zone, *Ficus exasperata* VAHL (with 3 plants representing 42.9% of the plants associated with cocoa) dominate in Zone A, *Pycnanthus angolensis* (with 35 plants representing 25% of the plants associated with cocoa) dominate in Zone B, *Macaranga monandra* MULL. ARG. (with 13 plants representing 25% of the plants associated with cocoa) dominate in Zone C and *Musanga cecropioides* (with 35 plants representing 47.94% of the plants associated with cocoa) dominate in Zone D. However, a large proportion of individuals (140 plants representing 51.47% of the plants associated with cocoa) of these suitable species for cocoa trees are concentrated in Zone B and a small proportion (seven plants representing 2.5% of the plants associated with cocoa) in Zone A.

Table 4. Abundance of suitable species for cocoa agronomy by cocoa agroforests in four zones in the Bengamisa-Yangambi landscape.

Species	Zone A (*n* = 6 Agroforests)	Zone B (*n* = 11 Agroforests)	Zone C (*n* = 3 Agroforests)	Zone D (*n* = 5 Agroforests)	Whole Region (*n* = 25 Agroforests)
Musanga cecropioides	2	10	10	35	57
Pycnanthus angolensis	0	35	3	0	38
Ficus exasperata	3	24	1	1	29
Petersianthus macrocarpus	0	17	6	4	27
Zanthoxylum gilletii	0	12	8	6	26
Macaranga monandra	0	2	13	4	19
Bridelia atroviridis	0	9	0	7	16
Macaranga spinosa	0	5	4	6	15
Albizia gummifera	0	10	1	1	12
Albizia adianthifolia	0	4	2	2	8
Alstonia boonei	1	4	0	1	6

Table 4. *Cont.*

Species	Zone A (*n* = 6 Agroforests)	Zone B (*n* = 11 Agroforests)	Zone C (*n* = 3 Agroforests)	Zone D (*n* = 5 Agroforests)	Whole Region (*n* = 25 Agroforests)
Albizia ferruginea	0	3	0	2	5
Canarium schweinfurthii	0	0	3	0	3
Croton haumanianus	0	2	0	0	2
Ficus elastica	0	1	1	0	2
Ficus mucuso	1	0	0	1	2
Ficus wildemaniana	0	1	0	1	2
Zanthoxylum lemairei	0	1	0	1	2
Harungana madagascariensis	0	0	0	1	1
Total of suitable plants	7	140	52	73	272
Total of all plants in study area	155	476	187	178	996
Percentage of suitable plants	4.52	29.41	27.81	41.01	27.31

Legend: Zone A (18–33 km from Kisangani city), Zone B (34–49 km from Kisangani city), Zone C (50–65 km from Kisangani city) and Zone D (66–81 km from Kisangani city).

3.3.2. Unsuitable Species for Cocoa Agronomy

Unsuitable species for cocoa production (Table 5) represented 13.35% of cocoa-associated plant stems recorded in the Bengamisa-Yangambi landscape. They are grouped in 22 species out of 90 species recorded (25% of the species of the whole region). *Pseudospondias microcarpa* (A. RICH.) ENGLER (medicinal species) (26 plants or 19.5%) and *Myrianthus arboreus* P. BEAUV. (edible fruit species) (16 plants or 12%), are the most abundant species in that category in cocoa agroforests of the study area. However, considering each zone, the highest representations are *Dacryodes edulis* (edible fruit species) with five plants or 71.43% in Zone A, *Pseudospondias microcarpa* (medicinal species) with 26 plants or 38.46% in Zone B, *Carapa procera* GILBERT (timber species) with 14 plants or 45.16% in Zone C, and both *Vernonia conferta* BENTHAM (medicinal species) and *Pterocarpus soyauxii* TAUB. (timber species) with six plants or 20% each in Zone D. However, a large proportion (48.87%) of individuals of these unsuitable species are concentrated in Zone B and are less represented (5.26%) in Zone A.

Table 5. Abundance of unsuitable species for cocoa agronomy by cocoa agroforests in four zones of the Bengamisa-Yangambi landscape.

Species	Zone A (*n* = 6 Agroforests)	Zone B (*n* = 11 Agroforests)	Zone C (*n* = 3 Agroforests)	Zone D (*n* = 5 Agroforests)	Whole Region (*n* = 25 Agroforests)
Pseudospondias microcarpa	1	25	0	0	26
Myrianthus arboreus	0	16	0	0	16
Carapa procera	0	1	14	0	15
Dacryodes edulis	5	3	1	3	12
Pterocarpus soyauxii	0	1	4	6	11
Rauvolfia vomitoria	0	4	1	3	8
Trichilia gilgiana	0	4	2	2	8
Vernonia conferta	0	0	0	6	6
Desplatsia dewevrei	0	4	1	0	5
Oncoba welwitschii	0	1	0	4	5
Blighia welwitschii	0	2	0	1	3
Synsepalum subcordatum	0	0	3	0	3
Uapaca guineensis	1	0	1	1	3
Barteria fistulosa	0	1	0	1	2
Canthium subcordatum	0	0	1	1	2

Table 5. *Cont.*

Species	Zone A (*n* = 6 Agroforests)	Zone B (*n* = 11 Agroforests)	Zone C (*n* = 3 Agroforests)	Zone D (*n* = 5 Agroforests)	Whole Region (*n* = 25 Agroforests)
Cola lateritia	0	2	0	0	2
Anonidium mannii	0	0	1	0	1
Cola marsupium	0	0	1	0	1
Drypetes gossweileri	0	0	1	0	1
Gilbertiodendron dewevrei	0	0	0	1	1
Homalium longistylum	0	1	0	0	1
Panda oleosa	0	0	0	1	1
Total of unsuitable plants	7	65	31	30	133
Total of all plants in study area	155	476	187	178	996
Percentage of unsuitable plants	4.52	13.66	16.58	16.85	13.35

Legend: Zone A (18–33 km from Kisangani city), Zone B (34–49 km from Kisangani city), Zone C (50–65 km from Kisangani city) and Zone D (66–81 km from Kisangani city).

3.4. Main Uses of Plants Associated with Cocoa

3.4.1. Main Uses of Plants Associated with Cocoa by Zone (i.e., Main Distance from Kisangani)

In the study area, we inventoried trees hosting caterpillars, trees with edible products, trees with medicinal properties, timber, and other trees with secondary or unknown uses (Table 6). Within these categories, trees with edible products were the most abundant (70.56 ± 50.4 trees per ha), especially in Zone A. The least abundant use category was timber (5.76 ± 7.1 individuals per ha) for the entire study area. Timber was most abundant in Zone C (13.33 ± 13.05 trees per ha) and Zone D (13 ± 4.69 trees per ha).

Table 6. Average number of tree species (±standard deviation) by cocoa agroforests in four zones of Yangambi-Bengamisa landscape according to their main uses.

Main Uses	Zone A (*n* = 6 Agroforests)	Zone B (*n* = 11 Agroforests)	Zone C (*n* = 3 Agroforests)	Zone D (*n* = 5 Agroforests)	Whole Region (*n* = 25 Agroforests)	*p*-Value
Edible	85.33 (±32.36)	78.91 (±49.48)	77.33 (±85.54)	30.4 (±39.76)	70.56 (±50.4)	0.265
Hosts for caterpillars	0.67 (±1.63)	12 (±14.86)	9.33 (±6.11)	12.8 (±10.35)	9.12 (±11.75)	0.07834
Medicinal	11.33 (±8.55)	38.91 (±37.23)	76 (±79.9)	25.6 (±17.57)	34.08 (±39.38)	0.115
Timber	1.33 (±2.07) a	13.45 (±12.93) a	53.33 (±52.2) b	52 (±18.76) b	23.04 (±28.38)	0.002034
Others (minor or no known uses)	4.67 (±7.34)	29.82 (±42.61)	33.33 (±34.02)	21.6 (±14.59)	22.56 (±31.91)	0.2558
Total	103.33 (±37.64) a	173.09 (±56.23) b	249.33 (±13.44) b	142.4 (±7.45) ab	159.36 (±71.96)	0.01405

Legend: Zone A (18–33 km from Kisangani city), Zone B (34–49 km from Kisangani city), Zone C (50–65 km from Kisangani city) and Zone D (66–81 km from Kisangani city). Zones not sharing a common letter (a and b) in a row are significantly different at *p* = 0.05.

3.4.2. Main Uses of Plants Associated with Cocoa by Cocoa Agroforest Models

Table 7 shows that trees for medicinal use are more abundant (48.8 ± 31.88 individuals per ha) in Model F, followed by tree species with secondary functions (fuelwood, construction wood, etc.) or unknown (37.6 ± 41.06 individuals per ha). In Model FP, individuals in the medicinal category have been most inventoried (52.8 ± 64.96 trees per ha), followed by those with individuals possessing certain edible products (50.4 ± 35.05 trees per ha). Finally, in Model P, individuals of species with edible products are more dominant (121.6 ± 24.6 trees per ha), followed by medicinal species (10 ± 12.82). There is a significant difference between the different models in each of the main uses (*p*-value < 0.05).

Table 7. Average number of tree species (±standard deviation) in three models of cocoa agroforests in Yangambi-Bengamisa landscape according to their main uses.

Main Uses	Model F (n = 10 Agroforests)	Model FP (n = 5 Agroforests)	Model P (n = 10 Agroforests)	Whole Region (n = 25 Agroforests)	p-Value
Edible	29.6 (±25.24) a	50.4 (±35.05) a	121.6 (±24.6) b	70.56 (±50.4)	0.0001507
Tree-hosting caterpillars	15.6 (±13.91) a	11.2 (±11.1) ab	1.6 (±2.8) b	9.12 (±11.75)	0.01258
Medicinal	48.8 (±31.88) a	52.8 (±64.96) ab	10 (±12.82) b	34.08 (±39.38)	0.007181
Timber	30.4 (±25.38) ab	39.2 (±43.58) a	7.6 (±13.79) b	23.04 (±28.38)	0.04679
Others (minor or no known uses)	37.6 (±41.06) a	28.8 (±25.2) ab	4.4 (±10.41) b	22.56 (±31.91)	0.0071
Whole region	162 (±77.98)	182.4 (±113.3)	145.2 (±38.69)	159.36 (±71.96)	0.653

Legend: Model F (agroforests in which companion plants [associated with cocoa trees] are dominated by forest species); Model FP (cocoa agroforests in which companion plants [associated with cocoa trees] are split equally between forest species and oil palms); and Model P (cocoa agroforests in which companion plants [associated with cocoa trees] are dominated by oil palms). Models not sharing a common letter (a and b) in a row are significantly different at p = 0.05.

3.5. Structure of Cocoa Agroforests

The average density and basal area of all species (cocoa trees and cocoa-associated plants) in the study area (Table 8) are 1048.16 trees/ha and 17.28 m^2/ha, respectively. Of these, cocoa- associated plants take up 15.20% of total density, but 55.84% of total basal area. The density and basal area of cocoa-associated plants in the cocoa agroforests of the study area increase with distance from Kisangani city, except in Zone D where they have decreased compared to the previous areas (Zone C and Zone B). The average density and basal area of cocoa-associated plants in the study area are, respectively, 159.36 trees/ha (p-value = 0.01405) and 9.65 m^2/ha (p-value = 0.273).

Table 8. Average density and basal area of cocoa-associated plants (±standard deviation) by cocoa agroforests in four zones of Yangambi-Bengamisa landscape.

Cocoa Agroforest Structure	Zone A (n = 6 Agroforests)	Zone B (n = 11 Agroforests)	Zone C (n =3 Agroforests)	Zone D (n = 5 Agroforests)	Whole Region (n = 25 Agroforests)	p-Value
Density of associated plants (n/ha)	103.33 (±37.64) a	173.09 (±56.23) b	249.33 (±113.44) b	142.4 (±7.45) ab	159.36 (±71.96)	0.01405
Basal area of associated plants (m^2/ha)	5.59 (±5.28)	10.95 (±9.4)	15.36 (±3.35)	8.26 (±4.66)	9.65 (±7.52)	0.273
Density of (n/ha) of cocoa	913.33 (±213.07)	959.27 (±154.07)	746.67 (±78.93)	789.6 (±92.81)	888.8 (±168.24)	0.107
Basal area of cocoa	10.91 (±2.63)	9.14 (±1.94)	8.2 (±2.04)	9.68 (±3.9)	9.56 (±2.58)	0.453
Density of whole region	1016.67 (±77.94)	1132.36(±187.62)	996 (±34.64)	932 (±77.82)	1048.16 (±70.36)	0.133
Basal area of whole region	13.13 (±3.04)	19.19 (±7.73)	17.28 (±2)	18.08 (±3.13)	17.28 (±5.91)	0.248

Legend: Zone A (18–33 km from Kisangani city), Zone B (34–49 km from Kisangani city), Zone C (50–65 km from Kisangani city) and Zone D (66–81 km from Kisangani city). Zones not sharing a common letter (a and b) in a row are significantly different at p = 0.05.

Table 9 shows that Model F had the highest density (1146 stems/ha) and basal area (22.99 m^2/ha), while Model P recorded the lowest density (971 trees/ha) and basal area (12.1 m^2/ha). The difference of density between the three models is not statistically significant, but the difference becomes significant when comparing the basal area (p-value = 9.62 × 10^{-7}).

Table 9. Average density and basal area of cocoa-associated plants (±standard deviation) in three models of cocoa agroforests in the Bengamisa-Yangambi landscape.

Cocoa Agroforest Structures	Model F ($n = 10$ Agroforests)	Model FP ($n = 5$ Agroforests)	Model P ($n = 10$ Agroforests)	Whole Region ($n = 25$ Agroforests)	*p*-Value
Density of associated plants (n/ha)	162 (±77.98)	182.4 (±113.3)	145.2 (±38.69)	159.36 (±71.96)	0.653
Basal area of associated plants (m^2/ha)	12.38 (±9.48)	9.87 (±4.83)	6.81 (±5.77)	9.65 (±7.52)	0.263
Density (n/ha) cocoa	984 (±200.55)	824 (±109.8)	826 (±114.53)	888.8 (±168.24)	0.0629
Basal area of cocoa	9.6 (±2.97)	9.81 (±3.45)	9.4 (±1.91)	9.56 (±2.58)	0.961
Density of whole region	1146 (±217.52)	1006.4 (±50.72)	971.2 (±103.67)	1048.16 (±70.36)	0.0525
Basal area of whole region	22.99 (±3.86) a	16.22 (±2.77) b	12.1 (±2.85) b	17.28 (±5.91)	9.62×10^{-7}

Legend: Model F (Agroforests in which companion plants [associated with cocoa trees] are dominated by forest species); Model FP (cocoa agroforests in which companion plants [associated with cocoa trees] are split equally between forest species and oil palms); and Model P (cocoa agroforests in which companion plants [associated with cocoa trees] are dominated by oil palms). Models not sharing a common letter (a and b) in a row are significantly different at $p = 0.05$.

3.6. Relationship between Density and Species Richness of Cocoa Agroforest

There is strong correlation ($r = 76\%$) between the species richness and the density of cocoa-associated plants in cocoa agroforests in the Bengamisa-Yangambi landscape (Figure 4) and in each zone of distance (p-value $= 9.4 \times 10^{-6}$, $R^2 = 58\%$). Moreover, the specific richness in the cocoa agroforests of the above-mentioned landscape is strongly correlated with the density of suitable associated plants ($r = 77\%$, p-value $= 6.537 \times 10^{-6}$, $R^2 = 57.6\%$) (Figure 4a) and with the density of unsuitable associated plants ($r = 84.8\%$, p-value $= 8.236 \times 10^{-8}$, $R^2 = 70.8\%$) to cocoa trees (Figure 4b).

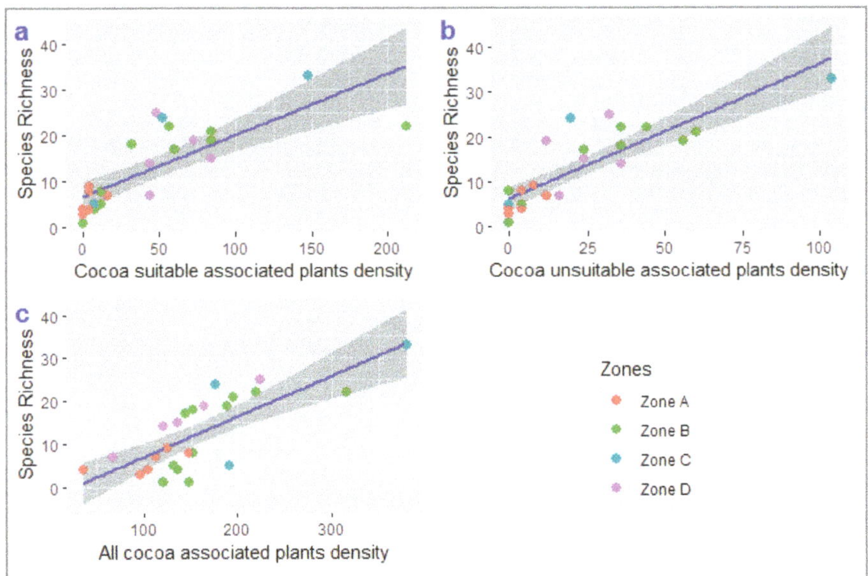

Figure 4. Correlation between species richness and plant density by cocoa agroforests in four zones of the Bengamisa-Yangambi landscape. Correlation between species richness and cocoa suitable associated plants density (**a**), cocoa suitable associated plants density (**b**) and all cocoa associated plants density (**c**). Legend: Zone A corresponds to 18–33 km from Kisangani city, Zone B corresponds to 34–49 km from Kisangani city, Zone C corresponds to 50–65 km from Kisangani city, Zone D corresponds to 66–81 km from Kisangani city and (•) corresponds to a cocoa agroforests.

4. Discussion

This study reveals that distance from Kisangani (The main city in the landscape) influences the plants composition (expressed here by species diversity and structure) of the cocoa agroforests. Market access associated to the proximity of this city and land-use intensity and the related disturbance impact the composition of plants associated with cocoa in the forest landscape.

4.1. Floristic Composition of Cocoa Agroforests

The cocoa agroforests of the Bengamisa-Yangambi landscape harbor substantial tree diversity. A total of 996 plants associated to cocoa belonging to 89 species, 77 genera, and 38 families were recorded in 25 cocoa agroforests across the Bengamisa-Yangambi landscape. An average of 13 species occurred per cocoa agroforest. Furthermore, these cocoa agroforests have a high tree diversity as is the case in other cocoa production systems in other parts of the tropics. For example, the authors of [55] inventoried 71 species and 32 families in the agroforestry systems in East Cameroon. Also, the authors of [56] inventoried 27 families and 62 species in cocoa farms in the southern region of Cameroon. However, these results are small compared to those of [57], who obtained 40 families, 112 genera, and 127 species in the Bajo Caguán zone in Colombia. These variabilities can be explained, among others, by the different ecosystems in which the cocoa agroforests were developed, the socio-economic of the landscapes, and the type of cocoa farming systems promoted [57] and sample size.

The high abundance of oil palms (366 of 996 total plants surveyed, i.e., 37%) in cocoa agroforests of the Bengamisa-Yangambi landscape may be explained by the fact that many cocoa trees were established under old oil palm plantations. Indeed, the Bengamisa-Yangambi landscape was subject to an intensification of oil palm cultivation during the colonial period [35]. Moreover, in that region, oil palm is the main source of oil consumed by local people. It is also the main source of beverage consumed (palm wine) in the region. These observations were made in other African countries like Côte d'Ivoire [50] and Cameroon [7,41,42].

The species rarefaction curve, which considers the number of individuals and specific richness (Figures 2 and 3), allows us to see that the number of species grows alongside the increase of cocoa agroforest. This suggests that farmers do not necessarily grow the same species. Since individual farmers have their own species interest [42,58] in separate cocoa agroforests, the combination of all their farms allows for a longer list of species in the cocoa landscape created between the city of Kisangani and the natural forest. The floristic composition of this cocoa landscape is a mixture of local forest species and exotic plants (avocado, etc.). Among these species, those introduced by farmers (oil palm, avocado, African pear) are more present in the cocoa agroforest near Kisangani. Conversely, native species are mainly found within cocoa agroforests close to the forest. However, the abundance of such introduced species indicates the degree of alteration of the cocoa agroforests compared to primary forest [41].

4.2. Specific Diversity in the Cocoa Agroforests of the Study Area

Even though the difference between zones is not statistically significant (one-way ANOVA), the diversity indices (Shannon-Wiener, Simpson's, Piélou equitability) revealed that diversity is increasing from Zone A (near Kisangani city) to Zone D (forest region). However, large agglomerations like the city of Kisangani exert strong pressure on the forest for satisfying their multiple needs (fuel wood, timber, NTFPs, etc.). This leads to forest fragmentation, followed by deforestation. Therefore, the pressure exerted on the forest also influences the species composition of the cocoa agroforests [41]. It is well established that the distance between the cocoa agroforests and the forest stands may affect the processes associated to forest tree species dissemination [42]. Moreover, when differences in forest coverage change (mainly his disturbance) were considered, plant species richness in cocoa agroforest decreased with increasing intensity of land-use [2], confirming previous studies [41] in the forest zone of southern Cameroon.

Among the three models of cocoa agroforests studied, Model F (agroforests in which companion plants (associated with cocoa trees) are dominated by forest species) is the most diversified. Model P (cocoa agroforests in which companion plants (associated with cocoa trees) are dominated by oil palms) is less diversified (one-way ANOVA). Indeed, the different models of cocoa agroforests displayed different levels of species diversity. These ranged from a critical reduction in species richness from the complex diversified multistate system (which is the most diverse) to the high density of perennial plants (like oil palm) models [57].

4.3. Structure of Cocoa Agroforests

The cocoa-associated plants represent a low density per hectare (15.20% of 1048.16 stems/ha) in the study area. However, they account for more than half of the average basal area in the study area (55.84% of 17.28 m^2/ha). Cocoa trees have small diameters compared to cocoa-associated plants, which are heterogeneous, ranging from small to large trees. Consequently, they have a greater influence on the average basal area, despite their low density per hectare. This has a direct impact on stored biomass. Many studies in the tropics have shown similar results, such as in Cameroon [59] and Indonesia [60].

The density of cocoa-associated plants increases significantly with distance from the city (one-way ANOVA, *p*-value = 0.0177). This difference shows the impact of anthropogenic factors on the vegetation composition in the study area [61]. The impact of anthropogenic footprints is also known to influence the relation between species richness and density [23] of plants associated with cocoa in the study area. Moreover, the management of plants associated with cocoa by smallholders is different between regions and strongly impacts cocoa landscapes [62].

We compared the average density and basal area in agroforestry landscape and natural forests. The density value is largely higher than 467 stems/ha in *Pericopsis elata* forest and 344 stems/ha in *Julbernardia seretii* forest obtained by [63] in the lowland forest of Uma in DRC. It is also higher than 412 and 343 stems/ha, respectively, in the mixed forest and monodominant *Gilbertiodendron dewevrei* forest obtained by [29] in the Yangambi forest in DRC.

The average basal area value is less than 29 and 24.5 m^2/ha obtained by [63] in forests of *Pericopsis elata* and *Julbernardia seretii*, respectively. It is less than 31.8 and 29.7 m^2/ha, respectively, in the mixed and monodominant *Gilbertiodendron dewevrei* forest obtained by [29]. It is also less than 23 m^2/ha in the *Gilbertiodendron dewevrei* forest and the 32.3 m^2/ha obtained by [64] in the mixed forest of the Rubi-Télé hunting domain in DRC. On the other hand, this basal area value is closer to the 19.21 m^2/ha obtained by [65] in a degraded forest in the north of Congo-Brazzaville.

However, cocoa agroforests are more comparable to secondary forests in terms of basal area and the high density of small, medium and (the few) large trees in contrast to primary forests (monodominant and mixed) [66]. Conversely, in tropical forests, the large proportion of basal area (biomass) is occupied by large trees. For this reason, even if cocoa agroforests have a high density of trees per hectare, they will not be able to replace primary forests. However, they offer opportunities for developing sustainable land-use systems within fragmented protected forest landscape (around the Yangambi forest reserve). This could help address land and environmental degradation problems, while ensuring provision of substantial household income to sustain livelihoods [67].

The strong positive correlation (*r* = 76%) between the species richness and the density of cocoa-associated plants in cocoa agroforests in the Bengamisa-Yangambi landscape is similar to the results obtained in tropical forests by several authors [68–70], demonstrating the importance of cocoa agroforests in floristic biodiversity conservation in forest landscape. However, this correlation depends on the structural characteristics of the vegetation in the landscape [71]. In the landscape where agronomy farming is introduced, the species proposed by extension services are not necessarily the farmer's priority [58]. Some of these trees retained are well known to be used (timber, medicine, and fuel wood) [72]. Thus, the management strategies in cocoa plantations affects species diversity and density at plantation and landscape level [72,73]. The strong correlation between species richness and plant density of associated unsuitable species suggest that cocoa bean production is not necessarily the aim of the farmers. And these retentions of species that are not suitable to cocoa agronomy may event, in some situations, be benefit to biodiversity conservation. Figure 4b clearly shows that the species diversity of these unsuitable species increases faster with the increase of their density than in the group of suitable associated species (Figure 4a). This finding suggests that some trade-off between cocoa agronomy and biodiversity conservation may need to be managed within cocoa agroforests.

4.4. Main Local Uses of Cocoa-Associated Plants

The results show that trees with edible products are more abundant and timber species are less abundant in the agroforests of our study area. In Models F and FP, trees for medicinal use are most abundant and in Model P edible species are dominant. This is sufficient proof that farmers are conserving more of the species they need in cocoa agroforests. These needs differ from one area to another. In the forest zone, farmers are more dependent on plants for their health care, whereas near the city they depend more on edible species (oil palm, avocado, African pear, etc.). Nowak et al. [25] obtained a similar result. Indeed, the presence or absence of certain species depends more on their interest to farmers [23]. Usefulness to the household may explain why farmers maintain certain species considered by extension services as potentially unsuitable to cocoa (Table 5) in their cocoa agroforests. Similar results have been obtained in other countries, such as Ghana [24], Cameroon [41,42], and Côte d'Ivoire [58].

4.5. Landscape Management Implication

Initially, cocoa was promoted for its beans. Gradually, however, cocoa agroforest became understood as potentially useful for biodiversity conservation and for climate change responses in forest landscapes of the tropics. Its biodiversity conservation function is mainly explained by the importance of plants associated with cocoa, and more specifically the forest species. Several authors have also recorded this observation in the tropics [1–3,41]. Thus, Model F (agroforests in which companion plants (associated with cocoa trees) are dominated by forest species) is the most appropriate if one wishes to approach specific richness and forest structure. The presence of associated plants also contributes to creating a microclimate favourable to development of cocoa trees [74]. On the other hand, several REDD+ projects have proposed cocoa agroforests as a response to deforestation and climate change. These include Mambasa Geographically Integrated REDD+ Pilot Project based on "green cocoa" in DRC [23,75], DRC Cocoa Partnership [20], the Ghana Cocoa Forest Programme [76], Mainstreaming Climate-smart Agricultural Practices in Cocoa Production in Ghana, Climate Cocoa Partnership for REDD+ Preparation [20], Zero Cocoa Deforestation [77], Initiative for Sustainable Landscapes in Cameroon, and the Climate Smart Cocoa Program in Côte d'Ivoire [20].

Without any intervention to reverse the trend, market access if intensified with the growing of Kisangani may contribute to the simplification of plants composition of the cocoa agroforests. This will gradually lead to replacement of forest/native species by the main consumed one, which many of them been exotic (i.e., introduced in DRC). The diversity of cocoa agroforest models offers a variety of options that can be used in the landscapes to search for balance between ecological conservation and farmers livelihoods.

4.6. Limitation of this Paper and Perspectives

This paper is one of the first to study the diversity of cocoa agroforest in DRC. However, further studies are still needed. These should examine better use of cocoa agroforests to support the Sustainable Development Goals (SDG) in rural areas. They should also assess how use of cocoa agroforest can help respond to the Convention on Biological Diversity (CBD) and the United Nations Framework Convention on Climate Change (UNFCCC). More importantly, consumers and chocolate industries are trying to import cocoa that was harvested sustainably without furthering deforestation. European governments are moving toward reducing/cancelling imports of crops that lead to deforestation, including cocoa. At the same time, the private sector (importers and distributors in Europe) wants its value chain to be free of links to deforestation. Plants associated with cocoa are the main carbon sinks [46,78,79] and key components for others ecosystems services. This creates expectations for cocoa agroforest and farming systems that should be explored through greater study. Therefore, the three models (F, PF, P) still require carbon stock studies to be used properly in REDD+ programs in Tshopo province. Other studies could integrate other ecosystem services (wildlife conservation, soil protection, etc.) and socio-economic concerns (improved economic conditions of farmers' households) related to these three models in Tshopo. Cocoa agroforests are generally recognized for their socio-economic and ecological importance. However, each farming system of the forest landscape needs to be studied better to generate information that will support the sustainable management of these rural landscapes.

5. Conclusions

Cocoa trees introduced in the Yangambi-Bengamisa landscape have been associated with other plants in different cocoa agroforests that are part of the current landscape. Although farmers were advised not to keep some trees in the same field as cocoa plants, they have maintained/introduced plants over the last decades that have some function (edible, hosting caterpillars, medicinal, etc.). A multitude of cocoa farming systems could be classified into three main groups: Model F (agroforests in which companion plants (associated with cocoa trees) are dominated by forest species), Model FP (cocoa agroforests in which companion plants (associated with cocoa trees) are split equally between forest species and oil palms), and Model P (cocoa agroforests in which companion plants (associated with cocoa trees) are dominated by oil palms). Such models thus present different options for strengthening the livelihood of farmers, improving biodiversity conservation, and/or defining an appropriate climate change response in Tshopo province and other parts of DRC. Based on the findings of this study, the diversity of cocoa agroforest models offers a variety of options that can be used in the landscapes to search for balance between ecological conservation and farmers livelihoods.

The distance from the city (Kisangani) is a determining factor in the floristic composition (species diversity and plant density) of the cocoa agroforests. The area close to the city is marked by an abundance of oil palms and some edible species, such as avocados and African pear. Areas near the forests are more abundant to forest species for medicinal use.

The findings provide scientific evidence that can be useful in harnessing cocoa agroforest to improve the livelihoods of farmers, conserve biodiversity, and respond to climate change in the Bengamisa-Yangambi landscape and other forest landscapes of DRC.

Author Contributions: D.J.S. conceptualized and supervised the study with J.-M.K. and J.E.; G.B. collected the data with the assistance of L.M. All authors have read and agreed to the published version of the manuscript.

Funding: This research was funded by EU (Contrat FED/2016/381-145) and NORAD Grant agreement code No: NOR114.

Acknowledgments: The authors would like to thank the European Union (EU) and CIFOR for funding through the Training and Research in Tshopo (FORETS: Formation, Recherche, Environment dans la Tshopo) project. We also thank the Resources & Synergies Development design office for its logistical support. We thank all cocoa farmers in the Bengamisa-Yangambi landscape for their support during the data collection period. Finally, we would like to thank all those who accompanied us in the field during this period (guides and botanical identifiers). This research was carried out to partially complete the requirements of a master's degree, but also as part of the CIFOR GCS-REDD+ project funded by the Norwegian Agency for Development Cooperation (NORAD).

Conflicts of Interest: The authors declare no conflict of interest.

References

1. Foley, J.A.; DeFries, R.; Asner, G.P.; Barford, C.; Bonan, G.; Carpenter, S.R.; Chapin, F.S.; Coe, M.T.; Daily, G.C.; Gibbs, H.K.; et al. Global consequences of land-use. *Science* **2005**, *309*, 570–574. [CrossRef] [PubMed]
2. Clough, Y.; Barkmann, J.; Juhrbandt, J.; Kessler, M.; Wanger, T.C.; Anshary, A.; Buchori, D.; Cicuzza, D.; Darras, K.; Dwi Putra, D.; et al. Combining high biodiversity with high yields in tropical agroforests. *Proc. Natl. Acad. Sci. USA* **2011**, *108*, 8311–8316. [CrossRef] [PubMed]

3. Rice, R.A.; Greenberg, R. Cacao cultivation and the conservation of biological diversity. *Ambio* **2000**, *29*, 167–173. [CrossRef]

4. Götz, S.; Harvey, C.A.; Grégoire, V. Agroforestry and biodiversity conservation in tropical landscapes. In *Agroforestry Systems*; Götz, S., da Fonseca, G., Harvey, C.A., Gascon, C., Vasconcelos, H., Izac, A.-M., Eds.; Island Press: Washington, DC, USA, 2004; Volume 68, pp. 247–249.

5. Bhagwat, S.A.; Willis, K.J.; Birks, H.J.B.; Whittaker, R.J. Agroforestry: A refuge for tropical biodiversity? *Trends Ecol. Evol.* **2008**, *23*, 261–267. [CrossRef] [PubMed]

6. Cicuzza, D.; Kessler, M.; Clough, Y.; Pitopang, R.; Leitner, D.; Tjitrosoedirdjo, S.S. Conservation value of cacao agroforestry systems for terrestrial herbaceous species in Central Sulawesi, Indonesia. *Biotropica* **2011**, *43*, 755–762. [CrossRef]

7. Sonwa, D.J.; Weise, S.F.; Schroth, G.; Janssens, M.J.J. Howard-Yana Shapiro Plant diversity management in cocoa agroforestry systems in West and Central Africa—Effects of markets and household needs. *Agrofor. Syst.* **2014**, *88*, 1021–1034. [CrossRef]

8. Blaser, W.J.; Oppong, J.; Yeboah, E.; Six, J. Shade trees have limited benefits for soil fertility in cocoa agroforests. *Agric. Ecosyst. Environ.* **2017**, *243*, 83–91. [CrossRef]

9. Schroth, G.; Harvey, C.A. Biodiversity conservation in cocoa production landscapes: An overview. *Biodivers. Conserv.* **2007**, *16*, 2237–2244. [CrossRef]

10. Toppo, P.; Raj, A. Role of agroforestry in climate change mitigation. *J. Pharmacogn. Phytochem.* **2018**, *7*, 241–243.

11. Vebrova, H.; Lojka, B.; Husband, T.P.; Zans, M.E.C.; Van Damme, P.; Rollo, A.; Kalousova, M. Tree diversity in cacao agroforests in San Alejandro, Peruvian Amazon. *Agrofor. Syst.* **2014**, *88*, 1101–1115. [CrossRef]

12. Sari, R.R.; Saputra, D.D.; Hairiah, K.; Rozendaal, D.M.A.; Roshetko, J.M.; van Noordwijk, M. Gendered species preferences link tree diversity and carbon stocks in Cacao agroforest in Southeast Sulawesi, Indonesia. *Land* **2020**, *9*, 108. [CrossRef]

13. Mbow, C.; Smith, P.; Skole, D.; Duguma, L.; Bustamante, M. Achieving mitigation and adaptation to climate change through sustainable agroforestry practices in africa. *Curr. Opin. Environ. Sustain.* **2014**, *6*, 8–14. [CrossRef]

14. Merijn, B.; Simone, S. Biodiversity conservation in cacao agroforestry systems. In *Biodiversity Conservation in Agroforestry Landscapes: Challenges and Opportunities*; Simonetti, J.A., Ed.; Editorial Universitaria: Santiago, Chile, 2013; pp. 61–76.

15. Daghela Bisseleua, H.B.; Fotio, D.; Yede; Missoup, A.D.; Vidal, S. Shade tree diversity, cocoa pest damage, yield compensating inputs and farmers' net returns in West Africa. *PLoS ONE* **2013**, *8*, e56115. [CrossRef]

16. Rajab, Y.; Leuschner, C.; Barus, H.; Tjoa, A.; Hertel, D. Cacao cultivation under diverse shade tree cover allows high carbon storage and sequestration without yield losses. *PLoS ONE* **2016**, *11*, e0149949. [CrossRef] [PubMed]

17. Ruf, F.O. The myth of complex cocoa agroforests: The case of Ghana. *Hum. Ecol.* **2011**, *39*, 373–388. [CrossRef]

18. Wessel, M.; Quist-Wessel, P.M.F. Cocoa production in West Africa, a review and analysis of recent developments. *NJAS Wagening. J. Life Sci.* **2015**, *74–75*, 1–7. [CrossRef]

19. Anderson Bitty, E.; Bi, S.G.; Bene, J.C.K.; Kouassi, P.K.; Scott McGraw, W. Cocoa farming and primate extirpation inside Cote d'Ivoire's protected areas. *Trop. Conserv. Sci.* **2015**, *8*, 95–113. [CrossRef]

20. Kroeger, A.; Bakhtary, H.; Haupt, F.; Streck, C. *Eliminating Deforestation from the Cocoa Supply Chain*; World Bank: Washington, DC, USA, 2017; pp. 1–61.

21. Bernard, F.; Minang, P.A. Community forestry and REDD+ in Cameroon: What future? *Ecol. Soc.* **2019**, *24*, 14. [CrossRef]

22. Downie, R. *Assessing the Growth Potential of Eastern Congo's Coffee and Cocoa Sectors*; Center for Strategic and International Studies: Washington, DC, USA, 2018; pp. 1–24.

23. Jassogne, L.; van Asten, P.; De Beule, H. *Cocoa: Driver of deforestation in the Democratic Republic of Congo?* CGIAR: Copenhagen, Denmark, 2014; pp. 1–31.

24. Schroth, G.; Jeusset, A.; da Silva Gomes, A.; Florence, C.T.; Coelho, N.A.P.; Faria, D.; Läderach, P. Climate friendliness of cocoa agroforests is compatible with productivity increase. *Mitig. Adapt. Strateg. Glob. Chang.* **2016**, *21*, 67–80. [CrossRef]

25. Nowak, A.; Rosenstock, T.S.; Hammond, J.; Degrande, A.; Smith, E. *Livelihoods of Households Living Near Yangambi Biosphere Reserve, Democratic Republic of Congo*; Center for International Forestry Research: Bogor, Indonesia, 2019; pp. 1–10.

26. Ngo Bieng, M.A.; Gidoin, C.; Avelino, J.; Cilas, C.; Deheuvels, O.; Wery, J. Diversity and spatial clustering of shade trees affect cacao yield and pathogen pressure in Costa Rican agroforests. *Basic Appl. Ecol.* **2013**, *14*, 329–336. [CrossRef]

27. Kyale Koy, J.; Wardell, D.A.; Mikwa, J.; Kabuanga, J.M.; Monga Ngonga, A.M.; Oszwald, J.; Doumenge, C. Dynamics of déforestation dans la Réserve de biosphère de Yangambi (République Démocratique du Congo): Variabilité spatiale et temporelle au cours des 30 dernières années. *Bois Forets Trop.* **2019**, *341*, 15. [CrossRef]

28. Moloba, Y.; Mobula, V.; Ntoto, R.; Mahungu, M. Dynamique socio-économique de l'adoption des variétés améliorées du manioc en République Démocratique du Congo (RDC): Cas des provinces de Kongo Central et la Tshopo. *Eur. Sci. J. ESJ* **2019**, *15*, 346–362.

29. Kearsley, E.; Verbeeck, H.; Hufkens, K.; Van de Perre, F.; Doetterl, S.; Baert, G.; Beeckman, H.; Boeckx, P.; Huygens, D. Functional community structure of African monodominant Gilbertiodendron dewevrei forest influenced by local environmental filtering. *Ecol. Evol.* **2017**, *7*, 295–304. [CrossRef]

30. Termote, C.; Van Damme, P.; Djailo, B.D. Eating from the wild: Turumbu, Mbole and Bali traditional knowledge on non-cultivated edible plants, District Tshopo, DR Congo. *Genet. Resour. Crop Evol.* **2011**, *58*, 585–618. [CrossRef]

31. Moonen, P.C.J.; Verbist, B.; Schaepherders, J.; Bwama Meyi, M.; Van Rompaey, A.; Muys, B. Actor-based identification of deforestation drivers paves the road to effective REDD+ in DR Congo. *Landuse Policy* **2016**, *58*, 123–132. [CrossRef]

32. Moonen, P.C.J.; Verbist, B.; Boyemba Bosela, F.; Norgrove, L.; Dondeyne, S.; Van Meerbeek, K.; Kearsley, E.; Verbeeck, H.; Vermeir, P.; Boeckx, P.; et al. Disentangling how management affects biomass stock and productivity of tropical secondary forests fallows. *Sci. Total Environ.* **2019**, *659*, 101–114. [CrossRef]

33. Akkermans, T.; Van Rompaey, A.; Van Lipzig, N.; Moonen, P.; Verbist, B. Quantifying successional land cover after clearing of tropical rainforest along forest frontiers in the Congo Basin. *Phys. Geogr.* **2013**, *34*, 417–440. [CrossRef]

34. Bamba, I.; Barima, Y.S.; Bogaert, J. Influence de la densité de la population sur la structure spatiale d'un paysage forestier dans le Bassin du congo en R. D. Congo. *Trop. Conserv. Sci.* **2010**, *3*, 31–44. [CrossRef]

35. Béguin, H. Densité de population, productivité et développement agricole. *Espace Géogr.* **1974**, *4*, 267–272. [CrossRef]

36. Ministère de l'Environnement Conservation de la Nature et Tourisme (MECNT). *Potentiel REDD + de la RDC*; MECNT: Kinshasa, Democratic Republic of the Congo, 2009; p. 66.

37. Torres, B.; Maza, O.J.; Aguirred, P.; Und, L.H.; Günter, S. Contribution of traditional agroforestry to climate change adaptation in the Ecuadorian Amazon: The Chakra System. In *Handbook of Climate Change Adaptation*; Filho, W.L., Ed.; Springer: Berlin/Heidleberg, Germany, 2014; Volume 1, pp. 1–19.

38. Zapfack, L.; Engwald, S.; Sonke, B.; Achoundong, G.; Madong, B.A. The impact of land conversion on plant biodiversity in the forest zone of Cameroon. *Biodivers. Conserv.* **2002**, *11*, 2047–2061. [CrossRef]

39. Lejoly, J.L.; Djele, M.N.; Eerinck, D.G. *Catalogue-Flore des Plantes Vasculaires des Districts de Kisangani et de la Tshopo (RD Congo)*; 4ème.; Taxonomania: Bruxelles, Belgium, 2012; pp. 1–313.

40. Oke, D.O.; Odebiyi, K.A. Traditional cocoa-based agroforestry and forest species conservation in Ondo State, Nigeria. *Agric. Ecosyst. Environ.* **2007**, *122*, 305–311. [CrossRef]

41. Sonwa, D.J.; Nkongmeneck, B.A.; Weise, S.F.; Tchatat, M.; Adesina, A.A.; Janssens, M.J.J. Diversity of plants in cocoa agroforests in the humid forest zone of Southern Cameroon. *Biodivers. Conserv.* **2007**, *16*, 2385–2400. [CrossRef]

42. Abada Mbolo, M.M.; Zekeng, J.C.; Mala, W.A.; Fobane, J.L.; Djomo Chimi, C.; Ngavounsia, T.; Nyako, C.M.; Menyene, L.F.E.; Tamanjong, Y.V. The role of cocoa agroforestry systems in conserving forest tree diversity in the Central region of Cameroon. *Agrofor. Syst.* **2016**, *90*, 577–590. [CrossRef]

43. Dawoe, K.; Asante, W.; Acheampong, E.; Bosu, P. Shade tree diversity and aboveground carbon stocks in Theobroma cacao agroforestry systems: Implications for REDD+ implementation in a West African cacao landscape. *Carbon Balance Manag.* **2016**, *11*, 1–13. [CrossRef] [PubMed]

44. Jadán, O.; Cifuentes, M.; Torres, B.; Selesi, D.; Veintimilla, D.; Günter, S. Influence of tree cover on diversity, carbon sequestration and productivity of cocoa systems in the Ecuadorian Amazon. *Bois Forets Trop.* **2015**, *325*, 35–47. [CrossRef]

45. Noiha, V.N.; Zapfack, L.; Mbade, L.F. Biodiversity management and plant dynamic in a Cocoa Agroforest (Cameroon). *Int. J. Plant Soil Sci.* **2015**, *6*, 101–108. [CrossRef]

46. Somarriba, E.; Cerda, R.; Orozco, L.; Cifuentes, M.; Dávila, H.; Espin, T.; Mavisoy, H.; Ávila, G.; Alvarado, E.; Poveda, V.; et al. Carbon stocks and cocoa yields in agroforestry systems of Central America. *Agric. Ecosyst. Environ.* **2013**, *173*, 46–57. [CrossRef]

47. Hervé, B.; Vidal, S. Plant biodiversity and vegetation structure in traditional cocoa forest gardens in southern Cameroon under different management. *Biodivers. Conserv.* **2008**, *17*, 1821–1835.

48. Sullivan, M.J.P.; Talbot, J.; Lewis, S.L.; Phillips, O.L.; Qie, L.; Begne, S.K.; Chave, J.; Cuni-sanchez, A.; Hubau, W.; Lopez, G.; et al. Diversity and carbon storage across the tropical forest biome. *Sci. Rep.* **2017**, *7–39102*, 1–12. [CrossRef]

49. Shahid, M.; Joshi, S.P. Relationship between tree species diversity and carbon stock density in moist deciduous forest of Western Himalayas, India. *J. For. Environ. Sci.* **2017**, *33*, 39–48. [CrossRef]

50. Grall, J.; Coic, N. *Synthèse des Méthodes D'évaluation de la Qualité du Benthos en Milieu Côtier Préliminaire*; Institut Universitaire Européen de la Mer–Université de Bretagne Occidentale Laboratoire des Sciences de l'Environnement MARin: Bretagne, France, 2006; pp. 1–90.

51. R Core Team. *R: A Language and Environment for Statistical Computing*; R Foundation for Statistical Computing: Vienna, Austria, 2020; pp. 1–3636.

52. Wickham, H. *ggplot2: Elegant Graphics for Data Analysis*; Springer: New York, NY, USA, 2016; ISBN 978-3-319-24277-4.

53. Kindt, R.; Coe, R. *Tree Diversity Analysis. A Manual and Software for Common Statistical Methods for Ecological and Biodiversity Studies*; World Agroforestry Centre (ICRAF): Nairobi, Kenya, 2005; ISBN 92-9059-179-X.

54. Oksanen, J.; Blanchet, F.G.; Friendly, M.; Kindt, R.; Legendre, P.; McGlinn, D.; Minchin, P.R.; O'Hara, R.B.; Simpson, G.L.; Solymos, P.; et al. Vegan: Community Ecology Package. R package version 2.5-6. Available online: https://CRAN.R-project.org/package=vegan (accessed on 18 July 2020).

55. Temgoua, L.F.; Dongmo, W.; Nguimdo, V.; Nguena, C. Diversité ligneuse et stock de carbone des systèmes agroforestiers à base de cacaoyers à l'Est Cameroun: Cas de la forêt d'enseignement et de recherche de l'Université de Dschang. *J. Appl. Biosci.* **2018**, *122*, 12274–12286. [CrossRef]

56. Zapfack, L.; Chimi Djomo, C.; Noiha Noumi, V.; Zekeng, J.C.; Meyan-ya Daghela, G.R.; Tabue Mbobda, R. Correlation between associated trees, cocoa trees and carbon stocks potential in cocoa agroforests of Southern Cameroon. *Sustain. Environ.* **2016**, *1*, 71–84.

57. Salazar, J.C.S.; Bieng, M.A.N.; Melgarejo, L.M.; Di Rienzo, J.A.; Casanoves, F. First typology of cacao (*Theobroma cacao* L.) systems in Colombian Amazonia, based on tree species richness, canopy structure and light availability. *PLoS ONE* **2018**, *13*, e0191003. [CrossRef] [PubMed]

58. Yao, C.Y.A.; Kpangui, K.B.; Vroh, B.T.A.; Ouattara, D. Pratiques culturales, valeurs d'usage et perception des paysans des espèces compagnes du cacaoyer dans des agroforêts traditionnelles au centre de la Côte d'Ivoire. *Rev. d'ethnoécologie* **2016**, *9*, 1–19. [CrossRef]

59. Sonwa, D.; Weise, L.; Chimi Djomo, C.; Kabelong Banoho, L.-P.; Forbi Preasious, F.; Tsopmejio Temfack, I.; Tajeukem Vice, C.; Ntonmen Yonkeu, A.; Tabue Mboba, R.; Nasang, J. Carbon storage potential of cacao agroforestry systems of different age and management intensity. *Clim. Dev.* **2018**, 1–12. [CrossRef]

60. Santhyami, S.; Basukriadi, A.; Patria, M.P.; Abdulhadi, R. The comparison of aboveground C-stock between cocoa-based agroforestry system and cocoa monoculture practice in West Sumatra, Indonesia. *Biodiversitas* **2018**, *19*, 472–479. [CrossRef]

61. Häger, A.; Fernández Otárola, M.; Stuhlmacher, M.F.; Acuña Castillo, R.; Contreras Arias, A. Effects of management and landscape composition on the diversity and structure of tree species assemblages in coffee agroforests. *Agric. Ecosyst. Environ.* **2014**, *199*, 43–51. [CrossRef]

62. Bisseleua, D.H.B.; Missoup, A.D.; Vidal, S. Biodiversity conservation, ecosystem functioning, and economic incentives under cocoa agroforestry intensification. *Conserv. Biol.* **2009**, *23*, 1176–1184. [CrossRef]

63. Omatoko, J.; Nshimba, H.; Bogaert, J.; Lejoly, J.; Shutsha, R.; Shaumba, J.P.; Asimonyio, J.; Ngbolua, K.N. Etudes floristique et structurale des peuplements sur sols argileux à Pericopsis elata et sableux à Julbernardia seretii dans la forêt de plaine d'UMA en République Démocratique du Congo. *Int. J. Innov. Appl. Stud.* **2015**, *13*, 452–463.

64. Kambale, J.K.; Asimonyio, J.A.; Shutsha, R.E.; Katembo, E.W.; Tsongo, J.M.; Kavira, P.K. Etudes floristique et structurale des forêts dans le domaine de chasse de Rubi-Télé (Province de Bas-Uélé, République Démocratique du Congo). *Int. J. Innov. Sci. Res.* **2016**, *24*, 309–321.

65. Koubouana, F.; Ifo, S.A.; Ndzai, S.F.; Stoffenne, B. Étude comparative d'une forêt primaire et d'une forêt dégradée au Nord de la République du Congo par référence à la structure des forêts tropicales humides. *Rev. Sci. Tech. Forêt Environ. du Bassin du Congo* **2018**, *11*, 11–25.

66. Sonwa, D.; Weise, S.F.; Tchatat, M.; Nkongmeneck, B.; Adesina, A.A.; Ndoye, O.; Gockowski, J. Rôle des agroforêts cacao dans la foresterie paysanne et communautaire au Sud-Cameroun. *Réseau For. Pour Développement Rural, document du réseau* **2001**, *25g*, 1–11.

67. Asare, R.; Asare, R.A.; Asante, W.A.; Markussen, B.; RÆbild, A. Influences of shading and fertilization on on-farm yields of cocoa in Ghana. *Exp. Agric.* **2017**, *53*, 416–431. [CrossRef]

68. Bohlman, S.A.; Bourg, N.A.; Brinks, J.; Bunyavejchewin, S.; Butt, N.; Chisholm, R.A.; Muller-Landau, H.C.; Rahman, K.A.; Bebber, D.P.; Bin, Y.; et al. Scale-dependent relationships between tree species richness and ecosystem function in forests. *J. Ecol.* **2013**, *101*, 1214–1224.

69. Sagar, R.; Singh, J.S. Tree density, basal area and species diversity in a disturbed dry tropical forest of northern India: Implications for conservation. *Environ. Conserv.* **2006**, *33*, 256–262. [CrossRef]

70. Blanc, L.; Florès, O.; Molino, J.F.; Gourlet-Fleury, S.; Sabatier, D. Diversité spécifique et regroupement d'espéces arborescentes en Forêt Guyanaise. *Rev. For. Fr.* **2003**, *55*, 131–146. [CrossRef]

71. Asase, A.; Ofori-frimpong, K.; Ekpe, P. Impact of cocoa farming on vegetation in an agricultural landscape in Ghana. *Afr. J. Ecol.* **2009**, *48*, 338–346. [CrossRef]

72. Middendorp, R.; Vanacker, V.; Lambin, E. Impacts of shaded agroforestry management on carbon sequestration, biodiversity and farmers income in cocoa production landscapes. *Landsc. Ecol.* **2018**, *33*, 1953–1974. [CrossRef]

73. Asigbaase, M.; Sjogersten, S.; Lomax, B.; Dawoe, E. Tree diversity and its ecological importance value in organic and conventional cocoa agroforests in Ghana. *PLoS ONE* **2019**, *14*, e0210557. [CrossRef]

74. Niether, W.; Armengot, L.; Andres, C.; Schneider, M.; Gerold, G. Shade trees and tree pruning alter throughfall and microclimate in cocoa (*Theobroma cacao* L.) production systems. *Ann. For. Sci.* **2018**, *75*, 38. [CrossRef]

75. Redd-RDC. *Mambasa Pays: République Démocratique Du Congo, Rapport D'évaluation du Projet*; Banque Africaine de Développement: Kinshasa, Democratic Republic of the Congo, 2010; pp. 1–18.

76. Mason, J.J.; Asare, R.; Cenamo, M.; Soares, P.; Carrero, G.; Murphy, A.J.; Bandari, C. *Ghana Cocoa REDD+ Programme: Draft Implementation Report*; Ghana cocoa Reed+ programme: Accra, Ghana, 2016; pp. 1–77.

77. Carodenuto, S.; Gromko, D.; Chia, E.L. Zero deforestation cocoa in Cameroon: Private sector engagement to support Reducing Emissions from Deforestation and forest Degradation (REDD+). *GIZ-ProPFE Policy Br.* **2017**, *4*, 1–3.

78. Fernandes, C.D.A.F.; Matsumoto, S.N.; Fernandes, V.S. Carbon stock in the development of different designs of biodiverse agroforestry systems. *Rev. Bras. Eng. Agrícola Ambient.* **2018**, *22*, 720–725. [CrossRef]

79. Sonwa, D.J.; Weise, S.F.; Bernard, A.; Nkongmeneck, M.T.; Janssens, M.J.J. Profiling carbon storage/stocks of cocoa agroforests in the forest landscape of southern Cameroon. In *Agroforestry: Anecdotal to Modern Science*; Dagar, J., Tewari, V., Eds.; Springer: Singapore, 2017; pp. 739–752.

Publisher's Note: MDPI stays neutral with regard to jurisdictional claims in published maps and institutional affiliations.

 *forests*

Article

The Agroforestry Heritage System of Sabana De Morro in El Salvador

Antonio Santoro [1,*], Ever Alexis Martinez Aguilar [2], Martina Venturi [1], Francesco Piras [1], Federica Corrieri [1], Juan Rosa Quintanilla [2] and Mauro Agnoletti [1]

[1] Department of Agriculture, Food, Environment and Forestry (DAGRI), University of Florence, Via San Bonaventura 13, 50145 Firenze, Italy; martina.venturi@unifi.it (M.V.); francesco.piras@unifi.it (F.P.); federica.corrieri@unifi.it (F.C.); mauro.agnoletti@unifi.it (M.A.)

[2] Facultad de Ciencias Agronómicas, Universidad de El Salvador, 1101 San Salvador, El Salvador; ever.martinez@ues.edu.sv (E.A.M.A.); juan.quintanilla@ues.edu.sv (J.R.Q.)

[*] Correspondence: antonio.santoro@unifi.it

Received: 4 June 2020; Accepted: 6 July 2020; Published: 9 July 2020

Abstract: Traditional agroforestry systems are recognized as having great importance for providing multiple benefits for local communities all over the world, especially in tropical countries. Thanks to their multifunctional role, they can support small farmers, contribute to hydrogeological risk reduction, water regulation, preservation of soil, agrobiodiversity and landscape, as well as being examples of mitigation and adaptation towards climate change. The Globally Important Agricultural Heritage Systems (GIAHS) programme of the Food and Agriculture Organization (FAO) aims to identify agricultural systems of global importance, preserving landscape, agrobiodiversity and traditional knowledge, through dynamic conservation principles. The Sabana de Morro is a traditional agroforestry system located in El Salvador based on cattle grazing in pastures with the presence of *Crescentia alata* and *Crescentia cujete* trees, locally called Morro or Jícaro. We documented the main characteristics of this system, that has never been deeply studied, in the Municipality of Dolores, in accordance with the five GIAHS criteria, and through detailed land use mapping, to assess the relations between landscape structure, agrobiodiversity and traditional silvopastoral practices. Sabana de Morro proved to be based on strong interactions between trees, cattle and farmers. The pulp of the Morro fruits is eaten by grazing cattle, completing their feeding and giving a peculiar taste to the locally produced cheese. Morro trees provide shade for the animals while cattle contribute by spreading their seeds that also take advantage of the manure. Results show that this agroforestry system contributes to the preservation of a rich agrobiodiversity and of the traditional landscape. At the same time, it supports local farmers' livelihood and is consistent with the aim of the GIAHS programme, even if further surveys and research are needed to assess the real possibility of the inclusion in this FAO programme.

Keywords: agricultural heritage; *Crescentia alata*; *morro*; GIAHS; El Salvador; traditional agroforestry

---

1. Introduction

In the last decades, traditional agroforestry systems have received increasing attention, being examples of sustainable systems able to provide multiple benefits for local communities all over the world, especially in tropical developing countries. In fact, these realities are often characterized by a multifunctional role [1–3]. The range of functions can be wide, depending on multiple factors as agroforestry systems can meet financial, social and environmental objectives [4]. Traditional agroforestry systems may potentially support livelihood improvement for small farmers and their families through simultaneous production of food, fodder and firewood [5], and they are often characterized by more stable levels of total production per unit area than high-input systems [6]. Moreover, they could contribute to hydrogeological risk reduction, water regulation, preservation of soil, agrobiodiversity and other natural resources [7]. Finally, traditional agroforestry systems might represent important tourist destinations for rural tourism, as well as examples of mitigation and adaptation towards climate change [8].

At the international level, the growing interest on topics related to rural areas' sustainable development, based on traditional agricultural systems, has contributed to the establishment of a FAO (Food and Agricultural Organization) programme called GIAHS (Globally Important Agricultural Heritage Systems). This programme, launched in 2002, has the aim of identifying and preserving agricultural systems of global importance with their landscapes, agro-biodiversity, traditional knowledge and associated culture. However, the aim is not only the preservation of these systems, but also to apply dynamic conservation principles in order to promote sustainable development of rural areas, with direct and indirect benefits for the community [9].

The Sabana de Morro, also known as Morrales, is a traditional agroforestry system located in El Salvador based on free cattle grazing in pastures characterized by the presence of *Crescentia alata* and *Crescentia cujete* trees, locally called Morro or Jícaro. The Sabana de Morro zone is distributed in different areas all over the country, as in the upper valley of the Lempa river, as well as in other regions such as Acajutla (Punta Remedios) in the plain of Ahuachapán, Santa Ana (Candelaria de la Frontera), Cabañas (Dolores), San Vicente (San Ildefonso) and Morazán (San Carlos). In some areas (Dolores, Chalatenango, San Miguel, Morazán and San Vicente), local farmers use the Morro fruit to integrate the cattle feeding because in the dry season there is shortage of green fodder, while in other parts of the country the use of the Morro fruits for cattle feeding is not traditionally applied by farmers [10]. The presence of *Crescentia* spp. trees in pastures is common also in other Central American countries and cattle are reported to frequently eat the pulp of the fruits [11–13].

The current research focused on Dolores Municipality where the tradition of cattle breeders and cheese makers is heritage-derived from the Spanish colonization and is supposed to have started with the foundation of the settlement in 1681 and still practiced today with very few changes. In this area, the Sabana de Morro represents a silvopastoral system based on strong interactions between trees, cattle and farmers. Morro trees provide shade to animals during the hottest hours of the day and the pulp of the fruits is eaten by the grazing cattle. Morro trees also take advantage from the cattle, as cows do not cause damage to young trees and contribute to spread the seeds that also take advantage of the manure to germinate. Farmers apply specific and traditional management of the pastures to ensure the renovation of Morro trees, as they believe that Morro fruits give a peculiar taste to the milk and to the locally produced cheese that is particularly appreciated and can guarantee a fair income to the families of the local farmers.

The main aim of the research was to investigate relationships between specific landscape structure, agrobiodiversity and the maintenance of traditional silvopastoral practices. Despite the fact that different studies in the American tropics mention the presence of *Crescentia* spp. trees in pastures, no specific studies have been carried out [14]. The research was based on detailed land use mapping, that might also represent a starting point for future monitoring, and on an accurate description of the system based on GIAHS criteria, since one of the aims was to verify if the Sabana de Morro has the characteristics to be proposed for the inscription in the GIAHS programme. Sabana de Morro, in fact, could represent an example of a traditional agroforestry system important for sustainable rural development, local farmers' life quality and preservation of related agrobiodiversity, traditional knowledge and landscape. This paper is also one of the output of the "GIAHS Building Capacity" project, funded by the Italian Agency for Development Cooperation (AICS) and by the Department of Agriculture, Food, Environment and Forestry (DAGRI) of the University of Florence, with the aim of spreading knowledge about GIAHS-related issues and for identifying potential GIAHS sites.

2. Materials and Methods

2.1. The Study Area

The site coveres 14,823 ha and corresponds to the Municipality of Ciudad Dolores (Department of Cabañas, El Salvador) (Figure 1). Dolores municipality is divided between the City (124 ha) and 6 Cantones: Cañafistula (4047 ha), Curarén (1511 ha), Chapelcoro (2459 ha), El Rincón (3194 ha), Niqueresque (590 ha), San Carlos (1656 ha). Dolores municipality has an approximate population of 6347 habitants (3064 males and 3283 females) with a density of population of 43 people/km^2. Rural activities are particularly important, as only 22.3% of the population is classified as urban.

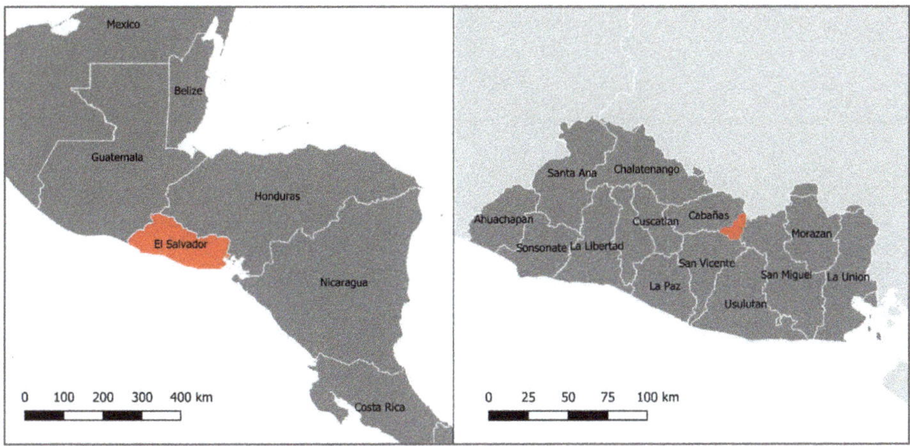

Figure 1. El Salvador is a small country in Central America (**left**). The Municipality of Dolores (13,581 ha) is in the south eastern part of Cabañas Department (**right**).

Dolores municipality has an elevation ranging from 33 to 577 m above sea level., with an average altitude of 195 m above sea level. However, most of the area is flat or with gentle slopes (half of the surface has a slope between 0% and 20%, and only 10% of the surface has a slope greater than 50%). There are higher and steeper slopes, especially in the western part of the area. The site is crossed by a rich hydrography. The eastern boundary is defined by the Rio Lempa and the southern one by the Rio Titihuapa, while smaller rivers cross the area, such as the Rio Gualpuca, the Rio Sisicua and the Rio Marcos.

According to the Köppen-Geiger climate classification, the local climate is Aw-Equatorial savannah with dry winter [15], with average rain per year of 1823 mm, and average temperature of 26.6 °C [16].

2.2. Methodology

The applied methodology was divided in two parts. The first was based on the creation of a detailed map of the Sabana de Morro agroforestry system. In fact, an exhaustive land use map represents a crucial instrument to study the transformations that could have affected the site in the past or could affect the site in the future. Moreover, a detailed land use map allows to deeply describe the landscape structure and the interrelation among the five GIAHS criteria. The land use map was realized through photointerpretation of Google Satellite images of November 2017 using Quantum GIS 3.10.3 and applying a minimum mapping surface of 250 m^2.

A Digital Terrain Model (DTM) was used to calculate slope classes. The DTM used had a 30 m resolution and was provided by the US Geological Service. It was processed using Quantum GIS 3.10.3, with support of the GRASS plug-in, to obtain the following informative layers: slopes (and slopes class), aspects and elevation class. From these layers, each characteristic was linked to each land use patch to obtain the corresponding value for each patch. At the end of this process, the attribute table of the land use layer included, for each patch, the land use categorization, the elevation class, the slope class and the dominant exposure.

Two indexes were applied to evaluate the land use structure. The first was the LSI (Landscape Shape Index), an index derived from the Edge Density, that was used to evaluate the degree of fragmentation for each land use through segmentation of edge based on the perimeter and the area [17]. The higher the *LSI* value, the greater the fragmentation. This index was calculated by the following formula:

$$LSI = \frac{p_i}{2\sqrt{\pi a_i}}$$

where p_i is the perimeter of each patch in meters and a_i is the area of the patch in hectares. After calculating the *LSI* for each patch, an average value for each land use was calculated.

The other index applied was Hill's Diversity Number that is used to obtain indications about the effective number of land uses that contribute to the diversity of the landscape [18]. It was calculated according to the following formula:

$$N_1 = e^{-\Sigma\left(\frac{n_1}{N}\right)*\ln\left(\frac{n_1}{N}\right)}$$

where n_1 is the total surface of the *i* land use class, and N is the total area of the study area.

The second part of the study was the description of the Sabana de Morro agroforestry system according to the five GIAHS criteria. Since one of the aims of the research was to verify if this agroforestry system could be a potential GIAHS site, it was important to carefully analyse the site peculiarities according to the concept of global importance and to the five GIAHS criteria, that are:

1. Food and livelihood security.
2. Agro-biodiversity.
3. Local and Traditional Knowledge systems.
4. Cultures, Value systems and Social Organisations.
5. Landscapes and Seascapes Features.

The concept of global importance is related to the contemporary relevance of the system, that is, according to FAO, "established by its present and future capacity to provide food and livelihood security, to contribute to human well-being and quality of life, and to generate other local, national and global economic and environmental goods and services to its community and wider society".

This second part of the paper is mainly based on bibliographic research, official statistical data, field surveys and observations made by the authors. The main source for statistical data was the IV Censo Agropecuniario [19]. Unfortunately, this survey refers to 2007 and there is no more recent reliable data, but it proved to be crucial for delineating the socioeconomic and productive framework.

3. Results

3.1. Land Use Structure

Detailed land use mapping refers to 2017 (Figure 2). Most of the analyzed area is occupied by forests (34.3%) and by Sabana de Morro (27.9%), followed by pastures (14.1%), shrublands (9.3%), cultivations (5%) and other land uses (9.4%) (Table 1). Different types of cultivations can be found in 5% of the surface, mainly located near the river where the soil is more fertile or near the main urban area, the city of Dolores. Urbanized areas account for 2.3% of the surface and mainly consist of Dolores city, while other urbanized areas are made of very small villages.

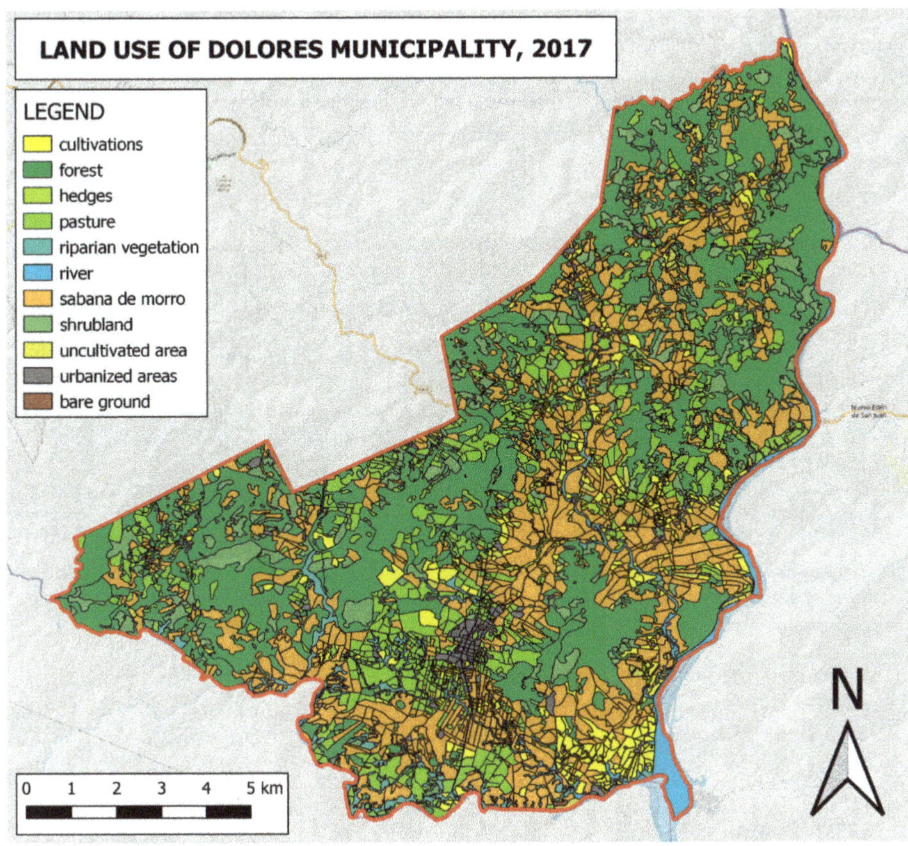

Figure 2. Land use map of 2017 for the municipality of Dolores, El Salvador.

Table 1. Different land uses, their surfaces and the Landscape Shape Index (LSI) for the Dolores municipality 2017.

Land Use	Surface (ha)	Surface (%)	LSI
bare ground	137	0.1	139.3
cultivations	745.6	5.0	136.2
forest	5080.7	34.3	211.0
hedges	137.8	0.9	247.4
pasture	2087.5	14.1	139.2
riparian vegetation	429.9	2.9	283.9
river	326.5	2.2	299.9
sabana de morro	4132.9	27.9	145.1
shrubland	1376.9	9.3	150.2
uncultivated area	151.0	1.0	143.8
urbanized areas	341.7	2.3	135.4
Total	14,824.3	100.0	

The total number of land uses was 11, while the Hill's Diversity Index was less than half [5.7], highlighting that the landscape is mainly characterized by a small number of land uses. In fact, the local landscape structure is based on big forested patches and smaller patches occupied by Sabana de Morro (average area of 3.9 ha) and cultivations (average area of 1.6 ha).

In Dolores, Sabana de Morro and cultivations are arranged in regular plots, as highlighted by the LSI values (Table 1), often divided by hedges, creating a complex mosaic at the lower altitudes (Figure 3). Despite the total small surface occupied, land uses such as hedges (0.9%) and riparian vegetation (2.9%), are particularly important for biodiversity, as they represent ecological networks, and for the preservation of a peculiar landscape structure. These land uses are also characterized by a high shape complexity since their LSI values are higher than those of other land uses, highlighting the ecological and landscape importance of these vegetal structures.

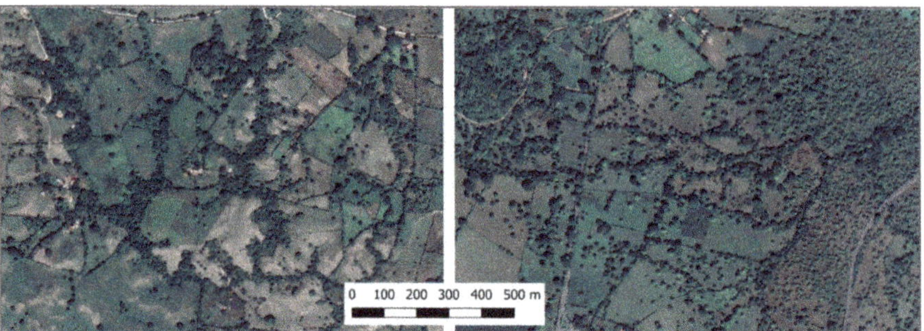

Figure 3. The presence of riparian vegetation and hedges of different width contributes in shaping a complex and aesthetically valuable landscape structure. These vegetal structures have also a great importance for biodiversity.

The density of the Morro trees varies from few trees to a high density where the crowns of the trees are almost connected to one another [20]. During our research, a survey of Morro trees density was done based on 10 different plots. The result confirmed the density variability, as detected Morro trees density varied from 8 trees/ha to 45 trees/ha (Figure 4).

Figure 4. The density of Morro trees is very variable, ranging from 8 trees/ha (**left**) to 45 trees/ha (**right**).

In addition, it was possible to identify a specific disposition of Sabana de Morro and forest patches, the two main land uses characterizing the landscape, both regarding altitude and slope class (Figure 5). Sabana de Morro is mainly located in areas characterized by an altitude lower than 200 m above sea level and a slope gradient between 0 and 20%. Forests can be mainly found at altitudes between 100 and 250 m, and on slopes in the range 30–50%. On the contrary, there seemed to be no correlation between the exposure of the slopes and land use distribution, except for a slightly greater presence of the forests on the slopes with a predominantly northern exposure.

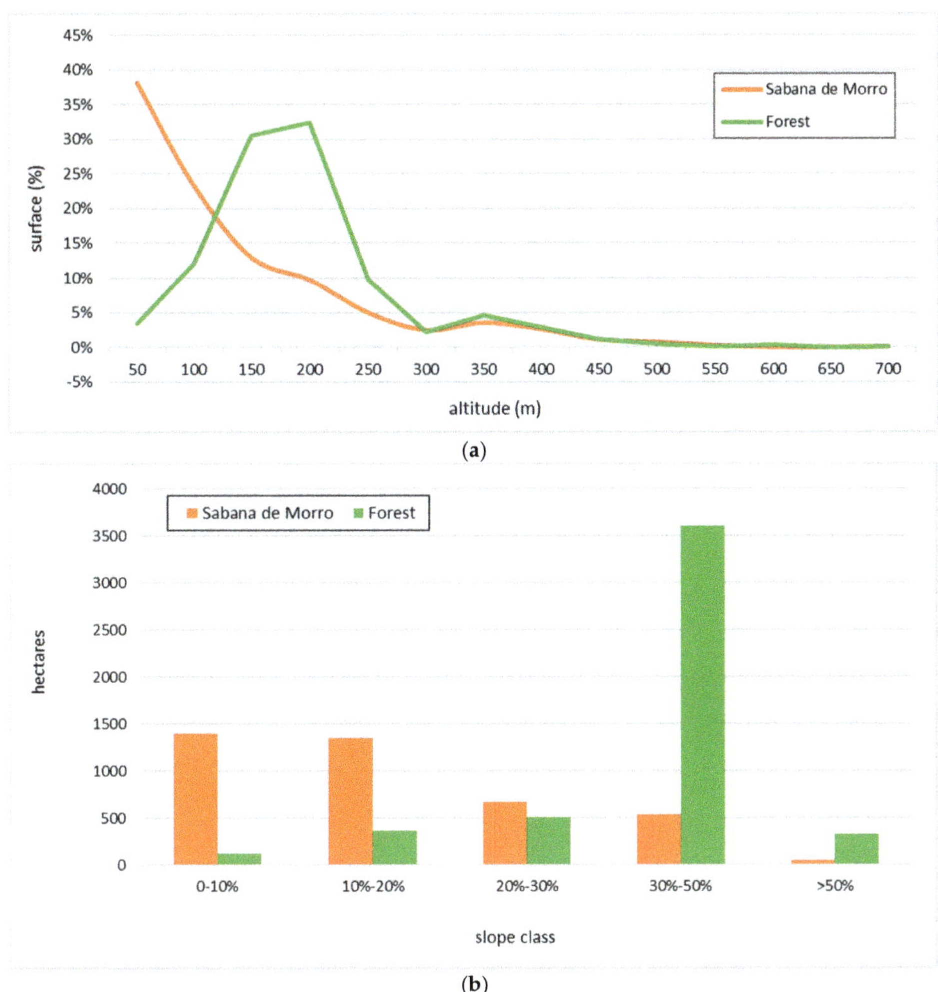

Figure 5. Distribution of Sabana de Morro and forests according to altitude (**a**) and slope class (**b**).

3.2. Description of the Agroforestry System according to the GIAHS Criteria

3.2.1. Food and Livelihood Security

Agricultural activity creates important job demands for families living in the area. Cattle breeding is one of the main agricultural activities here (Figure 6), and 260 breeders still use Morro fruits to integrate cattle feeding. The number of cattle, according to 2007 data for Dolores Municipality, amounts to 14,139, with 3888 calves, 2997 steers, 485 bulls, 64 oxen, 3064 cows not in production and 3642 cows under milk production [19]. The fact that local livelihood is based on cattle breeding and crops is also confirmed by a work of the World Food Program and the Government of El Salvador [21] that classified the municipality of Dolores in the class "western zone of basic grains and cattle".

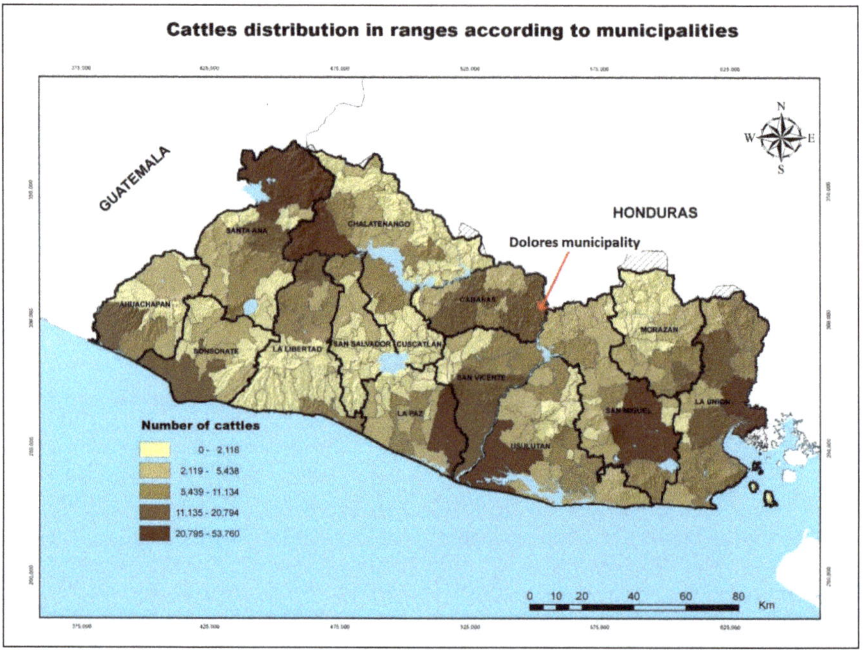

Figure 6. Dolores is one of the municipalities with a higher number of bovines in El Salvador (Source: Ministerio de Economia 2009, modified by the authors).

Average milk production, according to local farmers, is equal to 8 L/cow (11 bottles), in line with values of 4–8 L per day recorded by Ortéz et al. [22]. Almost all the local farmers sell their production of milk to processors of cheese. They say that in the past they used to produce Puebla Cheese on their own, but nowadays it is more practical and economically feasible to sell the entire milk production to big local cheese producers.

Beside cattle breeding, local farmers cultivate different crops, the main ones being maize (*Zea mays*), sorghum (*Sorghum vulgare*) and beans (*Phaseolus vulgaris*, Chaparrastique variety), with two production cycles per year (Table 2). Sorghum is used for feeding animals, and other forage species on smaller surfaces can be found in the areas such as *Hyparrhenia rufa*, *Digitaria swazilandesis*, *Digitaria decumbens*, *Brachiaria decumbens*, *Pennisetum purpureum*, *Cynodon* spp.

Table 2. Production (quintals) of maize, sorghum and beans in Dolores Municipality in 2017 [19].

	Human Consumption	Animal Consumption	For Seeds	To Sell
Maize	21,128	6802	17	14,713
Sorghum	-	8478	66	2379
Beans	464	-	8	83

Poultry and pigs that, respectively, are 858 and 514 in Dolores, are mainly raised for self-consumption or sold at the weekly local market that is an important place for farmers to sell animals and animal products. According to 2007 data, beside cattle, poultry and pigs, there were also 55 equines, 10 goats, 25 pelibueyes sheep and 33 beehives [19]. In the Department of Cabañas, migration of young people towards bigger cities is frequent, and the hiring of Honduran workers for agricultural activities is a usual practice.

Regarding the composition of Morro fruits, Benavides [23] found it had 17% of crude protein and 32% of in vitro digestibility of the dry matter. Zamora et al. [24] recommended giving 3.56–4.4 kg/animal/day of Morro fruit to adult cows and bulls after a period of adaptation. The quality of Morro fruit for cattle feeding is very good based on chemical analysis (Table 3), with 22.54% of crude protein in the pulp.

Table 3. Chemical analysis of Morro fruit. Values are expressed in percentages (Source: Department of Agricultural Chemistry, Faculty of Agronomic Science, UES).

	Humidity	Dry Matter	Ashes	Crude Protein	Ethereal Extract	Crude Fiber	Carbohydrates
Pulp	5.77	11.41	8.17	22.54	15.91	12.00	41.38
Peel	4.18	28.50	6.20	6.99	0.42	39.69	46.69

3.2.2. Agro-Biodiversity

The Sabana de Morro derives from the natural vegetation of El Salvador and represents a case of typical wooded savanna but can also adopt characteristics of steppe [12]. Sabana de Morro traditional management allows the growth of different herbaceous species on the pastures, resulting in a rich agrobiodiversity.

Typical plants of this agroforestry system are *Crescentia alata* and *Crescentia cujete*, locally called Morro and Jícaro, and can be recognized easily by the shape of their fruits. *Crescentia alata* is a small tree (4–8 m, max 18 m), with a diameter at chest height of 30 cm (max 60), slow growth, originally from Mexico and tolerant of temporary floods and soils with bad drainage. Fruits are round shaped with a diameter of 7–10 cm (max 15 cm), while *Crescentia cujete* is very similar but its fruits are bigger (15–30 cm). Morro trees start to produce after 15 years and, according to local farmers, can still produce after 50 years. Fruit production can reach up to 27 kg/tree/year from the 8th year, and varies between 10–200 (averages of 60–80) per tree [25].

Natural herbaceous species growing on Sabana de Morro pastures mainly belong to Gramineae, Leguminosae or Asteraceae families. Common herbaceous plants include *Cynodon dactylon* (barrenillo), *Echinocloa frumentacea*, *Echinocloa polystachya* (pasto alemán) and *Digitaria sanguinalis* (pata de gallina). Some of them are considered weeds, such as *Cyperus michelianus* and *C. difformis* (coyolillo) or Stylosanthes hamata. The Morrales are also rich in epiphytes: *Tillandsia caput-medusae* and *T. schiedeana* (both known as gallitos), and Orchids like *Laelia rubescens*. *Hylocereus undatus* (pitahaya) can also be found in the area. If the Sabana de Morro is not used for some years, spontaneous shrubs grow, starting an ecological succession towards Sabana de Arbustos Espinosos (thorny bushes savannah) that is formed by *Acacia farnesiana* and *Acacia cornigera* [26].

Regarding cattle breeding, local farmers used to cross different breeds, some of European origin (Holstein, Brown Swiss) that have good production of milk, with Brahman (*Bos indicus*) that is more tolerant to tropical conditions but produces less milk. An exception is the Gyr (*Bos indicus*) breed, recently introduced, with a high amount of milk production and good adaptation to the tropical climate. Pigs are also raised locally. Unfortunately, the number of traditional Creole breeds is decreasing with respect to the introduced Dalland breed. Local varieties of chickens, turkeys, goats, horses and sheep can be found in smaller numbers.

The different habitats that originate from the Sabana de Morro agroforestry systems are homes to a variety of wild animals including birds (*Polyborus plancus*, *Buteo nitidus*, *Icterus pectoralis*, *Columbina talpacoti*, *Eumomota superciliciosa*, *Zenaida asiatica*), reptiles (*Ctenosaura similis*, *Iguana iguana*, *Drymarchon melanurus*, *Boa constrictor*) and mammals (*Dasypus novemcinctus*, *Didelphis marsupialis*, *Canis latrans*, *Odocoileus virginianus*, *Sylvilagus floridanus*). Among the mammals, the coyote (*Canis latrans*) can cause significant economic losses for local farmers as it can attack pigs, poultry and cattle calves. Rivers and torrents are also rich in fish, shrimps and crabs.

3.2.3. Local and Traditional Knowledge Systems

The most common way for cattle to consume Morro fruits is to freely eat them when they fall on the ground. This system does not require a workforce or costs for the farmers, but it can be dangerous for the animals as entire fruits can remain stuck in their throats and even cause death by asphyxia (enmorradas). Farmers can intervene rapidly to avoid animal death by pushing the fruit down the throat of the animal with the thin stick obtained from the central part of the stem of a banana tree. Another way for farmers to supply the Morro fruit is by collecting the fruits from the trees, cutting them with a machete and leaving them on the ground. This method can cause problems too, as cutting the fruit can result in sharp edges of the hard fruit peel that can cause internal lacerations. The third way is to harvest the Morro fruits when they are ripe, or just before, cutting the fruit, extracting the pulp and putting it in the cattle feeder. The traditional animal-feeder is called a *canoa*, a long and rectangular shaped wooden container. In the last years, the use of Morro fruits for cattle feeding has been decreasing, as it is more profitable to sell the fruits for handicrafts ($ 40.00/1000 fruits) than pay workers to harvest and give the pulp to the cattle. However, some farmers still apply this traditional practice with the involvement of all the family members.

Cattle breeding is carried out according to traditional practices too. The management of pasture consists of letting the cattle graze in one plot until the grass lasts and then moving the cattle to another plot to allow vegetation growth. In recent years, farmers have not relied only on free grazing for milk producing cows but have usually integrated grazing with concentrated nutrition to get maximum milk production while the rest of the animals are fed only on pasture. In the dry season, farmers feed all the cattle with silage, which is prepared by farmers at the end of the rainy season with maize or sorghum that is cultivated by themselves. During the flowering of the Morro trees, after pollination when petals fall on the ground, farmers collect petals or send the animals on the pasture since Morro flowers can represent a source of food for the cattle. When a calf is born, the farmer lets the cow and the calf together for a week and during this first week the total milk production is used to feed the calf. After a week, the farmer starts manually milking with the calf tied to the leg of the cow, with three quarters of the milk for the farmer and one for the calf. After milking, the farmer releases both the cow and the calf and lets them stay together so that the calf learns to eat grass, separating them only during the night. This management lasts approximately five and a half months after the calf is born.

Even the transformation of milk into the traditional square-shaped cheese follows the same techniques of centuries ago, and is part of the cultural heritage of local farmers. After milk is collected it is deposited in a pot where rennet is added. Once the milk is curdled it is cut into cubes with a knife while buttermilk is drained from the cubes of cheese curd and put in a barrel, adding salt to prepare brine. The cubes of cheese are immersed in the brine for three days after which the cubes of cheese are put on a wooden square mold wrapped in tissue. Cheese curd is pressed using a wood press for two days to drain the remaining buttermilk and brine and to mature the curd. The result is a square Puebla cheese ready to consumption.

3.2.4. Cultures, Value Systems and Social Organizations

In Dolores municipality, according to 2007 data, there were 1005 farmers and about 2257 employed in agricultural activities [19]. These data mean that about 15.8% of the population was made of farmers and another 35.7% employed in agriculture, without taking into account that family members are often involved in farming activities too, so the social structure of Dolores was particularly based on rural activities. The importance of small farmers in Dolores was confirmed by 2007 official data [19], as only 290 farmers were classified as commercial farmers (18.6% women), while 714 were classified as small-producers (11.1% women), and 275 were family-farms (Figure 7). As these data show, the role of women as conductors of farms was, unfortunately, subordinate. Moreover, 56.7% of farms had a surface smaller than 1 ha, and 16% between 1 ha and 5 ha.

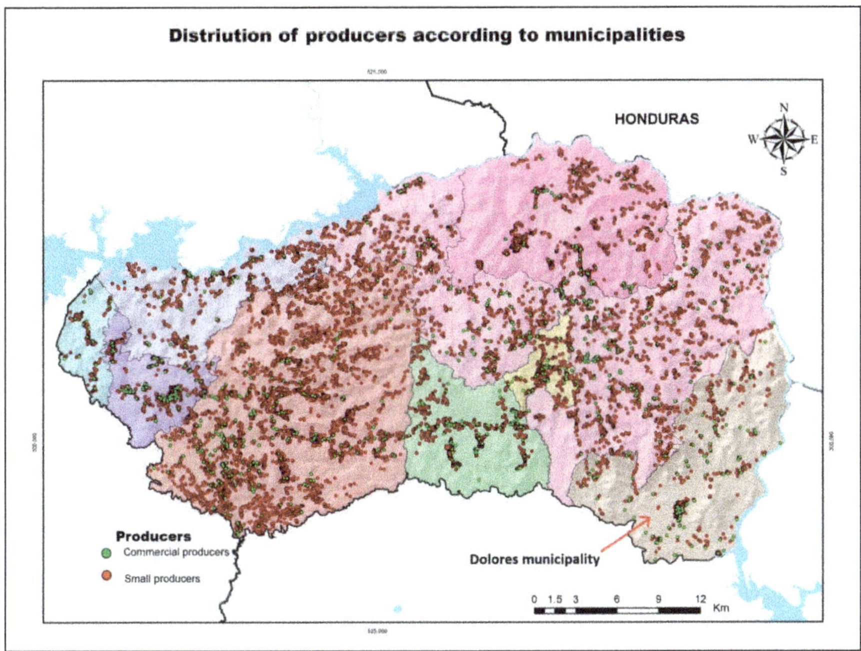

Figure 7. Official data about the presence of commercial farms (green) and small farms (red) in Cabanas department (Source: Ministerio de Economia 2009, modified by the authors) highlights the importance of small-producers and family-farming.

The main social organization related to cattle breeding in Dolores is a cooperative called ASOPUEBLA de RL, constituted by 27 breeders. Moreover, the government of El Salvador through the CENTA (Centro Nacional de Tecnologia Agropecuaria y Forestal) agency located in Guacotecti (Cabañas), gives technical support to at least 12 families of the municipality of Dolores in their agricultural and cattle breeding activities. According to 2007 data, 54 farmers requested and obtained subsidies for agricultural activities, and 27 for breeding activities, from public or private banks or from NGOs (Table 4).

Table 4. Number of farmers who requested subsidies for agricultural and breeding activities and financing institutions (Data source: Ministerio de Economia 2009).

Total Number of Subsidies Requests	Number of Farmers Who Asked and Received Subsidies for Agricultural Activities	Number of Farmers Who Asked and Received Subsidies for Breeding Activities	Private Banks	National Bank	ONGs	Local Lender
69	54	27	8	42	5	16

3.2.5. Landscapes and Seascapes Features

The Sabana de Morro is not only an ecosystem or an ecological classification of tropical dry forest, but also represents a very defined landscape (Figure 8) characterized by specific vegetation and plain or smooth slope morphology. The main features of the landscape structure are reported in Section 3.1. The landscape is mainly made of Sabana de Morro patches surrounded by bigger forested patches. Local forests are classified as Tropical Dry Forest typical of the climate Tropical Arid Low-Hot Land (below 700 m above sea level.) with 4–7 months rain-free dry season. These forests are characterized by the fall of leaves every year during the dry season and represent the potential forest type of El Salvador at lower altitudes that has been largely cut in the past to obtain land for agricultural activities. The most common species are *Byrsonima crassifolia* (Nance), *Apeiba tibourbou* (Peine de mico), *Curatelia americana* (Chaparro), *Stemmadenia donnell-smithii* (Cojon de puerco),

Bursera simaruba (Jiote), *Gliricidia sepium* (Madre cacao), *Luehea candida* (Cabo de hacha), *Cochlospermum vitifolium* (Tecomasuche), *Guazuma ulmifolia* (Caulote), *Tabebuia chrysantha* (Cortez), *Ipomoea arborescens* (Siete pellejos). All these species are not commonly used to obtain high-quality timber but for firewood or for fence poles, like *Gliricidia sepium* and *Bursera simaruba* [12,26–31].

Figure 8. Pictures of Sabana de Morro in Dolores Municipality, El Salvador.

Beside rural features, the landscape is also characterized by small settlements. The biggest one is the city of Dolores with the main church, a remnant from the time of the Spanish colony that over the centuries has been renovated several times due to damages caused by earthquakes. Some houses of Dolores still preserve the traditional design of colonial houses with a shingle roof and a porch in front of the entrance to create a cool and shaded place.

4. Discussion

According to the results of this study, Sabana de Morro seems to be a well-preserved traditional agroforestry system driven by a rural economy mainly focused on free cattle grazing. Sabana de Morro probably originated through the incremental development of an adaptive strategy, as farmers decided to leave Morro trees in the pastures because they provided multiple services, and therefore they have incidentally become engaged in a type of agroforestry and have also incidentally changed the landscape [14]. Local communities intervened on the environment, selecting and preserving useful ecosystem services with positive effects on the environment and on the production system, being based on a mutual exchange between the Morro trees, the cattle and the farmers.

At the national level, this agroforestry system has undergone an important decrease while it has been maintained in this area constituting one of the main sources for local sustenance. In fact, at the national level, this plant formation covers 1.38% of the territory [32] which, compared to the 2.82% reported by Flores [33] for 1978, shows a significative decrease. In twenty-two years about 29,321 ha of Sabana de Morro have disappeared in the country, with an average of 1332 ha/year, mainly due to abandonment of grazing activities and transformation into sugar cane cultivation in suitable soil conditions. In Dolores Municipality, soils are of poor quality because the high amount of clay can lead to water stagnation during the rainy season, and to dehydration and to the formation of cracks in the dry season. Therefore, transformation of Sabana de Morro into cultivations, or in improved pastures by sowing productive forage species to integrate with natural pastures, is not possible, and the Sabana de Morro agroforestry system is the most efficient way to provide an income to local farmers, contributing to its maintenance. The ecology of Morro trees contributes to the preservation of the agroforestry system, since they do not require specific management, their seeds are efficiently dispersed and fertilized by cattle and the young trees are not damaged by the animals.

These are the main reasons that have contributed to the maintenance of this agroforestry system in Dolores, where the main traditional features seem to be well-preserved. This is important considering that the conversion of traditional forest formations is the most important driver of tropical biodiversity loss and associated ecosystem services in the tropics [7]. The comparison of our land use mapping of 2017 with the data of the Censo Agropecuario 2007 [19] showed little changes compared to the rest of the country. Cultivated areas showed a decrease in the last ten years from 1025 ha to 724 ha, while different types of pastures decreased from 6600 ha to 6041 ha.

Silvopastoral systems with the presence of *Crescentia alata* or *C. cujete*, can also be found in other Central America countries. In Mexico, leaves and fruits are used to integrate the alimentation of lambs during the dry season, with evidence of good fiber supply for the diet of the animals and good growth of young lambs [34]. In the Caribbean region of Colombia 83% of local farmers include *Crescentia cujete* among the four trees on which they are "highly dependent" during periods of pasture shortage [35]. For Costa Rica, Janzen [36] reported that horses break the hard, ripe fruits of *Crescentia alata* with their incisors and swallow the small seeds embedded in the sugar-rich fruit pulp. According to the same study, Morro indehiscent fruits fall to the ground and rot without being touched by dispersal agents so that seeds usually die without germinating. Horses, therefore, become the main dispersing agents, and the seeds also take advantage of the manure to germinate, as happens in the Dolores pastures thanks to the presence of cattle. Bass [14] described in great detail the origin and use of *Crescentia* spp. trees in Honduran pastures, stating that farmers protect and encourage their presence in semiarid pastures where they serve as food for cattle when the dry season drastically reduces herbaceous forage.

As with other traditional silvopastoral systems, the Sabana de Morro represents an option for productive and sustainable landscapes [37]. In many tropical landscapes, agroforestry systems are the major ecosystems that resemble natural forest [38–40]. As these systems potentially have a high biodiversity conservation value [41,42], protection of pristine habitat needs to be combined with such environmentally friendly and sustainable land-use systems [7]. As demonstrated for other traditional agricultural models in Latin America [6], and in other tropical countries [3], the Sabana de Morro agroforestry system represents a promising example for other areas as it promotes biodiversity, supporting, at the same time, the livelihood of local farmers. The tree component, if appropriately managed as happens in Sabana de Morro, can enhance nutrient cycling, benefit pastures, provide complementary tree products in the form of fodder, timber, firewood and other tree products, and improve animal productivity [43]. While conventional cattle ranching has proved to be a major source of greenhouse gas emissions, traditional silvopastoral systems can have a crucial role as examples of adaptation and mitigation of climate change. Sabana de Morro can be compared to similar silvopastural systems for cattle breeding in Latin America where the presence of trees on pastures can increase carbon sequestration above and below ground, with values of carbon stock (2.43–74 Mg/ha) and of carbon sequestration (0.49–4.93 Mg/ha/year) on average higher for traditional pastures with natural trees compared to pastures with planted trees [42]. At the same time, temperatures can be 2–5 °C lower under the tree canopy compared to temperatures measured outside the tree canopy [44]. The shade effect of the trees is particularly important. Braun et al. [37] reported that the shade provided by trees on pastures can improve animal welfare, increasing milk production from 12 to 15% and reducing the number of veterinary services.

Sabana de Morro can also be compared to other traditional silvopastoral systems in Europe, such as the Dehesa in Spain, the Montado in Portugal and pastures with carob trees in southern Italy. European silvopastoral systems, beside their importance for environmental-related issues, are considered a legacy of traditional land use and areas where traditional practices and their associated cultural values still exist [45]. Moreover, they can be examples, if adequately supported and promoted through public policies, of a significative economic resource thanks to their recreational use and high-quality products [46].

5. Conclusions

The main vulnerability of this agroforestry system is linked to the risk of abandonment of cattle breeding due to the aging of farmers and to emigration of young people, while the main cost for the farmers relates to the harvesting of Morro fruits and extraction of the pulp to feed animals. Even though the habitants of Dolores have a strong sense of place, and all the interviewed farmers were born in

Dolores, the socioeconomic situation of the country can lead to a decrease in the number of farmers, especially among the younger generations. It must be considered that the Municipality of Dolores is one of the most backward in the country, where illiteracy reaches 32.9%, twice compared to the national rate (16%) in a country where 42.2% of people in the range 15–29 years seriously consider leaving the country [47].

The programme promoted by FAO focuses on sites that can be considered of global importance, a concept exemplified by the five criteria requested for inscription. This agroforestry system is the result of interaction between trees, cattle and farmers and has a long history rooted in the country. The cattle breeders' and cheese makers' traditions derive from the Spanish colonization and are still practiced today with very few changes. The maintenance of the Sabana de Morro system not only guarantees the protection of local traditions, but is also a sustainable way to provide good livelihood conditions while respecting the surrounding environment. Moreover, the diversity of land use, the landscape structure, the presence of hedges and of scattered trees in the pastures, play a fundamental role in agrobiodiversity protection and richness.

Results of this research proved that this agroforestry system has some important characteristics, in accordance to the five GIAHS criteria. The main findings are related to the effects of Sabana de Morro in preserving the traditional landscape and high levels of agrobiodiversity. Despite the maintenance of important environmental, socioeconomic and cultural services, further surveys and researches are needed to assess the effective potentiality of this agroforestry system to be included in the GIAHS programme, especially to precisely evaluate the contribution of this system to the local livelihoods and the problems related to a scarce generational turnover. In addition, it should be necessary to update the 2007 statistical data of the Censo Agropecuario to assess the recent evolution of the local socioeconomic and productive sector.

Author Contributions: Conceptualization, A.S., E.A.M.A. and J.R.Q.; methodology, A.S., F.P., F.C., M.V. and E.A.M.A.; software and analysis, A.S., F.P., F.C. and M.V.; writing, A.S., F.P., F.C., M.V. and E.A.M.A.; supervision, M.A. and A.S. All authors have read and agreed to the published version of the manuscript.

Funding: This research is part of the "GIAHS Building Capacity" project, funded by the Italian Agency for Development Cooperation (AICS) and by the Department of Agriculture, Food, Environment and Forestry (DAGRI) of the University of Florence.

Conflicts of Interest: The authors declare no conflict of interest.

References

1. Jose, S. Agroforestry for ecosystem services and environmental benefits: An overview. *Agrofor. Syst.* **2009**, *76*, 1–10. [CrossRef]
2. Roshetko, J.M.; Bertomeu, M. Multi-species and multifunctional smallholder tree farming systems in Southeast Asia: Timber, NTFPs, plus environmental benefits. *Ann. Silvic. Res.* **2015**, *39*, 62–69. [CrossRef]
3. Kumar, V. Multifunctional agroforestry systems in tropics region. *Nat. Environ. Pollut. Technol.* **2016**, *15*, 365.
4. Rahman, S.A.; Foli, S.; Pavel, M.A.; Mamun, M.A.; Sunderland, T. Forest, trees and agroforestry: Better livelihoods and ecosystem services from multifunctional landscapes. *Int. J. Dev. Sustain.* **2015**, *4*, 479–491.
5. Pandey, D.N. Multifunctional agroforestry systems in India. *Curr. Sci.* **2007**, 455–463.
6. Altieri, M.A. Enhancing the productivity and multifunctionality of traditional farming in Latin America. *Int. J. Sustain. Dev. World Ecol.* **2000**, *7*, 50–61. [CrossRef]
7. Tscharntke, T.; Clough, Y.; Bhagwat, S.A.; Buchori, D.; Faust, H.; Hertel, D.; Holscher, D.; Juhrbandt, J.; Kessler, M.; Perfecto, I.; et al. Multifunctional shade-tree management in tropical agroforestry landscapes—A review. *J. Appl. Ecol.* **2011**, *48*, 619–629. [CrossRef]
8. Van Noordwijk, M.; Hoang, M.H.; Neufeldt, H.; Öborn, I.; Yatich, T. *How Trees and People Can Co-Adapt to Climate Change: Reducing Vulnerability Through Multifunctional Agroforestry Landscapes*; World Agroforestry Centre (ICRAF): Nairobi, Kenya, 2011.
9. Koohafkan, P.; Altieri, M.A. Globally Important agricultural heritage systems. In *A Legacy for the Future*; Food and Agriculture Organization of the United Nations: Rome, Italy, 2011.

10. Aguilar Ramírez, J.I.; Carrillo Serrano, S.F.; Ramírez, N.R. Evaluación de la Harina de Fruto de Morro (Crescentia alata) en la Alimentación de Cabro en Desarrollo, en la Época Seca, en la Comunidad "San José", Departamento de Morazán. Universidad de El Salvador. 1994. Available online: http://ri.ues.edu.sv/id/eprint/8994 (accessed on 7 July 2020).
11. Hernandez, F. *Historia de las Pldntas de Nueva Espana*; Imprenta Universitaria: Albacete, Mexico, 1943.
12. Lötschert, W. La Sabana de Morros en El Salvador, con una vista conjunta de la vegetación en el país. *Comunicaciones* **1953**, *2*, 122–128.
13. Tapia, M.E. *Cultivos Andinos Subexplotados y su Aporte a la Alimentacion*; Oficina Regional de la FAO para América Latina y el Caribe: Santiago, Chile, 2000.
14. Bass, J. Incidental Agroforestry in Honduras: The "jícaro" tree ("*Crescentia*" spp.) and pasture land use. *J. Lat. Am. Geogr.* **2004**, *3*, 67–80. [CrossRef]
15. Kottek, M.; Grieser, J.; Beck, C.; Rudolf, B.; Rubel, F. World Map of the Köppen-Geiger climate classification updated. *Meteorol. Z* **2006**, *15*, 259–263. [CrossRef]
16. Centro de Meteorología e Hidrología. *Tablas de Datos Climatológicos*; Ministerio de Agricultura y Ganadería (MAG): San Salvador, El Salvador, 1993.
17. Tang, J.; Wang, L.; Yao, Z. Analyses of urban landscape dynamics using multi-temporal satellite images: A comparison of two petroleum-oriented cities. *Landsc. Urban Plan.* **2008**, *87*, 269–278. [CrossRef]
18. Hill, M.O. Diversity and evenness: Unifying notation and its consequences. *Ecology* **1973**, *54*, 427–432. [CrossRef]
19. Ministerio de Economia. *IV Censo Agropecuniario*; Resultados Nacionales: Ministerio de Economia, El Salvador, 2009.
20. Dirección General de Economía Agropecuaria (DGEA). *Posibilidades de Incremento de Producción e Industrialización del Morro en El Salvador*; Ministerio de Agricultura y Ganaderia (MAG): San Salvador, El Salvador, 1983.
21. World Food Program (WFP)-Government of El Salvador (GOES). *Mapa de Medios de Vida, Cabañas*; World Food Program: Rome, Italy, 2018.
22. Ortez, O.; Flores, H.; Alemán, S.; Osorio, M.; Solorzano, S. *El Salvador: Informe Nacional Sobre el Estado de la Biodiversidad Para la Alimentación y la Agricultura*; MAG (Ministerio de Agricultura y Ganadería, SV)-CENTA (Centro Nacional de Tecnología Agropecuaria y Forestal "Enrique Álvarez Córdova"): San Salvador, El Salvador, 2016.
23. Benavides, J. Árboles y arbustos forrajeros en América Central. CATIE. *Ser. Técnica Inf. Técnico* **1994**, *236*, 3–28.
24. Zamora, S.; García, J.; Bonilla, G.; Aguilar, H.; Harvey, C.; Ibrahim, M. ¿Cómo utilizar los frutos de Guanacaste (*Enterolobium cyclocarpum*), Guacimo (*Guazuma ulmifolia*), Genizaro (*Pithecellobium saman*) y jicaro (*Crescentia alata*) en alimentación animal? *Agroforestería De Las Américas* **2001**, *8*, 45–49.
25. Reyna de Aguilar, M.L. *El Morro (Crescentia alata Kunth.)*; Pankia 10; Jardín Botánico La Laguna: Pasaje Privado D, El Salvador, 1991; pp. 3–6.
26. Lagos, J.A.; Mejía, L.P.; Rosales, R. *Compendio de Botánica Sistemática*; Concultura: San Salvador, El Salvador, 1997.
27. Janzen, D. Tropical Dry Forest. In *Biodiversity*; Wilson, E., Peter, F., Eds.; National Academy of Sciences: Washington, DC, USA, 1988; pp. 130–144.
28. Dickey, D.R.; Van Rossem, A.J. *The Birds of El Salvador*; Field Museum of Natural History: Chigago, IL, USA, 1938; 231p.
29. Kovar, P.A. Idea general de la vegetación de El Salvador. In *Plants and Plant Science in Latin America*; The Chronica Botanica Company: Waltham, MA, USA, 1945; pp. 56–57.
30. Mertens, R. *Die Amphibien und Reptilien van El Salvador*; Abhdlgn. Senckenb, Naturf: Frankfurt Main, Germany, 1952; 487p.
31. Baiza Avelar, V.; Martinez Funes, R.; Medrano, L. *Alternativas del Manejo Integral del Area Boscosa en la Cooperativa Santa Anita, Mercedes Umaña, Usulutan*; Thesis in Ingeniería Agronómica, Universidad de El Salvador: San Salvador, El Salvador, 1998.
32. Ventura, N.; VIllacorta, R. *Mapeo de Vegetación Natural de Ecosistemas Terrestres y Acuáticos de Centro América*; Ministerio de Medio Ambiente y Recursos Naturales (MARN): San Salvador, El Salvador, 2000.
33. Flores, J.S. *Curso Fundamental de Ecología*; Department Biología, Fac. CC y HH, Universidad de El Salvador: San Salvador, El Salvador, 1978; 208p.
34. Rojas-Hernandez, S.; Olivares-Perez, J.; Aviles-Nova, F.; Villa-Mancera, A.; Reynoso-Palomar, A.; Camacho-Díaz, L.M. Productive response of lambs fed Crescentia alata and Guazuma ulmifolia fruits in a tropical region of Mexico. *Trop. Anim. Health Prod.* **2015**, *47*, 1431–1436. [CrossRef] [PubMed]

35. Cajas-Giron, Y.S.; Sinclair, F.L. Characterization of multistrata silvopastoral systems on seasonally dry pastures in the Caribbean Region of Colombia. *Agrofor. Syst.* **2001**, *53*, 215–225. [CrossRef]

36. Janzen, D.H. How and why horses open Crescentia alata fruits. *Biotropica* **1982**, 149–152. [CrossRef]

37. Braun, A.; Van Dijk, S.; Grulke, M. Upscaling Silvopastoral Systems in South America; Inter American Investment Corporation (IIC)-Inter American Development Bank (IDB). 2016. Available online: https://publications.iadb.org/en/publication/17180/upscaling-silvopastoral-systems-south-america (accessed on 7 July 2020).

38. Schroth, G.; Harvey, C. Biodiversity conservation in cocoa production landscapes. *Biodivers. Conserv.* **2007**, *16*, 2237–2244. [CrossRef]

39. Schroth, G.; Fonseca, G.A.B.; Harvey, C.A.; Gascon, C.; Vasconcelos, H.L.; Izac, A.N. *Agroforestry and Biodiversity Conservation in Tropical Landscapes*; Island Press: Washington, DC, USA, 2004.

40. Bhagwat, S.A.; Willis, K.J.; Birks, H.J.B.; Whittaker, R.J. Agroforestry: A refuge for tropical biodiversity? *Trends Ecol. Evol.* **2008**, *23*, 261–267. [CrossRef] [PubMed]

41. Jose, S. Agroforestry for conserving and enhancing biodiversity. *Agrofor. Syst.* **2012**, *85*, 1–8. [CrossRef]

42. Montagnini, F.; Ibrahim, M.; Murgueitio, E. Silvopastoral systems and climate change mitigation in Latin America. *Bois Et Des Trop.* **2013**, *316*, 3–16. [CrossRef]

43. Muschler, R.G. Agroforestry: Essential for sustainable and climate-smart land use. In *Tropical Forestry Handbook*; Pancel, L., Köhl, M., Eds.; Springer: Berlin/Heidelberg, Germany, 2015. [CrossRef]

44. Murgueitio, E.; Calle, Z.; Uribe, F.; Calle, A.; Solorio, B. Native trees and shrubs for the productive rehabilitation of tropical cattle ranching lands. *For. Ecol. Manag.* **2011**, *261*, 1654–1663. [CrossRef]

45. Den Herder, M.; Moreno, G.; Mosquera-Losada, R.M.; Palma, J.H.; Sidiropoulou, A.; Freijanes, J.J.S.; Papanastasis, V.P. Current extent and stratification of agroforestry in the European Union. *Agric. Ecosyst. Environ.* **2017**, *241*, 121–132. [CrossRef]

46. Moreno, G.; Franca, A.; Pinto-Correia, T.; Godinho, S. Multifunctionality and dynamics of silvopastoral systems. *Options Méditerranéennes* **2014**, 421–436. Available online: https://agris.fao.org/agris-search/search.do?recordID=QC2017600093 (accessed on 7 July 2020).

47. Fundación Salvadoreña para el Desarrollo Económico y Social (FUSADES). Informe de Coyuntura Social 2017–2018. Available online: http://fusades.org/sites/default/files/Coyuntura%20Social%20Sept.%202018%20color_1.pdf (accessed on 7 July 2020).

 *forests*

Article

Vertical Distribution and Elevation Preference for the Breeding of Fairy Pittas on Jeju Island, Korea

Eun-Mi Kim [1,2], Chang-Wan Kang [2,3], Chang-Yong Choi [2,4,*], Jung-Hwa Chun [5] and Hyun-Young Nam [2,6]

1   Warm Temperate and Subtropical Forest Research Center, National Institute of Forest Science, Jeju 63582, Korea; kptta@naver.com
2   Jeju Nature Park, Seowipo, Jeju 63630, Korea; jejubirds@daum.net (C.-W.K.); stern23@snu.ac.kr (H.-Y.N.)
3   The Korea Association for Bird Protection Jeju Branch, Seogwipo, Jeju 63541, Korea
4   Research Institute of Agriculture and Life Sciences, Seoul National University, Seoul 08826, Korea
5   Research Planning and Coordination Division, National Institute of Forest Science, Seoul 02455, Korea; chunjh69@korea.kr
6   School of Biological Sciences, Seoul National University, Seoul 08826, Korea
*   Correspondence: sub95@snu.ac.kr; Tel.: +82-2-880-4766

Received: 16 September 2019; Accepted: 7 November 2019; Published: 12 November 2019

Abstract: Elevation often becomes an important component in the breeding site selection of forest birds because it may affect individual fitness. To understand how the threatened fairy pitta (*Pitta nympha* Temminck & Schlegel) selects a particular elevation for breeding and whether the pitta achieves better reproductive performance in its preferred elevation, we surveyed for the presence of breeding pittas and recorded their reproductive performances at six different elevation zones on Mount Halla in Jeju Island, Korea. We expected that preference for breeding sites and reproductive performance would gradually decrease with increasing elevation. In fact, 73 presence and 78 absence records indicated no preference for breeding sites at elevations lower than 400 m. However, forest habitats between 400 and 600 m were strongly preferred, while locations above 800 m were clearly avoided. The egg-laying date was significantly earlier at lower elevations, but other measures of reproductive performance did not differ between the preferred and non-preferred elevations. Contrary to our expectations, this indicates that there was no clear advantage to a pitta's reproductive performance based on elevation preference. Our findings suggest that the inherent optimal selection for the best reproductive outcomes was not a key driving factor for the pattern of elevation preference observed, and that the pitta's preference might be a response to external and environmental factors such as climate conditions. The reduction of anthropogenic impacts by mitigating human–pitta conflicts at lower elevations, such as accidental non-reproductive mortality and forest loss, would help enhance the conservation of the fairy pitta on Jeju Island, a stronghold of this threatened species in Korea.

Keywords: elevation; *Pitta nympha*; preference; reproductive performance; selection; conservation

---

1. Introduction

The habitat selection of birds, especially breeding site selection, is often regarded as a non-random process of natural selection because it may affect individual fitness [1,2]. In particular, breeding or nesting site selection is often closely linked with individual fitness, including survival and reproductive performance; therefore, such choices are under diverse and strong selection pressures [2–4]. Higher elevation is often characterized by lower temperatures, delayed plant phenology, unstable climates, reduced food supply and habitat quality, and lower nest site availability, which affect many aspects of avian ecology [5–7]. Therefore, elevation often becomes an important component in the breeding site selection of forest birds because it influences many reproductive aspects like the clutch size and egg size [6], the timing of breeding and reproductive investments [7], and predation risks [8].

The fairy pitta (*Pitta nympha* Temminck & Schlegel) is an obligate migrant breeding in forest environments in the Korea (hereafter Korea), China, Taiwan, and Japan, and it is believed to spend winter in Southeast Asian countries [9]. Its breeding habitats in forests have been largely disturbed and reduced by diverse anthropogenic factors such as deforestation, forest fires for slash and burn farming, unsustainable forest management, the logging and harvesting of forest products, residential and commercial developments, the construction of roads and dams, and illegal trapping [9,10] as well as physical collisions with man-made structures [11]. In addition to the small global population of only 1500–7000 mature individuals and their limited geographic distribution, confined to East Asia, these threats have resulted in a trend of decline for the pitta, such that it is now a vulnerable species on the IUCN (International Union for Conservation of Nature) Red List [9]. Although the general biology of breeding fairy pittas has been described [9–13], detailed information is still limited due to its elusive behavior at breeding sites, which are often associated with dense forests and forested valleys [12,13].

The fairy pitta is known as a breeder in lowland forests, and its upper limit of elevation seems to be 1200 m [9,10]. More specifically, the pitta in Taiwan was observed at elevations from sea level to 980 m, but more than 75% of them were observed at elevations lower than 280 m [14]. In Korea, Kim et al. [15] reported that the vertical distribution of breeding pairs on Jeju Island ranged from 50 to 800 m, and about 91% of them were confirmed in forests at less than 600 m. However, this report in Korea contradicts previous older records documented in the 1960s that the pittas were distributed between 1200 and 1600 m [16] and bred between 1000 and 1200 m [17] on the same island. During the 1980s, this species was reported in forested areas at 500–600 m [18,19]. This inconsistency among the previous reports of the pitta's occurrence along the elevation gradient on Mount Halla, which supports the largest breeding population of the pitta in Korea [20], may lead to great uncertainty in species conservation and habitat management. Kim et al. [21] explained that the inconsistency in the vertical occurrence of breeding pittas on Jeju Island is caused by long-term changes in habitat, vegetation cover, and forest landscapes based on reviews of satellite images, ground-based landscape photos, and the literature. They judged that the forest landscapes of Jeju Island had been improved both in quantity and quality, suggesting increased potential for breeding habitats for the pitta [21]. However, the review study did not examine any direct linkage between the breeding habits of the pitta and the changing forest landscapes in the key breeding area on Jeju Island.

Understanding a pattern for how available habitat differs from breeding habitat is often the first step toward discovering the process of natural selection for breeding-site selection [2]. Given the unknown pattern of vertical selection of breeding sites by the threatened fairy pitta, the aim of this study was to understand the breeding habitat preference along the elevation gradient of forests on Mount Halla and to test the hypothesis that the pitta selectively prefers a particular elevation to achieve better reproductive performance (e.g., the egg-laying date, clutch size, number of hatchlings, and number of fledglings). Specifically, because the common breeding habitats of pittas were known to be lower elevation forests (<600 m) [14,15], we expected that the breeding site preference and reproductive performance would gradually decrease with increase in elevation. In this paper, we discuss our findings as the vertical distributions of pittas may relate to the potential influences of forest habitat change in the context of the pitta conservation on Jeju Island, the Korea.

2. Materials and Methods

2.1. Study Area

Our study area was Mount Halla (or Hallasan; 33°21′ N, 126°32′ E) on Jeju Island, Jeju Province, in the Korea. Jeju Island is a volcanic island located in the southernmost part of Korea and belongs to a semitropical climate zone [22]. During the breeding season of fairy pittas (May to August), the island has a mean temperature of 22.7 °C and average precipitation of 228.3 mm [22]. Mount Halla is a shield volcano on Jeju Island and is the highest mountain in Korea with its summit about 1950 m above sea level. The gradual change in elevation from the coastline to the summit on Mount Halla forms diverse microclimates and microhabitats for wildlife.

The fairy pitta was known as a rare breeding bird on Jeju Island, but Kim [20] confirmed that more than 60 pairs breed annually on the island. This finding indicated that the island supports the largest known breeding population in Korea and that the Jeju population seems to be the stronghold of this threatened species in Korea.

2.2. Field Survey

Based on previous studies on the habitat use of fairy pittas [15], we identified potential breeding habitats of the species that should include broad-leaved and evergreen forests larger than 3 ha in size. We also considered that the potential habitats were dominated by tall trees (>10 m) with no dense vegetation on the forest floors and were well-shaded, humid areas (average humidity of 70%–80%) with a mean temperature of 20–25 °C in June and July [15]. After identification of the potential breeding habitats of fairy pittas, we grouped them into six elevation zones from sea level to 1200 m above sea level at 200 m intervals. A total of 151 survey sites were selected and surveyed between 2002 and 2017: 27 sites were at 0–200 m, 39 were at 200–400 m, 49 were at 400–600 m, 20 were at 600–800 m, 9 were at 800–1000 m, and 7 sites were at 1000–1200 m. Despite records in the older literature [16], we did not consider higher elevations (>1200 m) as potential habitat for the pitta because there is no confirmed breeding record of this species there since 2000. Because the fairy pittas are present between May and September in Jeju [11], we carried out playback surveys using recordings of the pitta's song between 20 May and the end of August to confirm the occurrence of breeding pairs at 151 sites: 103 sites between 2002 and 2003, 30 in 2011, and 18 in 2017. The sites were repeatedly visited and the playback survey was performed twice a week in May (arrival, territory establishment, and nest building stages), and three times a week between June to August (nest-building, egg-laying, brooding, and fledging stages) in 2002, 2003, and 2011 [15], while the 18 sites in 2017 were surveyed once or twice in June. Two or more observers remained stationary while playing back recordings at a randomly-selected position in each site. One cycle of the survey included the playback of a recorded pitta song for 30 seconds, and then pausing for 10 seconds to wait for pitta responses; we repeated three cycles in each site to detect any response of the pittas [15]. We used different positions for the playbacks at the same site but also visited the same position where the pittas previously responded. When one or more pittas were vocally or visually detected responding to our playbacks in a study site, we confirmed their presence and also searched for their active nests to collect more detailed breeding information at the site. All playback surveys were conducted between sunrise and sunset under affordable weather conditions with good visibility, no precipitation, and light winds.

2.3. Preference for a Particular Breeding Elevation

To describe the vertical distribution and elevation preference of the pittas, we used three methods considering their characteristics in resource selection studies: a generalized linear model (GLM), and Jacobs' preference and Manly's selection indices. The GLM can investigate the overall influence of elevation on presence/absence of pittas, and it can predict the pittas' vertical distribution based on the predicted chance of presence (or encounter rate) in each elevation zone [23]. The Jacobs' and Manly's indices are both useful in studies examining the relationship of habitat selection with other variables using the proportions of available and used habitats [24,25]. We used the Jacobs' index to test where there was preference or avoidance in each elevation zone because it can provide values in finite symmetrical scales—negative values (0 to −1) for avoidance and positive values (0 to 1) for preference [24]. Manly's index with Bonferroni adjustment can provide relative selection values (rather than preference or avoidance) that come with conservative confidence intervals for each elevation zone; we provide here, the standardized selection ratios to interpret the unadjusted selection ratios in asymmetrical scales [25].

We tested a hypothesis that elevation had an influence on the probability that a fairy pitta was present on a site using a presence/absence analysis [23]. First, we built a contingency or cross-tabulation table of pitta presence (true = 1) and absence (false = 0) with elevation categories for a chi-square test [23]. Then, we used a generalized linear model (GLM) with a binomial logit link to examine the hypothesis.

In addition, the Jacobs' preference index [24] was used to estimate the preference and avoidance of breeding elevations on Mount Halla (Equation (1)):

$$D_i = (r_i - p_i)/(r_i + p_i - 2r_ip_i),\tag{1}$$

where p_i is the proportion of available habitats and r_i is that of occupied habitats in each elevation zone i. Because this index ranging from −1.0 (avoidance) to 1.0 (preference) provided a single D_i for each elevation zone, and we used a bootstrapping method, that is, the easiest and commonly-used method, to estimate the variances of preference indices [25]. To test where there was preference or avoidance at each elevation, and whether preferences were different across the six elevation zones, we performed a bootstrapping analysis through 100 random resamplings. Because the bootstrapping resamples did not pass normality tests, we used one-sample, signed-rank tests to examine whether the mean of the preference was different from zero (which means no preference nor avoidance). The differences in preference among elevations were examined using the Kruskal–Wallis test and all pairwise multiple comparison procedures (Tukey's test).

We also calculated Manly's selection ratio (w; Equation (2)) and standardized selection ratio (B; Equation (3)) [25] as below:

$$w_i = r_i/p_i\tag{2}$$

$$B_i = (r_i/p_i)/(\sum r_i/p_i),\tag{3}$$

where r_i and p_i are the proportions of used and potential breeding habitats, respectively, that are in category i, and u_i is the number of occupied sites in elevation category i. The standardized selection ratio, ranging from 0 to 1, is a probability that a category i habitat unit would be the next one selected if it was possible to make each of the types of habitat unit equally available [25]. Bonferroni corrected confidence intervals (CI) for proportions used (r) and selection ratios (w) of six elevation groups were estimated with $\alpha; = 0.05/6 = 0.0083$, and their standard errors were calculated, as shown below (Equations (4) and (5)):

$$r_i \pm Z_{\alpha/2}\sqrt{r_i(1-r_i)/\sum u_i}\tag{4}$$

$$w_i \pm Z_{\alpha/2}\sqrt{r_i(1-r_i)/(\sum u_i \times p_i^2)}.\tag{5}$$

2.4. Reproductive Performance

Whenever nests of the pittas that responded to our playbacks were found, we recorded breeding parameters such as date of first egg-laying, clutch size, and numbers of nestlings and fledglings for each of the nests found. If not directly observed, the first egg-laying date was estimated based on the number of eggs laid in the egg-laying period or by taking the known incubation period from the hatching date once the clutch size was determined. If at least one chick fledged from a nest, we considered it a successful nest and documented the number of fledglings per successful nest by elevation. We compared the breeding parameters between different elevations using ANOVA and the Kruskal–Wallis test followed by a Tukey test for an all pairwise multiple comparisons when necessary.

We used R software version 3.1.1 [26] for all statistical analyses, and all values are given as means ± standard deviations (SD) unless otherwise noted.

2.5. Changes in the Area of Forests

To assess forest areas by elevations and changes in those over time, we used Landsat satellite images and output raster maps processed in Kim et al. [21]. The detailed procedures were as follows.

Landsat MSS images (March 1975) and Landsat 7 images (March 2002) were selected to produce landcover maps for two periods. A surface terrain model for rectification was generated by acquiring and mosaicking together topographical maps (1:5000 scale) produced by the National Geographic Information Institute that covered the study area. Both of the images were geometrically rectified to the transverse Mercator coordinate, utilizing ground control points identifiable in the imagery. Radiometric calibration and subsequent transformation to at-satellite reflectance was performed by utilizing standard procedures by Markham and Barker [27]. Then, C-correction [28,29] was applied to the images to correct the reflectance of sun-facing and anti-sun facing slopes utilizing the DEM (digital elevation model) extracted from the topographical maps.

Both the unsupervised and supervised approaches were applied to the classification of the preprocessed images. Original images were converted to PCA (principal component analysis) and NDVI (normalized difference vegetation index) raster data to stack into single, multilayer image files including original bands except for thermal layers.

In order to recognize the spectral characteristics of the diverse ground objects represented in the satellite imagery, an unsupervised training algorithm (isodata) was run on the stacked images to generate appropriate spectral clusters and corresponding signatures.

Field surveys were also conducted over the study period to gather data on the pre-clustered spectral classes. A compass and laser distance-measuring devices were used together with real-time differential GPS (HDOP < 1.5) to identify the locations of certain characteristics. Survey points and the site description data were constructed into a spatial database and utilized for combining the isodata classes and supervised classification.

Based on the unsupervised classification results and ground truth data, a training dataset was created as a group of seeds for extracting the signature of each class for maximum likelihood supervised classification. After each classification process, the classified area and boundaries were visually evaluated based on the ground truth data. If the results were not acceptable, the training dataset was adjusted and the classification process was iterated. Finally, classification results were developed into maps showing six categories of cover, including deciduous forest, conifer forest, cultivated land, barren land, urban area, and water bodies. In this study, only deciduous and conifer forest categories were extracted and their changes assessed over time. The classification accuracy of the forest category was 95% based on field survey results and aerial photos.

3. Results

3.1. Preference for Breeding Elevation

We confirmed the occurrence of breeding fairy pittas at 73 out of 151 survey sites (67 of 133 from 2002 to 2011 and six of 18 in 2017), and the breeding pairs occurred in the potential habitats between 50 and 800 m on Jeju Island, Korea.

Based on the contingency tables developed by the 73 presence and 78 absence records, the proportions of occupied habitats by breeding pittas were significantly different from those of unused habitats across the elevation categories ($\chi^2 = 18.754$, $df = 5$, $p = 0.002$). Binomial GLM with a logit link also confirmed that the predicted chance of pitta presence differed by elevation (AIC = 198.87, $\chi^2 = 22.293$, $df = 5$, $p < 0.001$). The predicted chance for pitta presence in the range 400–600 m was the highest (0.65, 95% CI: 0.51–0.77) and was significantly different from the probability at 800–1000 m (0.11, 95% CI: 0.02–0.50) and above, whereas there were wide overlaps between the other elevations (Figure 1a).

Jacobs' preference indices (D) calculated from 100 random resamplings also confirmed the difference in preference by elevation ($df = 5$, $H = 512.579$, $p < 0.001$) (Figure 1b). There was no preference or avoidance at 0–200 m, but preference was documented between 200 and 600 m, while avoidance was found between 600–1200 m. The breeding pittas strongly preferred 400–600 m ($t = 27.820$, $df = 99$, $p < 0.001$), while they avoided higher elevations (600–800 m: $t = -13.222$, $df = 99$, $p < 0.001$; 800–1000 m: $t = -27.064$, $df = 99$, $p < 0.001$).

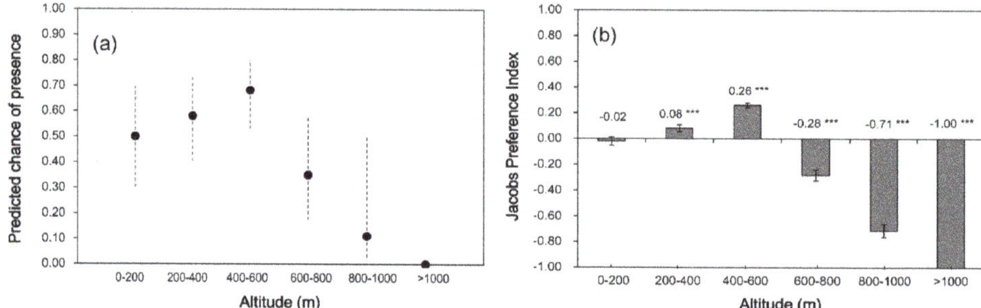

Figure 1. Occurrences and habitat preferences of fairy pittas on Jeju Island, Korea. (**a**) Predictions for the presence (the 95% confidence intervals (95% CI)) of fairy pittas for each elevation category. (**b**) Changes in the mean Jacob preference index for breeding habitat selection by elevation based on 100 resamplings; vertical bars denote standard deviations. Numeral letters and vertical lines on bars represent the mean preference indices and 95% confidence limits after 100 bootstrap resamples. Asterisks denote statistically significant differences (*** $p < 0.001$) from zero based on a Student's *t*-test for one sample (no preference or avoidance).

We estimated Manly's selection indices and confidence intervals for the proportions of used habitats compared with the available proportions (Table 1). Selection indices for forests above 800 m were significantly different from those of lower habitats by being smaller. Standardized selection indices indicated that forests in the range 400–600 m were 1.87 and 5.88 times more selected than habitats at 600–800 and 800–1000 m.

Table 1. Estimated selection indices and Bonferroni confidence intervals for the presence of fairy pittas (*Pitta nympha*) at different elevations on Jeju Island, Korea.

	Available Habitats		Used Habitats		Selection Index (w) with Bonferroni Confidence Limits	Standardized Selection Index (B)
Elevation	**Counts (A)**	**Proportion (p)**	**Counts (u)**	**Proportion (r) with Bonferroni Confidence Limits**		
0–200 m	27	0.179	14	0.192 (0.065–0.319)	1.073 (0.793–1.352)	0.245
200–400 m	39	0.258	19	0.260 (0.119–0.402)	1.008 (0.767–1.248)	0.230
400–600 m	49	0.325	32	0.438 (0.000–0.598)	1.351 (1.106–1.596)	0.308
600–800 m	20	0.132	7	0.096 (0.001–0.191)	0.724 (0.513–0.935)	0.165
800–1000 m	9	0.060	1	0.014 (0.000–0.051)	0.230 (0.157–0.303)	0.052
1000–1200 m	7	0.046	0	0.000	0.000	0.000
Total	151	1.000	73	1.000	4.384	1.000

3.2. Reproductive Performance

According to the 29 nests found with known breeding parameters, the first egg-laying date was after 174.9 ± 14.1 Julian days ($n = 26$) ranging from 157 at 395 m to 200 at 618 m. There was a general positive relationship between the date and the nesting elevation (coefficient ± standard error: 0.046 ± 0.0163; $r = 0.495$, $p = 0.010$) (Figure 2), indicating delayed egg-laying at higher elevations, and in particular, breeding in nests at <400 m elevation was 17–20 days earlier than that in nests at higher elevations (Table 2). However, no influence of elevation on clutch size was detected, nor on the number of hatchlings and the number of fledglings, even between the preferred (400–600 m) and avoided elevations (600–800 m) (Table 2).

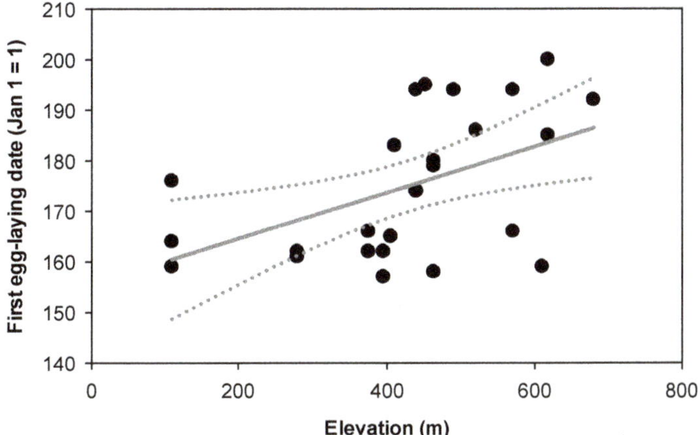

Figure 2. Changes in the first egg-laying dates of fairy pittas by elevation on Jeju Island, Korea. The linear relationship is shown as a solid line (gray) with 95% confidence intervals (dotted gray).

Table 2. Reproductive parameters (mean ± SD) of the fairy pitta (*Pitta nympha*) at three elevations with different preference on Jeju Island, Korea. Different superscript letters indicate significant differences ($p < 0.05$).

Parameter	Elevation with No Preference (0–400 m, $n = 12$)	Preferred Elevation (400–600 m, $n = 13$)	Avoided Elevation (600–800 m, $n = 4$)	df	F	H	p
First egg-laying date	163.2 ± 5.4 [a]	180.2 ± 12.4 [b]	184.0 ± 17.8 [b]	2	7.303		0.004
Clutch size	4.7 ± 0.7	4.5 ± 0.8	4.8 ± 1.0	2		0.395	0.821
No. hatchlings	3.4 ± 2.1	3.6 ± 1.8	3.8 ± 2.6	2		0.373	0.830
No. fledglings	3.4 ± 2.1	3.2 ± 2.0	3.5 ± 2.4	2		0.257	0.879

3.3. Changes in Forested Areas

The overall size of forested areas on Jeju Island remained nearly the same (2% increase in size) between 1975 and 2002 (Table 3). However, the area of preferred forest habitats at 400–600 m increased from 80 to 128 km², while coastal zones were deforested during the three decades (Table 3, Figure 3). Constructed forest maps also show deforestation and fragmentation of forested areas at lower elevations on the coast, but reforestation at the 400–600 m elevation zone (Figure 3).

Table 3. Changes in forested areas on Jeju Island between 1975 and 2002 in three different elevations with preference for breeding elevations.

Elevation by Habitat Preference	Forested Area (km²)		Difference (Change in %)
	1975	2002	
<400 m (elevation with no preference)	592.1	555.5	−36.6 (−6.2%)
400–600 m (preferred elevation)	79.8	127.8	48.0 (60.3%)
>600 m (avoided elevation)	226.1	232.3	6.2 (2.7%)
Total	898.0	915.6	17.6 (2.0%)

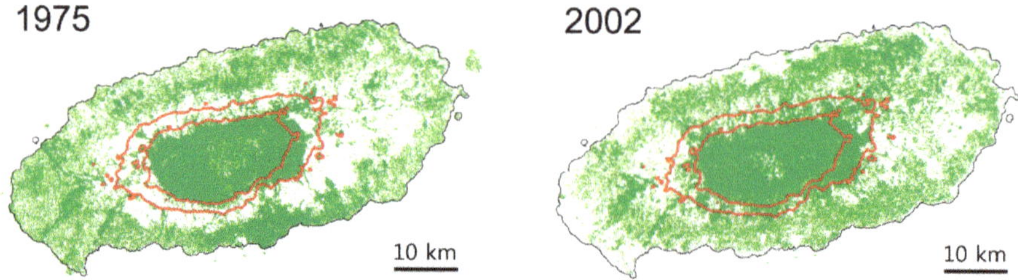

Figure 3. Changes in forested areas (marked green) between 1975 (**left**) and 2002 (**right**) on Jeju Island, Korea. The two bold red lines around Mount Halla in the center of the island indicate 400 and 600 m contours, and the areas between the two contours are at the breeding elevations preferred by the fairy pitta.

4. Discussion

The results of this study indicated that there was clear elevation preference and avoidance for breeding by the threatened fairy pitta on Jeju Island, the stronghold of the national population. However, contrary to our expectation, no preference of the pitta for elevations below 400 m for breeding habitat was detected. Instead, forests at 400–600 m were strongly preferred and selected as the breeding elevation, whereas the higher elevations were clearly avoided. This pattern of elevation preference was supported by three different methods: the predicted chance of presence [23], Jacob's preference index [24], and Manly's selection index [25].

Out of four parameters examined for reproductive performance, only the first egg-laying date showed a significant difference among the three different elevation categories, and the timing of breeding initiation was obviously delayed at elevations above 400 m. This vertical difference in breeding phenology is often caused by weather conditions such as temperature and humidity, as well as by ecological conditions such as prey availability [7,30]. In terms of prey availability, earthworms were the most important food item for nestlings of the pitta and accounted for >73% of the food delivered during chick provisioning periods [13]. According to a previous study on prey availability for breeding pittas, the earthworm density in forest floors around the pitta breeding habitat at 100 m elevation was higher than at 380 m elevation [31], suggesting a decline in prey availability along the elevation gradient. The distribution, growth, and activity of earthworms are also largely affected by non-biotic factors, such as humidity and temperature [32]. Because Jeju is a very humid island due to the high amount of average precipitation (2309 mm/year) [22,33,34], temperature more strongly affects the vegetation structure of Mount Halla than humidity does [35]. For example, the temperature range most suitable for raising earthworms is known to be 20–25 °C [36–38], and the forest floor on Mount Halla below 250 m elevation reaches that temperature range in June and July [35]. However, this temperature range is achieved a month later (July and August) in forests at 500–700 m elevation [35], implying that the occurrence and activity of the main prey for breeding pittas is delayed at higher elevations. The lower temperature and delayed food availability at a higher elevation may result in a delayed start of fairy pitta breeding on Mount Halla. Other studies on forest birds also revealed that the egg-laying date was delayed more in upland areas than in lowland areas for forest birds such as great tits (*Parus major* L.) [39], varied tits (*Poecile varius* (Temminck & Schlegel)) [8], and Tickell's leaf-warblers (*Phylloscopus affinis* (Tickell)) [40].

On the other hand, the remaining three parameters (clutch size, number of hatchlings, and number of fledglings) were not different among preferred, non-preferred, and avoided elevations. The change in clutch size and its related parameters along the elevation gradient may not have a certain trend when climatic and environmental factors are not actually deterministic for selection, or not predictable in relation to elevation [41]. Similarly, our findings indicated that there was no clear benefit in the pitta's breeding performance due to preference in elevation. As habitat selection patterns may not always indicate the optimal choices of a species [2], the result suggests that the pitta's inherent selection for the best reproductive performance was not a key driving factor of the elevation preference observed, and that environmental and external factors may affect the vertical distribution and elevation preference of the breeding pittas.

In many cases, the effects of habitat selection on avian breeding success or fitness are altered or deterred by other biological factors, such as nest predation. In particular, high nest predation rates may place a strong selection pressure on decisions affecting the selection of breeding sites [5,42,43]. Though the selection of a breeding habitat and nest site may often affect breeding success [3], the breeding success of five passerine species in South America, that demonstrated preference in nesting plants, was largely determined by the presence of predators rather than by nesting site preference [4]. Breeding performances of members of the family Laniidae in France were also affected by predation, rather than by nesting selection or preference [44]. As an important process reducing avian breeding success [45,46], nest predation of fairy pittas by corvids on Jeju Island was the most significant natural threat to the breeding pittas [11]. Large-billed crows (*Corvus macrorhynchos* Wagler) are fairly common nest predators of the pitta and cause 66% of their nest failures (Kim, unpublished data). The crows are widely distributed throughout the island but are often attracted even to the higher elevations by the high volume of tourists and food waste along trails [47]. The invasive Eurasian magpies (*Pica pica* (L.)) are also known to be a nest predator of the pitta [11], are increasing in number, and their distribution is expanding at lower elevations, particularly at less than 600 m, on Jeju Island [48]. Therefore, though the breeding performances observed under the existing predation risk remained the same up to 800 m in this study, the expected breeding performances would be better, especially at lower elevations, if no predation by invasive corvids were to happen.

In addition to the higher predator density at lower elevations, negative human impacts on pitta survival, such as car accidents and window strikes also occur at lower elevations [11]. This kind of mortality not directly related to reproduction was not considered in our reproductive performance measurements. Because the overall proportion of anthropogenic causes of non-reproductive mortality could be higher than that of natural causes in fairy pittas on Jeju Island [11], the breeding pittas might be more exposed to the greater human footprint in deforested areas at lower elevations.

While the overall size of forested areas remained nearly the same and coastal areas were deforested between 1975 and 2002, the area of forests in the range of elevations pittas prefer increased from 80 to 128 km^2. In addition, evergreen broadleaved forest, which is the preferred breeding environment for pittas [15], is currently dominant up to 600 m on Mount Halla [19]. Kim et al. [21] described that this increase of potential breeding habitat and the development of evergreen broadleaved forest are largely the result of reforestation in the mid-mountain areas between 200 and 600 m after the 1980s. The reforestation on Jeju Island has probably improved breeding habitats, but the Jeju Dwarf Bamboo (*Sasa quelpaertensis* Nakai) at higher elevations reduces the potential breeding and feeding habitats for the pittas by covering the forest floors [49]. As reforestation proceeds on Mount Halla, the dwarf bamboo colonies are also expanding into reforested areas at lower elevations [34], resulting in the potential for degradation of breeding habitat for the pittas.

5. Conclusions

As previously reported [14,21], fairy pittas clearly avoided higher mountain areas (above 600 m) for breeding habitat on Jeju Island. This might have been caused by the lower temperature, a delayed breeding season, lower prey abundance, and less broad-leaved forest areas, along with expanding Jeju dwarf bamboo on forest floors. On the other hand, forested areas lower than 400 m are expected to be advantageous for the breeding pittas because of higher temperatures and higher prey abundances (represented by earthworms) that clearly allowed the pittas to attempt earlier breeding. Despite the advantages at the lower elevations, significant preference was not detected and better reproductive outcomes were not guaranteed. That may be because of increasing environmental pressures including deforestation, habitat degradation, and human-related, non-reproductive mortality in the coastal zone.

Consequently, this study identified that the elevation range 400–600 m provided the currently preferred breeding habitat for the threatened fairy pittas on Jeju Island. This observed pattern in the preferred elevations might be a response to external and environmental factors—possible tradeoffs between upward shifts by human-related impacts (i.e., habitat loss and degradation, non-reproductive mortality, and higher predator density) and downward shifts by vertically graduated environmental parameters (i.e., temperature, vegetation, prey abundance)—rather than an inherent optimal selection for the best reproductive outcomes.

Despite no preference for them during breeding habitat selection by elevation, forested areas at lower elevation still provide critical breeding habitats for the threatened fairy pittas because the low forests have greater potential to host more breeding pairs in their four-times-larger area (555.5 km^2) than in the forest area at preferred elevations (127.8 km^2). Therefore, it is important to maintain good breeding populations in the lower elevation forests for the conservation of the overall population on Jeju Island. Increasing the preferred breeding habitat in size may benefit declining and threatened avian populations, and higher temperatures and prey abundance are still positive factors for increasing habitat preference and reproductive outcomes of the pitta at lower elevations. However, human impacts and development pressure on the mid-mountain region and coastal areas of Jeju Island are rapidly increasing [50]. Therefore, increasing human–pitta conflicts, habitat loss and degradation by on-going coastal developments, and anthropogenic threats and disturbance at lower elevations, should be effectively mitigated for the successful conservation of the threatened pittas on Jeju Island.

Author Contributions: Conceptualization, E.-M.K., C.-W.K., and C.-Y.C.; data curation, E.-M.K.; formal analysis, E.-M.K., C.-W.K., C.-Y.C., and J.-H.C.; funding acquisition, E.-M.K. and C.-W.K.; investigation, E.-M.K., C.-W.K., C.-Y.C., and H.-Y.N.; methodology, C.-Y.C., J.-H.C., and H.-Y.N.; project administration, E.-M.K. and C.-W.K.; software, J.-H.C.; supervision, C.-Y.C.; visualization, C.-Y.C. and J.-H.C.; writing—original draft, E.-M.K., C.-Y.C., and H.-Y.N.; writing—review and editing, C.-W.K., C.-Y.C., J.-H.C., and H.-Y.N.

Funding: The main parts of the fieldwork were funded by the Korea Association for Bird Protection Jeju Branch with the support of the Warm Temperate and Subtropical Forest Research Center, National Institute of Forest Sciences. Field surveys in 2017 were also supported by the endangered species monitoring program of the National Institute of Biological Resources.

Acknowledgments: We thank Mirae Oh as well as the staff of the Jeju Wildlife Rescue and Rehabilitation Center who helped our fieldwork.

Conflicts of Interest: The authors declare no conflict of interest.

References

1. Southwood, T.R.E. Habitat, the templet for ecological strategies? *J. Anim. Ecol.* **1977**, *46*, 337–366. [CrossRef]
2. Clark, R.G.; Shutler, D. Avian habitat selection: Pattern from process in nest-site use by ducks? *Ecology* **1999**, *80*, 272–287. [CrossRef]
3. Avilés, J.M.; Sánchez, J.M.; Parejo, D. Nest-site selection and breeding success in the Roller (*Coracias garrulus*) in the Southwest of the Iberian Peninsula. *J. Ornithol.* **2000**, *141*, 345–350. [CrossRef]
4. Mezquida, E.T. Nest site selection and nesting success of five species of passerines in a South American open *Prosopis* Woodland. *J. Ornithol.* **2004**, *145*, 16–22. [CrossRef]
5. Li, P.; Martin, T.E. Nest-site selection and nesting success of cavity-nesting birds in high elevation forest drainages. *Auk* **1991**, *108*, 405–418.
6. Johnson, L.S.; Ostlind, E.; Brubaker, J.L.; Balenger, S.L.; Johnson, B.G.P.; Golden, H. Changes in egg size and clutch size with elevation in a Wyoming population of Mountain Bluebirds. *Condor* **2006**, *108*, 591–600. [CrossRef]
7. Lee, J.K.; Chung, O.K.; Lee, W.S. Altitudinal variation in parental provisioning of nestling Varied Tits (*Poecile varius*). *Wilson J. Ornithol.* **2011**, *123*, 283–288. [CrossRef]
8. Boyle, W.A. Can variation in risk of nest predation explain altitudinal migration in tropical birds? *Oecologia* **2008**, *155*, 397–403. [CrossRef] [PubMed]
9. BirdLife International. *Threatened Birds of Asia: The BirdLife International Red Data Book*; BirdLife International: Cambridge, UK, 2001.

10. BirdLife International. Pitta Nympha (amended version of 2016 assessment). In *The IUCN Red List of Threatened Species 2017: E.T22698684A116880779*; BirdLife International: Cambridge, UK, 2017. [CrossRef]

11. Kim, E.M.; Choi, C.Y.; Kang, C.W. Causes of injury and mortality of Fairy Pitta *Pitta nympha* on Jeju Island, Korea. *Forktail* **2013**, *29*, 145–148.

12. Lambert, F.; Woodcock, M. *Pittas, Broadbills and Asities*; Pica Press: Sussex, UK, 1996.

13. Lin, R.S.; Yao, C.T.; Lee, P.F. The diet of Fairy Pitta *Pitta nympha* nestling in Taiwan as revealed by videotaping. *Zool. Stud.* **2007**, *46*, 355–361.

14. Ko, C.Y.; Lee, P.F.; Bai, M.L.; Lin, R.S. A rule-based species predictive model for the vulnerable Fairy Pitta (*Pitta nympha*) in Taiwan. *Taiwania* **2009**, *54*, 28–36.

15. Kim, E.M.; Oh, H.S.; Kim, S.B.; Kim, W.T. The distribution and habitat environment of Fairy Pitta (*Pitta nympha* Temminck & Schlegel) on Jeju Island, Korea. *Korean J. Ornithol.* **2003**, *10*, 77–86.

16. Kim, H.K. The ecology of Fairy Pitta. *Bull. Korea Cult. Res. Inst.* **1964**, *5*, 235–240.

17. Won, P.O. *Report of the Academic Surveys on Mt. Hallasan and Hongdo Island*; Ministry of Culture and Information: Seoul, Korea, 1968.

18. Park, H.S.; Kim, W.T. Forest bird surveys in Jeju Island. *Cheju National University Journal.* **1981**, *13*, 151–165.

19. Park, H.S. A study on the community structure of forest birds in the northern slope on Mt. Halla. *Cheju National University Journal.* **1984**, *19*, 171–183.

20. Kim, E.M. The distribution and breeding ecology of Fairy Pitta (*Pitta nympha*) on Mt. Halla. In *Report of Survey and Study of Hallasan Natural Reserve*; Ko, Y.J., Ko, J.G., Eds.; Research Institute for Mt. Halla, Jeju Special Self-governing Province: Jeju, Korea, 2006; pp. 533–545.

21. Kim, E.M.; Kwon, J.O.; Kang, C.W.; Chun, J.H. Causes of the difference of inhabited altitudes above sea level of Fairy Pitta (*Pitta nympha*) on Jeju Island followed by forest landscape through the comparison of Landsat images and the literature review. *J. Korean Assoc. Geogr. Inf. Stud.* **2013**, *16*, 79–90. [CrossRef]

22. Korea Meteorological Administration. *The Understanding of Climate Change VI: The Climate Change of Jeju Island*; Jeju Regional Meteorological Office, National Institute of Meteorological Research: Jeju, Korea, 2010.

23. Kindt, R.; Coe, R. *Tree Diversity Analysis. A manual and Software for Common Statistical Methods for Ecological and Biodiversity Studies*; World Agroforestry Centre (ICRAF): Nairobi, Kenya, 2005.

24. Jacobs, J. Quantitative measurement of food selection: A modification of the forage ratio and Ivlev's electivity index. *Oecologia* **1974**, *14*, 413–417. [CrossRef] [PubMed]

25. Manly, B.F.; McDonald, L.L.; Thomas, D.; McDonald, T.L.; Erickson, W.P. *Resource Selection by Animals: Statistical Design and Analysis for Field Studies*; Kluwer Academic Publishers: Boston, MA, USA, 2002.

26. R Development Core Team. *R: A Language and Environment for Statistical Computing*; R Foundation for Statistical Computing: Vienna, Austria, 2014; Available online: http://www.r-project.org (accessed on 1 June 2015).

27. Markham, B.L.; Barker, J.L. Landsat MSS and TM post-calibration dynamic ranges, exoatmospheric reflectances and at-satellite temperatures. *EOSAT Landsat Tech. Notes* **1986**, *1*, 3–8.

28. Teillet, P.M.; Guindon, B.; Goodenough, D.G. On the slope-aspect correction of multispectral scanner data. *Can. J. Remote Sens.* **1982**, *8*, 84–106. [CrossRef]

29. Meyer, P.; Itten, K.I.; Kellenberger, T.; Sandmeier, S.; Sandmeier, R. Radiometric corrections of topographically induced effects on Landsat TM data in an alpine environment. *ISPRS J. Photogram. Remote Sens.* **1993**, *48*, 17–28. [CrossRef]

30. Badyaev, A.V.; Ghalambor, C.K. Evolution of life histories along elevational gradients: Trade-off between parental care and fecundity. *Ecology* **2001**, *82*, 2948–2960. [CrossRef]

31. Kim, E.M.; Choi, H.S.; Kang, C.W.; Min, D.W.; Yang, E.J.; Oh, M.R. Comparative studies on earthworm density by breeding place characteristics of Fairy Pitta on Jeju Island. *J. Korean Environ. Res. Tech.* **2014**, *17*, 43–49. [CrossRef]

32. Martin, S.; Lavelle, P. A simulation model of the vertical movements of an earthworm population (*Millsonia anomala*, Omodeo, Megascolecidae) in an African savanna (Lamto, Ivory Coast). *Soil Biol. Biochem.* **1992**, *24*, 1419–1424. [CrossRef]

33. Kim, K.D.; Lee, S.H. The characteristics of folk house related to climate in Cheju Island. *J. Korean Assoc. Region. Geogr.* **2001**, *7*, 29–43.

34. Koh, J.G. Global warming and vegetation of Mt. Halla. *Research Report on Mt. Halla.* **2007**, *6*, 3–17.

35. Song, C.K.; Park, Y.M.; Cho, N.K.; Ko, Y.W.; Kang, D.I. Growth responses of some medicinal plants in different altitudes of Mountain Halla. *Korean J. Med. Crop Sci.* **2000**, *8*, 134–145.

36. Edwards, C.A. Breakdown of animal, vegetable, and industrial organic wastes by earthworms. In *Earthworms in Waste and Environmental Management*; Edwards, C.A., Neuhauser, E.F., Eds.; SPB Academic Publishing BV: The Hague, Netherlands, 1988; pp. 21–31.

37. Pawlett, M.; Hopkins, D.W.; Moffett, B.F.; Harris, J.A. The effect of earthworms and liming on soil microbial communities. *Biol. Fertil. Soils* **2009**, *45*, 361–369. [CrossRef]

38. Correia, F.V.; Moreira, J.C. Effects of glyphosate and 2, 4-D on earthworms (*Eisenia foetida*) in laboratory tests. *Bull. Environ. Contam. Toxicol.* **2010**, *85*, 264–268. [CrossRef] [PubMed]

39. Beldal, E.J.; Barba, E.; Gil-delgado, J.A.; Iglesias, D.J.; Lopez, G.M.; Monros, J.S. Laying date and clutch size of Great Tits (*Parus major*) in the Mediterranean region: A comparison of four habitat types. *J. für Ornithol.* **1998**, *139*, 269–276. [CrossRef]

40. Lu, X. Breeding ecology of an old world high-altitude warbler, *Phylloscopus affinis*. *J. Ornithol.* **2008**, *149*, 41–47. [CrossRef]

41. Krementz, D.G.; Handford, P. Does avian clutch size increase with altitude? *Oikos* **1984**, *43*, 256–259. [CrossRef]

42. Storch, I. Habitat fragmentation, nest selection, and nest predation risk in Capercaillie. *Ornis Scand.* **1991**, *22*, 213–217. [CrossRef]

43. Ortego, J. Consequences of Eagle Owl nest-site habitat preference for breeding performance and territory stability. *Ornis Fenn.* **2007**, *84*, 78–90.

44. Isenmann, P.; Fradet, G. Nest site, laying period, and breeding success of the Woodchat shrike (*Lanius senator*) in Mediterranean France. *J. für Ornithol.* **1988**, *139*, 49–54. [CrossRef]

45. Ricklefs, R.E. An analysis of nesting mortality in birds. *Smithson. Contrib. Zool.* **1969**, *9*, 1–48. [CrossRef]

46. Mayfield, H. Brown-headed cowbird: Agent of extermination. *Am. Birds* **1977**, *31*, 107–113.

47. Park, C.R.; Kim, E.M.; Kang, C.W. Seasonal change of altitudinal occurrence of birds at Mt. Hallasan: Escalator effects? *Int. For. Rev.* **2010**, *12*, 24–25.

48. Park, J.Y.; Kim, B.S.; Oh, H.S. A Study on the breeding density and diet of magpie *Pica pica* in Jeju Island. *Korean J. Environ. Ecol.* **2008**, *22*, 648–657.

49. Kim, C.S. Outline of plants of Hallasan Natural Reserve. In *Report of Survey and Study of Hallasan Natural Reserve*; Ko, Y.J., Ko, J.G., Eds.; Research Institute for Mt. Halla, Jeju Special Self-governing Province: Jeju, Korea, 2006; pp. 109–137.

50. Nam, H.Y.; Kim, E.M.; Choi, C.Y.; Kang, C.W. Avifauna of Gungdae Oreum and its seasonal changes in the Jeju Eastern Oreum Group in Jeju Island, Korea. *J. Asia-Pac. Biodiver* **2019**, in press. [CrossRef]

Review

Forest Climax Phenomenon: An Invariance of Scale

Raimundas Petrokas

Institute of Forestry, Lithuanian Research Centre for Agriculture and Forestry, LT-53101 Kaunas distr., Lithuania; raimundas.petrokas@lammc.lt

Received: 6 December 2019; Accepted: 28 December 2019; Published: 2 January 2020

Abstract: We can think of forests as multiscale multispecies networks, constantly evolving toward a climax or potential natural community—the successional process-pattern of natural regeneration that exhibits sensitivity to initial conditions. This is why I look into forest succession in light of the Red Queen hypothesis and focus on the key aspects of ecological self-organisation: dynamical criticality, evolvability and intransitivity. The idea of the review is that forest climax should be associated with habitat dynamics driven by a large continuum of ecologically equivalent time scales, so that the same ecological conclusions could be drawn statistically from any scale. A synthesis of the literature is undertaken in order to (1) present the framework for assessing habitat dynamics and (2) present the types of successional trajectories based on tree regeneration mode in forest gaps. In general, there are four types of successional trajectories within the process-pattern of forest regeneration that exhibits sensitivity to initial conditions: advance reproduction specialists, advance reproduction generalists, early reproduction generalists and early reproduction specialists. A successional trajectory is an expression of a fractal connectivity among certain patterns of natural regeneration in the multiscale multispecies networks of landscape habitats. Theoretically, the organically derived measures of pattern diversity, integrity and complexity, determined by the rates of recruitment, growth and mortality of forest tree species, are the means to test the efficacy of specific interventions to avert the disturbance-related decline in forest regeneration. That is of relevance to the emerging field of biocomplexity research.

Keywords: habitat; regeneration; succession; dynamics

---

1. Background

New forestry practices combine technical efficiency, economic and environmental performance, resistance and resilience to disturbances; however, forests demonstrate a greater vulnerability to anthropogenic impacts and gradual climatic changes—the two major forms of disturbance occurring today—than to large infrequent disturbances [1,2]. "Current views of succession emphasise ongoing process rather than the climax community as the stable end point or product" [3], and "the sharp distinction between successional and climax forests is widely applied today, with major implications for conservation practices and land-use policy" [4]. Forests are intensely harvested for timber and biofuel and are never allowed to recover the natural climax state; the water regulation potential of such forests is low, while their susceptibility to fires and pests is high [5]. Therefore, we have to ask here: Are not forest resistance and resilience to disturbances the result of ecological invariance? Because the essence of ecological patterns and processes is invariance [6]. Ecological invariants are known as "scaling" and "power laws" that describe power relationships across species or across ecological systems [7]. "Scale invariance, also called scaling, or scale-free dynamics, implies that the phenomenical or phenomenological dynamics are driven by a large continuum of equally important time scales, rather than by a small number of characteristic scales" [8,9]. "By scale invariance in ecology, we mean that scales are ecologically equivalent so that the same ecological conclusions may be drawn from any scale statistically" [10]. A characteristic indicator of scale invariance is when the frequency distribution of the events of self-organisation decays only as a power law, reflecting the self-similarity of the critical state [11–13]. For instance, Manrubia and Solé [14] performed an extensive study of a real rainforest in Barro Colorado Island, Panama, and found strong evidence of a self-organised critical state in the power laws followed by the magnitudes of the system, both in space (fractality, correlation function, clearings and tree size distributions) and time (biomass fluctuations). In general, common probability patterns arise from simple invariances; invariance defines scaling relations and probability patterns [15]. Nottale [16] offers fractal space–time as a method for establishing an invariance of scale. What is more, the fundamental principle underlying the theory of invariance is that the laws of nature always have the same form for all observers [17]. This leads to a reconsideration of the traditional approach to forests focused on long-term dynamics in favour of a successional approach [18–22], which has emerged from the recognition that "even intense natural disturbances leave biological legacies and spatial heterogeneity in the new forest, which contrasts with the simple and homogeneous environment that is often the outcome of traditional harvesting practices, particularly clear-cutting" [23].

The goals of the review were to (1) present the framework for assessing habitat dynamics and (2) present the types of successional trajectories based on tree regeneration mode in forest gaps. A synthesis of the literature was undertaken within the context of the fractal fragmentation and connectivity of landscape habitats. "Fractals are dynamic process-structures that etch time into space; by illuminating fractals, we self-reflexively illuminate the observer in the observed in nothing short of nature herself" [24]. However, it is necessary to clarify that, contrary to ideal fractals, landscape patterns can be considered fractals only for limited scale intervals; this is due to the dimensionality of successional elements, determined by their regeneration response to disturbance [25]. Watt [26] was the first to link space and time at the landscape scale. There are two main models of forest dynamics, developed from Watt's seminal idea of patch dynamics, i.e., the patch–mosaic model and the gap–phase model [27]. The patch hierarchy approach, based on the hierarchical patch dynamics paradigm [28] that integrates hierarchy theory with the patch dynamics perspective, has proven useful in scaling landscape patterns and processes [29]. Scale is a main concept in landscape ecology that focuses on the influence exerted by spatio-temporal patterns on the organisation of, and interaction among, functionally integrated multispecies ecosystems [30].

2. Fractal Forest

In mathematics, symmetries have the peculiar status of being both invariant and invariant-preserving transformations, which is why a fractal, being a highly nontrivial representation of the two fundamental symmetries of nature, dilation ($r \rightarrow ar$) and translation ($r \rightarrow r + b$), exhibits self-similarity or pattern integrity—the retention of copies of itself on a hierarchy of scales [19,31–35]. In other words, a fractal is known as expanding symmetry or evolving symmetry [36]. By virtue of occupying the exact portion of the geometrical space that it occupies, a fractal has a non-integer dimension that is less than, or equal to, the Euclidean dimension of the space it occupies [37,38]. If a fractal space in which a dynamic process takes place becomes a Euclidean space with integer dimension, this means that the process has left its strange attractor (i.e., an attractor of fractal dimension) and tends toward, or already is in, the state with a lower number of possible directions of further evolution [39]. For instance, Palmer [22] demonstrated that increasing fractal dimension (decreasing spatial dependence in a landscape) allowed more species to exist per microsite and per landscape. It must be noted, nevertheless, that, when dealing with fractals, the fundamental characteristic of being differentiable is missing. Therefore, it is a challenging problem to define operators on fractal sets [40]. "Strict fractal objects require infinite power–law scaling, which fails to address the limited range of scale invariance observed in nature" [41]. So, in the end, the question is whether an ecological system manifests a physically meaningful degree of fractal connectivity between its subunits [10,42,43].

With the discovery that a set of symbols has been used by nature to encode the information for the construction and maintenance of all living things, fractals—the invariant sets of chaotic systems—serve paradoxical functions as physical boundary keepers, both to separate and connect various subsystems and levels of being [24,44–46]. The key notion of a fractal is that it possesses structures on a hierarchy of scales generated by the reiteration of a mathematical formula [35], which is "a form of feedback, where the answer to the formula recycles into the original formula to generate the next solution" [47]. Therefore, the complexity of dynamic process-structures is measured by their fractal dimension [24,48]. For instance, the number $N(x)$ of objects with a characteristic linear dimension greater than x can be given by $N(x) \sim x^{-D}$, where $N(x)$ is a number measure corresponding to the scale unit x and D is the fractal dimension [8]. The value of the scaling exponent of the number–size relationship may vary widely, and the power–law scaling only holds over a finite range of time scales in real landscapes [49,50]. A power–law distribution of the probability density function of the pieces of an object in space or the parts of a process in time is evidenced in a straight line on a plot of log (number) vs. log (size) [51]. To estimate the habitat patch fractals, an alternative power–law distribution can be written: $N(m) = Cm^{-b}$, where $N(m)$ is the number of habitat patches with a biomass greater than m, C is a constant and b is a scaling exponent; noting that $m \sim x^{-3}$, we can find from a comparison with the number–size fractal distribution that $D = 3b$ [8]. From a forest management perspective, the patchwork of habitats in varying successional stages of recovery may correspond to forest stand development patterns, and the biomass to the stand volume. The power–law distribution of fractal fragments can be used as an indicator of the fragmentation of the landscape habitat into patches and landscape connectivity change [8]. Fractal fragmentation is often a scale-invariant process, but nevertheless most ecological patterns and processes show scaling thresholds at which abrupt changes in scaling relationships occur, corresponding to shifts in underlying mechanisms [52].

3. Coexistence

The basic components of a network of elements whose creation, evolution, destruction and interaction cause the emergence of a particular behaviour or feature that cannot be reduced to the properties of an individual system's components are called coexisting attractors [32,53,54]. An attractor—a region in state space that a system can enter but not leave—is a mathematical model of causal closure [55]. Closure usually results from the nonlinear, feedback nature of interactions [46,56]. We can generalise an attractor as any state toward which a dynamical system tends to evolve. Within the framework of chaos theory, the overloaded vegetation climax is considered a "strange attractor": the smallest invariant set of the events of self-organisation that exhibit exponential sensitivity to initial conditions [13,57–61]—"sensitivity in fact to any numerical rounding at any calculational step, not necessarily at the initial time" [62]. According to Anand [63], "this attractor itself is moving on a deterministic path imposed on the process by a highest order environmental constraint, the long term evolution of the Earth's climate". More than that, the final state toward which an ecosystem tends to evolve usually depends on the initial conditions involving several coexisting attractors [64–69]. As Allesina and Levine [70] put it, "just a handful of limiting factors can generate the coexistence of many species, a feature of intransitive networks". Climax forest is an excellent example of such an ecosystem in which climate, landscape, vegetation and fauna are closely interconnected: when one of these components is destroyed, whether partially or completely, the other components undergo an equally violent change [71]. Therefore, it follows that forest climax should be associated with habitat dynamics driven by a large continuum of ecologically equivalent time scales: the frequency of occurrence of an event of a given magnitude x is inversely proportional to some power α of its magnitude, $f(x) \sim x^{-\alpha}$ [33,72]. $f(x) \sim x^{-1}$—a critical dependence—is often associated with a feedback dynamic that creates a stable equilibrium at a critical point. It provides a general mechanism for the emergence of scale-free networks with the power–law degree distribution [16,73–78]. However, real-world networks do not follow power–law degree distribution over the whole range of a degree. In many real-world networks, scale-free property coexists with a hierarchy of nodes, low node separation and high clustering [79]. The hierarchy always follows a pair of exponential laws and a power law; it appears if a certain pattern is added at each time unit into the network [79,80]. "Networks which display scale-free properties are the most hierarchical" [81]. Hierarchical network structure promotes a dynamical robustness, the origins of which are in the understanding of the impact of node failures on the integrity of a network [82]. This is important for many disciplines, as many real-world networks are organised into many small, highly connected modules that combine in a hierarchical manner into larger units, with their number and degree of clustering following a power law [83,84]. Power–law distributions can be produced by endogenous processes like feedback loops, self-organisation, network effects, etc., so the key problem is to understand why nature gives rise to the wide diversity of degree structures found in real-world networks and why scale-free networks are rare [10,85].

We can think of forests as multiscale multispecies networks, constantly evolving toward a climax or potential natural community—the successional process-pattern of natural regeneration [3,12,74,86–92]. Known to ecologists as secondary forest succession, natural regeneration is the regrowth and reestablishment of the forests, recovering from natural or human disturbance [4,91–94]. Natural regeneration, however, is not achieved or accomplished—it is lived and evolved—and this is especially true when there is a focus on ecological self-organisation: organisms connected in communities transform the ecosystem while transforming themselves, and the chemical outputs of organisms, self-produced through feedback loops, are used by other organisms to facilitate their own self-reproduction [95,96]. To put it simply, the biology of self-replication is self-referential, as embodied by nucleic acid replication mechanisms; self-reference is "the hinge upon which levels of serial inclusiveness intercross" [24]—a critical scale of a phenomenon [12,47,97–101]. Furthermore, that is why I look into forest succession in light of the Red Queen hypothesis: life has evolved in order to stay extant, or else go extinct. The Red Queen hypothesis, as formulated by van Valen [102], is similar to that of a system obeying a self-organised criticality, which means that a given Red Queen phenomenon is caused by the system that organises its critical state by itself [103,104].

Within biology, the developmental process of organisms as well as their metabolisms, growth and learning have been identified as self-organising processes; nevertheless, there is yet no unique theory of self-organisation [105]. One of the objectives of the present article was, therefore, to give prominence to the key aspects of ecological self-organisation: dynamical criticality, evolvability and intransitivity. "Dynamical criticality, a central property for the functioning of a living organism, naturally emerges as a consequence of evolution that favours evolvability" [106]. Dynamical criticality explains evolution by reference to the broad internal disposition of a population to evolve in order to stay extant, rather than any actual evolutionary trajectory of populations by capturing the influence that the internal features of populations can have upon the outcomes of evolution [107]. Taking into account the above, I propose the conceptual framework for assessing habitat dynamics at a network–system–trajectory interface (Table 1): The network, because nature can be viewed in terms of multilevel, multidimensional hierarchies of inter-related event clusters that form a metaheterarchy, or a heterogeneous general hierarchy [24]; the system, because the experience of events can be viewed as a summary of the facts through which the events took place—a fact pattern [108,109]; and the trajectory—an expression of a relation among certain fact patterns in the network [79,110].

Table 1. The conceptual framework for assessing habitat dynamics.

Aspects	Concepts	Verifiers	
		Pattern	**Process**
Intransitivity	Network	Diversity [70,111]	Robustness [82,83,112,113]
Criticality	System	Integrity [114]	Fitness [115–121]
Evolvability	Trajectory	Complexity [22,107]	Inclusiveness [24,101,122]

4. Successional Species Turnover

The view of succession was developed from the patchwork pattern of habitats in varying stages of recovery from human disturbances. However, a compositional shift, in which post-disturbance stands dominated by fast growing shade-intolerant tree species are eventually replaced by late seral, shade-tolerant species, is not a simple unidirectional sequence of stages, but rather a complex model subject to differential species responses to factors such as physical site conditions, initial stand composition and intermediate disturbance effects [123]. Furthermore, "although successional stages are defined by characteristics of a forest stand, successional trajectories are fundamentally determined by rates of recruitment, growth, and mortality of populations of the component tree species" [124]. "Trees define the communities that they inhabit, are host to a multiplicity of other organisms and can determine the ecological dynamics of other plants and animals" [114]. In woodland habitat modelling, tree species form the basis of the parameters used to represent a range of component species within a particular process-pattern of the events of self-organisation and to derive a network. Moreover, unlike the action of seasons and natural disasters, long-term change in the composition of communities is brought about by the activities of living organisms which themselves inhabit the environment [19]: "Many of our rarest species are associated with ancient trees and only occur where there has been a continuous cover of old trees back through time on the site" [125]. For these reasons, the use of measures that account for the Red Queen dynamics of interacting populations of the component species that form the continuous cover of trees through time on the site may provide new ways to monitor succession and test the efficacy of specific interventions to modify the disturbance-related changes in successional process properties: robustness, fitness and inclusiveness (Table 1) [21,120,122,126–132]. "It has been suggested that Red Queen dynamics underlie a large number of important biological processes, some of which are still poorly understood, such as genetic recombination and sexual reproduction" [120].

The shift in dominance by shade-intolerant tree species to shade-tolerant tree species is the most generalisable and predictable feature of successional pathways [4]. The most distinctive difference between shade-intolerant and shade-tolerant tree species is that the former are incapable of establishing themselves in a forest understorey (Table 2), because their seeds do not accumulate in a long-lived seed pool. Instead, they germinate immediately upon dispersal or soon thereafter [133]. Shade-intolerant tree species, colonising from a refuge site, fast-growing, having low wood density, branching with axial differentiation, short-lived, rapidly establishing on disturbed sites, showing early reproduction, high fecundity and large dispersal, often bear numerous small seeds, annually produced and wind- or animal-dispersed [4,134–137]. Shade-tolerant tree species, advance reproduction-dependent, slow growing, branching without axial differentiation, long-lived, gradually replacing intolerants in the absence of disturbance, often bear few seeds, larger in size, sometimes masting, sometimes dispersed only locally and by diverse dispersal agents. Still, in the context of forest dynamics, the low light survival/high light growth tradeoff is "only one of the many possible strategies for trees to differentiate along a disturbance gradient", and "it is unlikely to function as an important mechanism for the stable coexistence of several tree species" [134]. This is not to deny the importance of the tradeoff in determining the successional status of species, yet "successional species turnover, in which pioneer species are being replaced by shade-tolerant species, already starts at the very early years of succession" [138]. The model of initial floristic composition postulates that most late successional species (like shrubs and trees) that will later dominate the community are already available at the onset of succession: "They are either part of the soil seed bank or present with vegetative propagules, rhizomes, or a sapling bank" [116]. So, "in actual practice, the distinction between a 'successional' and a 'climax' forest is subjective; there is no magical moment when a forest stops undergoing succession" [4].

Table 2. The types of successional trajectories based on tree regeneration mode in forest gaps. Chazdon's et al. [124] successional trajectories (Clark and Clark's [139] species groups A–D) correspond roughly to Whitmore's [140] species groups 1–4, having increasing "pioneer index", Yamamoto's [141] four major types of tree regeneration mode in gaps (numbered I–IV) and Petrere's et al. [90] four community types. Modified from Franklin [142].

Growth	Establishment	
	Forest	**Gaps**
	Old-growth specialists (A)	Successional generalists (C)
Forest	Establish and grow in dark forest; shade-tolerant species. Low potential and average growth rates, especially as juveniles. (1)	Establish in gaps, grow best in gaps, but can survive as saplings in closed forest. Higher juvenile growth potential than groups A or B. (3)
	Advance regeneration/gap filler/understorey tree (III)	Gap coloniser/gap filler/canopy tree/gap maker (II)
	Old-growth generalists (B)	Successional specialists (D)
Gaps	Establish in shade but show increased association with gaps as saplings. Growth rates as low as group A as juveniles, increasing with size. (2)	Establish and grow best in gaps at all juvenile stages; shade-intolerant species. Highest growth potential, especially as juveniles. (4)
	Advance regeneration/gap filler/canopy tree/gap maker (I)	Gap coloniser/canopy tree/gap maker (IV)

5. Conclusions

Is the Red Queen hypothesis similar to that of the system obeying a power–law sensitivity to initial conditions? Well, yes it is—and that is all about the time scale of self-reference. All climax communities exhibit sensitivity to initial conditions at any time scale of self-reference. There are four types of successional trajectories within the process-pattern of forest regeneration which exhibits sensitivity to initial conditions: advance reproduction specialists—gap fillers/understorey trees—establishing and growing in a dark forest; advance reproduction generalists—gap fillers/canopy trees/gap makers—establishing in the shade but showing increased association with gaps as saplings; early reproduction generalists—gap colonisers/gap fillers/canopy trees/gap makers—establishing in gaps, growing best in gaps, but surviving as saplings in a closed forest; early reproduction specialists—gap colonisers/canopy trees/gap makers—establishing and growing best in gaps at all juvenile stages. Recognising that dynamical criticality, a central property for the evolvability and intransitivity of living organisms, naturally emerges as a consequence of ecological self-organisation, forest climax should be associated with habitat dynamics driven by a large continuum of observer-invariant time scales. For this reason, the organically derived measures of pattern diversity, integrity and complexity, determined by rates of recruitment, growth and mortality of forest tree species that form a climax or potential natural community, are the means to test the efficacy of specific interventions to modify the disturbance-related changes in successional process properties: robustness, fitness and inclusiveness.

Funding: This research did not receive any specific grant from funding agencies in the public, commercial or not-for-profit sectors.

Acknowledgments: I am especially grateful to David Smart, who, as an English native speaker, willingly assessed the text of my review. This study was supported by the long-term research programme "Sustainable forestry and global changes", implemented by the Lithuanian Research Centre for Agriculture and Forestry.

Conflicts of Interest: The author declares no conflicts of interest.

References

1. Seynave, I.; Bailly, A.; Balandier, P.; Bontemps, J.-D.; Cailly, P.; Cordonnier, T.; Deleuze, C.; Dhôte, J.-F.; Ginisty, C.; Lebourgeois, F.; et al. GIS Coop: networks of silvicultural trials for supporting forest management under changing environment. *Ann. For. Sci.* **2018**, *75*, 48. [CrossRef]

2. Cole, L.E. How Quickly do Tropical Forests Recover from Disturbance? 2014. Available online: https://www.kew.org/blogs/kew-science/how-quickly-do-tropical-forests-recover-from-disturbance (accessed on 16 May 2018).

3. Yearsley, K.; Parminter, J. Seral Stages Across Forested Landscapes: Relationships to Biodiversity. Res. Prog. Extension Note 18; B.C. Ministry of Forests, 8p. Available online: https://www.for.gov.bc.ca/hfd/pubs/docs/En/En18.pdf (accessed on 28 March 2018).

4. Chazdon, R.L. *Second Growth: The Promise of Tropical Forest Regeneration in an Age of Deforestation*; University of Chicago Press: Chicago, IL, USA, 2014.

5. Makarieva, A.M.; Gorshkov, V.G.; Li, B.-L. Revisiting forest impact on atmospheric water vapor transport and precipitation. *Theor. Appl. Climatol.* **2013**, *111*, 79–96. [CrossRef]

6. Frank, S.A.; Bascompte, J. Invariance in ecological pattern. *bioRxiv* **2019**, 673590. [CrossRef]

7. Pulliam, H.R.; Waser, N.M. Ecological invariance and the search for generality in ecology. In *The Ecology of Place: Contributions of Place-Based Research to Ecological Understanding*; Billick, I., Price, M.V., Eds.; University of Chicago Press: Chicago, IL, USA, 2010; pp. 69–92.

8. Wendt, H.; Didier, G.; Combrexelle, S.; Abry, P. Multivariate Hadamard self-similarity: Testing fractal connectivity. *Physica D* **2017**, *356–357*, 1–36. [CrossRef]

9. Combrexelle, S.; Wendt, H.; Didier, G.; Abry, P. Multivariate scale-free dynamics: Testing fractal connectivity. In Proceedings of the 42nd IEEE international conference on acoustics, speech and signal processing (ICASSP), New Orleans, LA, USA, 5–9 March 2017; pp. 3984–3988. [CrossRef]

10. Li, B.-L. Fractal geometry applications in description and analysis of patch patterns and patch dynamics. *Ecol. Model.* **2000**, *132*, 33–50. [CrossRef]

11. Parrott, L.; Lange, H. An introduction to complexity science. In *Managing World Forests as Complex Adaptive Systems*; Messier, C., Puettmann, K.J., Coates, K.D., Eds.; Routledge: London, UK, 2013; pp. 17–32.

12. Kurakin, A. The self-organizing fractal theory as a universal discovery method: The phenomenon of life. *Theor. Biol. Med. Model.* **2011**, *8*, 4. [CrossRef]

13. Carlson, J.M.; Doyle, J. Complexity and robustness. *Proc. Natl. Acad. Sci. USA* **2002**, *99*, 2538–2545. [CrossRef]

14. Manrubia, S.C.; Solé, R.V. Self-organized criticality in rainforest dynamics. *Chaos Solitons Fractals* **1996**, *7*, 523–541. [CrossRef]

15. Frank, S.A. Common probability patterns arise from simple invariances. *arXiv* **2016**, arXiv:1602.03559v2.

16. Nottale, L. Lecture 19. Scale relativity. In *Scale Invariance and Beyond. Les Houches Workshop, 10–14 March 1997*; EDP Sciences; Dubrulle, B., Graner, F., Sornette, D., Eds.; Springer: Berlin/Heidelberg, Germany, 2013; pp. 249–261. [CrossRef]

17. Rusbult, C. Einstein's Theory of Relativity is a Theory of Invariance-Constancy. 2007. Available online: https://www.asa3.org/ASA/education/views/invariance.htm (accessed on 27 October 2019).

18. Morin, X.; Fahse, L.; Jactel, H.; Scherer-Lorenzen, M.; García-Valdés, R.; Bugmann, H. Long-term response of forest productivity to climate change is mostly driven by change in tree species composition. *Sci. Rep.* **2018**, *8*, 5627. [CrossRef]

19. Petrokas, R.; Baliuckas, V. Self-sustaining forest. *Appl. Ecol. Environ. Res.* **2017**, *15*, 409–426. [CrossRef]

20. Kirschbaum, M.; Fischlin, A. Climate change impacts on forests. In *Climate Change 1995—Impacts, Adaptations and Mitigation of Climate Change: Scientific-Technical Analysis*; Watson, R., Zinyowera, M.C., Moss, R.H., Eds.; Contribution of Working Group II to the Second Assessment Report of the Intergovernmental Panel of Climate Change (IPCC); Cambridge University Press: Cambridge, UK, 1996; pp. 95–129.

21. MCPFE Resolution H1: General Guidelines for the Sustainable Management of Forests in Europe. In Proceedings of the Second Ministerial Conference on the Protection of Forests in Europe (MCPFE), Helsinki, Finland, 16–17 June 1993.

22. Palmer, M.W. The coexistence of species in fractal landscapes. *Am. Nat.* **1992**, *139*, 375–397. [CrossRef]

23. Gustafsson, L.; Baker, S.C.; Bauhus, J.; Beese, W.J.; Brodie, A.; Kouki, J.; Lindenmayer, D.B.; Lõhmus, A.; Martínez Pastur, G.; Messier, C.; et al. Retention forestry to maintain multifunctional forests: A world perspective. *BioScience* **2012**, *62*, 633–645. [CrossRef]

24. Marks-Tarlow, T. Fractal entanglement between observer and observed. *Int. J. Semiot. Vis. Rhetor.* **2018**, *2*, 1–14. [CrossRef]

25. Andronache, I.; Marin, M.; Fischer, R.; Ahammer, H.; Radulovic, M.; Ciobotaru, A.M.; Jelinek, H.F.; Di Ieva, A.; Pintilii, R.D.; Drăghici, C.C.; et al. Dynamics of forest fragmentation and connectivity using particle and fractal analysis. *Sci. Rep.* **2019**, *9*, 12228. [CrossRef] [PubMed]

26. Watt, A.S. Pattern and process in the plant community. *J. Ecol.* **1947**, *35*, 1–22. [CrossRef]

27. Král, K.; Shue, J.; Vrška, T.; Gonzalez-Akre, E.B.; Parker, G.G.; McShea, W.J.; McMahon, S.M. Fine-scale patch mosaic of developmental stages in Northeast American secondary temperate forests: The European perspective. *Eur. J. For. Res.* **2016**, *135*, 981–996. [CrossRef]

28. Wu, J.; Loucks, O.L. From balance-of-nature to hierarchical patch dynamics: A paradigm shift in ecology. *Q. Rev. Biol.* **1995**, *70*, 439–466. [CrossRef]

29. Wu, J.; Li, H.; Jones, K.B.; Loucks, O.L. (Eds.) Scaling with known uncertainty: A synthesis. In *Scaling and Uncertainty Analysis in Ecology: Methods and Applications*; Springer: Dordrecht, The Netherlands, 2006; pp. 329–346.

30. Urban, D.L.; O'Neill, R.V.; Shugart, H.H., Jr. Landscape ecology: A hierarchical perspective can help scientists understand spatial patterns. *BioScience* **1987**, *37*, 119–127. [CrossRef]

31. Huynh, H.N.; Pradana, A.; Chew, L.Y. The complexity of sequences generated by the arc-fractal system. *PLoS ONE* **2015**, *10*, e0117365. [CrossRef]

32. Drake, J.; Fuller, M.; Zimmermann, C.; Gamarra, J. Emergence in ecological systems. In *From Energetics to Ecosystems: The Dynamics and Structure of Ecological Systems*; Rooney, N., McCann, K., Noakes, D., Eds.; Springer: New York, NY, USA, 2007; pp. 157–184.

33. Komulainen, T. *Self-Similarity and Power Laws*; Report 145/2004, 109–122; Helsinki University of Technology, Control Engineering Laboratory: Helsinki, Finland, 2004.

34. Li, B.-L. Fractal dimensions. In *Encyclopedia of Environmetrics*; El-Shaarawi, A.H., Piegorsch, W.W., Eds.; John Wiley & Sons Ltd.: Chichester, UK, 2002; Volume 2, pp. 821–825.

35. Salingaros, N.A.; West, B.J. A universal rule for the distribution of sizes. *Environ. Plan. B Plan. Des.* **1999**, *26*, 909–923. [CrossRef]

36. Kumar, D.K.; Arjunan, S.P.; Aliahmad, B. *Fractals: Applications in Biological Signalling and Image Processing*; CRC Press: Boca Raton, FL, USA, 2017.

37. Barnsley, M.F. *Fractals Everywhere*; Academic Press: New York, NY, USA, 1988.

38. Mandelbrot, B.B. *The Fractal Geometry of Nature*, 2nd ed.; Macmillan: London, UK, 1983.

39. Devaney, R.L. *An Introduction to Chaotic Dynamical Systems*; Benjamin/Cummings: Menlo Park, CA, USA, 1986; pp. 176–202.

40. Cattani, C. Fractal and fractional. *Fractal Fract.* **2017**, *1*, 1. [CrossRef]

41. Halley, J.D.; Winkler, D.A. Critical-like self-organization and natural selection: Two facets of a single evolutionary process? *BioSystems* **2008**, *92*, 148–158. [CrossRef] [PubMed]

42. Rickles, D.; Hawe, P.; Shiell, A. A simple guide to chaos and complexity. *J. Epidemiol. Community Health* **2007**, *61*, 933–937. [CrossRef]

43. Oldershaw, R.L. A Fractal Universe? 2003. Available online: http://www3.amherst.edu/~{}rloldershaw/NOF.HTM (accessed on 4 February 2019).

44. Brier, S. Biosemiotics. In *Encyclopedia of Language and Linguistics*, 2nd ed.; Elsevier Science: Oxford, UK, 2006; Volume 2, pp. 31–40.

45. Mindell, A. *Quantum Mind: The Edge between Physics and Psychology*; Lao Tse Press: Portland, OR, USA, 2000.

46. Gyr, J. Psychophysics: The self-referent holonomic observer-observed relation. In *Origins: Brain and Self Organization. Proceedings of 2nd Appalachian Conference on Behavioral Neurodynamics, Radford, Virginia, October*; Pribram, K.H., Ed.; Lawrence Erlbaum: Hillsdale, NJ, USA, 1994; pp. 102–120.

47. Welch, K. A Fractal Topology of Time: Implications for Consciousness and Cosmology. Ph.D. Thesis, California Institute of Integral Studies, San Francisco, CA, USA, 2010.

48. Joosten, J.J.; Soler-Toscano, F.; Zenil, H. Fractal dimension versus process complexity. *Adv. Math. Phys.* **2016**, *2016*, 5030593. [CrossRef]

49. Crawley, M.J.; Harral, J.E. Scale dependence in plant biodiversity. *Science* **2001**, *291*, 864–868. [CrossRef]
50. Schoener, T.W.; Spiller, D.A.; Losos, J.B. Natural restoration of the species–area relation for a lizard after a hurricane. *Science* **2001**, *294*, 1525–1528. [CrossRef]
51. Liebovitch, L.S.; Shehadeh, L.A. Introduction to fractals. In *Tutorials in Contemporary Nonlinear Methods for the Behavioral Sciences*; Web Book; Riley, M.A., Orden, G.V., Eds.; National Science Foundation, Directorate for Social, Behavorial and Economic Sciences: Alexandria, VA, USA, 2005; pp. 178–266.
52. Turcotte, D.L. Fractals and fragmentation. *J. Geophys. Res.* **1986**, *91*, 1921–1926. [CrossRef]
53. Tilebein, M. A complex adaptive systems approach to efficiency and innovation. *Kybernetes* **2006**, *35*, 1087–1099. [CrossRef]
54. Camazine, S.; Deneubourg, J.L.; Franks, N.R.; Sneyd, J.; Théraulaz, G.; Bonabeau, E. *Self-Organization in Biological Systems*, 2nd ed.; Princeton University Press: Princeton, NJ, USA, 2003.
55. Falconer, K.J. *Fractals: A Very Short Introduction*; Oxford University Press: Oxford, UK, 2013. [CrossRef]
56. Heylighen, F. The science of self-organization and adaptivity. In *Knowledge Management, Organizational Intelligence and Learning, and Complexity*; Kiel, L.D., Ed.; Encyclopedia of Life Support Systems (EOLSS); Eolss Publishers: Oxford, UK, 2001; Volume III, pp. 849–874.
57. Boeing, G. Visual analysis of nonlinear dynamical systems: Chaos, fractals, self-similarity and the limits of prediction. *Systems* **2016**, *4*, 37. [CrossRef]
58. Bartone, S. Strange Attractors: Queers, Chaos, and Evolution. 2014. Available online: https://journals.uvic.ca/index.php/adcs/article/view/17821/0 (accessed on 27 March 2018).
59. Rietkerk, M.; van de Koppel, J. Regular pattern formation in real ecosystems. *Trends Ecol. Evol.* **2008**, *23*, 169–175. [CrossRef] [PubMed]
60. Decocq, G. Determinism, chaos and stochasticity in plant community successions: Consequences for phytosociology and conservation ecology. In *Nature Conservation: Concepts and Practice*; Gafta, D., Akeroyd, J.R., Eds.; Springer: Berlin, Germany, 2006; pp. 254–266.
61. Eckmann, J.P.; Ruelle, D. Ergodic theory of chaos and strange attractors. *Rev. Mod. Phys.* **1985**, *57*, 617. [CrossRef]
62. Tsallis, C.; Plastino, A.R.; Zheng, W.M. Power-law sensitivity to initial conditions—new entropic representation. *Chaos Solitons Fractals* **1997**, *8*, 885–891. [CrossRef]
63. Anand, M. Towards a Unifying Theory of Vegetation Dynamics. Ph.D. Thesis, The University of Western Ontario, London, ON, Canada, 1997.
64. Klimenko, A.Y. Intransitivity in theory and in the real world. *Entropy* **2015**, *17*, 4364–4412. [CrossRef]
65. Freire, J.G.; Bonatto, C.; DaCamara, C.C.; Gallas, J.A.C. Multistability, phase diagrams, and intransitivity in the Lorenz-84 low-order atmospheric circulation model. *Chaos* **2008**, *18*, 033121. [CrossRef] [PubMed]
66. Gleick, J. *Chaos: Making a New Science*, 20th ed.; Penguin Books: New York, NY, USA, 2008.
67. Lorenz, E.N. Can chaos and intransitivity lead to interannual variability? *Tellus* **1990**, *42*, 378–389. [CrossRef]
68. Crutchfield, J.P.; Farmer, J.D.; Packard, N.H. Shaw RS Chaos. *Sci. Am.* **1986**, *254*, 46–57. [CrossRef]
69. McDonald, S.W.; Grebogi, C.; Ott, E.; Yorke, J.A. Fractal basin boundaries. *Physica D* **1985**, *17*, 125–153. [CrossRef]
70. Allesina, S.; Levine, J.M. A competitive network theory of species diversity. *Proc. Natl. Acad. Sci. USA* **2011**, *108*, 5638–5642. [CrossRef]
71. Richards, P.W. *The Tropical Rain Forest: An Ecological Study*; Cambridge University Press: New York, NY, USA, 1952.
72. Rhodes, C.J.; Anderson, R.M. Power laws governing epidemics in isolated populations. *Nature* **1996**, *381*, 600–602. [CrossRef]
73. Haimovici, A.; Marsili, M. Criticality of mostly informative samples: A Bayesian model selection approach. *J. Stat. Mech. Theory E* **2015**, *2015*, P10013. [CrossRef]
74. Graham, B. *Nature's Patterns: Exploring Her Tangled Web*, 2nd ed.; FreshVista: New York, NY, USA, 2014.
75. Laurienti, P.J.; Joyce, K.E.; Telesford, Q.K.; Burdette, J.H.; Hayasaka, S. Universal fractal scaling of self-organized networks. *Physica A* **2011**, *390*, 3608–3613. [CrossRef] [PubMed]
76. Messier, C.; Puettmann, K.J. Forests as complex adaptive systems: Implications for forest management and modelling. *L'Italia For. Mont.* **2011**, *66*, 249–258. [CrossRef]
77. Turcotte, D.L. Self-organized criticality. *Rep. Prog. Phys.* **1999**, *62*, 1377–1429. [CrossRef]

78. Bak, P.; Chen, K.; Creutz, M. Self-organized criticality in the 'Game of Life'. *Nature* **1989**, *342*, 780–782. [CrossRef]

79. Náther, P.; Markošová, M.; Rudolf, B. Hierarchy in the growing scale-free network with local rules. *Phys. A Stat. Mech. Appl.* **2009**, *388*, 5036–5044. [CrossRef]

80. Chen, Y.G. Zipf's law, 1/f noise, and fractal hierarchy. *Chaos Solitons Fractals* **2012**, *45*, 63–73. [CrossRef]

81. Mishkovski, I. Hierarchy and vulnerability of complex networks. In *ICT Innovations 2013. Advances in Intelligent Systems and Computing (AISC)*; Trajkovik, V., Anastas, M., Eds.; Springer: Heidelberg, Germany, 2014; Volume 231, pp. 273–281.

82. Smith, C.G.; Puzio, R.S.; Bergman, A. Hierarchical network structure promotes dynamical robustness. *arXiv* **2015**, arXiv:1412.0709v2.

83. Barabási, A.-L.; Pósfai, M. *Network Science*; Cambridge University Press: Cambridge, UK, 2016.

84. Jeong, H.; Tombor, B.; Albert, R.; Oltvai, Z.N.; Barabási, A.-L. The large-scale organization of metabolic networks. *Nature* **2000**, *407*, 651–654. [CrossRef]

85. Broido, A.D.; Clauset, A. Scale-free networks are rare. *Nat. Commun.* **2019**, *10*, 1017. [CrossRef]

86. Donaldson, L.; Wilson, R.J. Maclean IMD Old concepts, new challenges: Adapting landscape-scale conservation to the twenty-first century. *Biodivers. Conserv.* **2017**, *26*, 527–552. [CrossRef]

87. Duan, R.; Huang, M.; Kong, X.; Wang, Z.; Fan, W. Ecophysiological responses to different forest patch type of two codominant tree seedlings. *Ecol. Evol.* **2015**, *5*, 265–274. [CrossRef] [PubMed]

88. Hopkins, D. Resilience and Regime Change in Ecosystems: Bifurcation and Perturbation Analysis. 2009. Available online: https://math.dartmouth.edu/archive/m53f09/public_html/ (accessed on 22 March 2017).

89. Girvetz, E.H.; Greco, S.E. How to define a patch: A spatial model for hierarchically delineating organism-specific habitat patches. *Landsc. Ecol.* **2007**, *22*, 1131–1142. [CrossRef]

90. Petrere, M., Jr.; Giordano, L.C.; De Marco, P., Jr. Empirical diversity indices applied to forest communities in different successional stages. *Braz. J. Biol.* **2004**, *64*, 841–851. [CrossRef] [PubMed]

91. Borman, M.M.; Pyke, D.A. Successional theory and the desired plant community approach. *Rangelands* **1994**, *16*, 82–84.

92. Clements, F.E. Nature and structure of the climax. *J. Ecol.* **1936**, *24*, 252–284. [CrossRef]

93. Barbour, M.G.; Billings, W.D. (Eds.) *North American Terrestrial Vegetation*, 2nd ed.; Cambridge University Press: Cambridge, UK, 2000.

94. Richards, P.W. The secondary succession in the tropical rain forest. *Sci. Prog.* **1955**, *43*, 45–57.

95. Longo, G.; Montevil, M. Extended criticality, phase spaces and enablement in biology. *Chaos Solitons Fractals* **2013**, *55*, 64–79. [CrossRef]

96. Padgett, J. Autocatalysis in Chemistry and the Origin of Life. 2011. Unpublished Paper. Available online: http://home.uchicago.edu/~{}jpadgett/unpub.html (accessed on 13 June 2014).

97. Wu, J.; Li, H. Concepts of scale and scaling. In *Scaling and Uncertainty Analysis in Ecology: Methods and Applications*; Wu, J., Jones, K.B., Li, H., Loucks, O.L., Eds.; Springer: Dordrecht, The Netherlands, 2006; pp. 3–15.

98. Earl, D.J.; Deem, M.W. Evolvability is a selectable trait. *Proc. Natl. Acad. Sci. USA* **2004**, *101*, 11531–11536. [CrossRef]

99. Stauffer, D.; Aharony, A. *Introduction to Percolation Theory*, 2nd ed.; Taylor & Francis: London, UK, 1994.

100. Varela, F. *Principles of Biological Autonomy*; North Holland: New York, NY, USA, 1979.

101. Varela, F. A calculus for self-reference. *Int. J. Gen. Syst.* **1975**, *2*, 5–24.

102. Van Valen, L. A new evolutionary law. *Evol. Theory* **1973**, *1*, 1–30.

103. Pruessner, G. *Self-Organised Criticality: Theory, Models and Characterisation*; Cambridge University Press: Cambridge, UK, 2012.

104. Adami, C. Self-organized criticality in living systems. *Phys. Lett. A* **1995**, *203*, 29–32. [CrossRef]

105. Banzhaf, W. Self-organizing systems. In *Encyclopedia of Complexity and Systems Science*; Meyers, R., Ed.; Springer: New York, NY, USA, 2009; pp. 8040–8050.

106. Torres-Sosa, C.; Huang, S.; Aldana, M. Criticality is an emergent property of genetic networks that exhibit evolvability. *PLoS Comput. Biol.* **2012**, *8*, e1002669. [CrossRef] [PubMed]

107. Brown, R.L. What evolvability really is. *Br. J. Philos. Sci.* **2014**, *65*, 549–572. [CrossRef]

108. Brier, S. Cybersemiotics: An evolutionary world view going beyond entropy and information into the question of meaning. *Entropy* **2010**, *12*, 1902–1920. [CrossRef]

109. Peirce, C.S. *The Essential Peirce: Selected Philosophical Writings*; Indiana University Press: Bloomington, IL, USA, 1998.

110. Li, A.; Wang, L.; Schweitzer, F. The optimal trajectory to control complex networks. *arXiv* **2018**, arXiv:1806.04229.

111. Briers, R. Habitat Networks—Reviewing the Evidence Base: Final Report. Contract Report to Scottish Natural Heritage, Contract Number 29752. 2011. Available online: https://www.nature.scot/habitat-networks-reviewing-evidence-base-final-report (accessed on 4 February 2019).

112. Ortiz-Barrientos, D.; Baack, E.J. Species integrity in trees. *Mol. Ecol.* **2014**, *23*, 4188–4191. [CrossRef]

113. Kaiser-Bunbury, C.N.; Blüthgen, N. Integrating network ecology with applied conservation: A synthesis and guide to implementation. *AoB Plants* **2015**, *7*, plv076. [CrossRef]

114. Landi, P.; Minoarivelo, H.O.; Brännström, Å.; Hui, C.; Dieckmann, U. Complexity and stability of ecological networks: A review of the theory. *Popul. Ecol.* **2018**, *60*, 319–345. [CrossRef]

115. Ramiadantsoa, T.; Ovaskainen, O.; Rybicki, J.; Hanski, I. Large-scale habitat corridors for biodiversity conservation: A forest corridor in Madagascar. *PLoS ONE* **2015**, *10*, e0132126. [CrossRef]

116. Goetze, D.; Karlowski, U.; Porembski, S.; Tockner, K.; Watve, A.; Riede, K. Spatial and temporal dimensions of biodiversity dynamics. In *Biodiversity: Structure and Function. Encyclopedia of Life Support Systems (EOLSS)*; Barthlott, W., Linsenmair, K.E., Porembski, S., Eds.; Developed under the Auspices of the UNESCO; Eolss Publishers: Oxford, UK, 2014; pp. 166–208.

117. Kärenlampi, P.P. Symmetry of interactions rules in incompletely connected random replicator ecosystems. *Eur. Phys. J. E* **2014**, *37*, 56. [CrossRef] [PubMed]

118. Hidalgo, J.; Grilli, J.; Suweis, S.; Muñoz, M.A.; Banavar, J.R.; Maritan, A. Information-based fitness and the emergence of criticality in living systems. *Proc. Natl. Acad. Sci. USA* **2014**, *111*, 10095–10100. [CrossRef] [PubMed]

119. Robert, A. Find the weakest link. A comparison between demographic, genetic and demo-genetic metapopulation extinction times. *BMC Evol. Biol.* **2011**, *11*, 260. [CrossRef]

120. Dercole, F.; Ferriere, R.; Rinaldi, S. Chaotic Red Queen coevolution in three-species food chains. *Proc. R. Soc. B* **2010**, *277*, 2321–2330. [CrossRef] [PubMed]

121. Dieckmann, U.; Ferrière, R. Adaptive dynamics and evolving biodiversity. In *Evolutionary Conservation Biology*; Ferrière, R., Dieckmann, U., Couvet, D., Eds.; Cambridge University Press: Cambridge, UK, 2004; pp. 188–224.

122. Coren, R.L. *The Evolutionary Trajezctory: The Growth of Information in the History and Future of Earth*; The World Futures General Evolution Studies Series 11; CRC Press: London, UK, 1998. [CrossRef]

123. Taylor, A.R. Concepts, Theories and Models of Succession in the Boreal Forest of Central Canada. Ph.D. Thesis, Lakehead University, Thunder Bay, ON, Canada, 2009.

124. Chazdon, R.L.; Chao, A.; Colwell, R.K.; Lin, S.-Y.; Norden, N.; Letcher, S.G.; Clark, D.B.; Finegan, B.; Arroyo, J.P. A novel statistical method for classifying habitat generalists and specialists. *Ecology* **2011**, *92*, 1332–1343. [CrossRef] [PubMed]

125. Rose, B. Tree Ecology. 2005. Available online: http://www.monkey-do.net/sites/default/files/Tree%20ecology.pdf (accessed on 1 September 2015).

126. Cannon, J.B.; Peterson, C.J.; O'Brien, J.J.; Brewer, J.S. A review and classification of interactions between forest disturbance from wind and fire. *For. Ecol. Manag.* **2017**, *406*, 381–390. [CrossRef]

127. Saura, S.; Martín-Queller, E.; Hunter, M.L., Jr. Forest landscape change and biodiversity conservation. In *Forest Landscapes and Global Change: Challenges for Research and Management*; Azevedo, J.C., Pereram, A.H., Pinto, M.A., Eds.; Springer: New York, NY, USA, 2014; pp. 167–198. [CrossRef]

128. Desrochers, R.E.; Anand, M. Quantifying the components of biocomplexity along ecological perturbation gradients. *Biodivers. Conserv.* **2005**, *14*, 3437–3455. [CrossRef]

129. McCleary, K.; Mowat, G. Using forest structural diversity to inventory habitat diversity of forest-dwelling wildlife in the West Kootenay region of British Columbia. *BC J. Ecosyst. Manag.* **2002**, *2*, 1–13.

130. Michener, W.K.; Baerwald, T.J.; Firth, P.; Palmer, M.A.; Rosenberger, J.L.; Sandlin, E.A.; Zimmerman, H. Defining and unraveling biocomplexity. *BioScience* **2001**, *51*, 1018–1023. [CrossRef]

131. McEvoy, T.J. *Introduction to Forest Ecology and Silviculture*; The University of Vermont: Burlington, VT, USA, 1995.

132. Lipsitz, L.A.; Goldberger, A.L. Loss of 'complexity' and aging: Potential applications of fractals and chaos theory to senescence. *JAMA* **1992**, *267*, 1806–1809. [CrossRef]

133. Sousa, W.P. The role of disturbance in natural communities. *Ann. Rev. Ecol. Syst.* **1984**, *15*, 353–391. [CrossRef]

134. Gravel, D.; Canham, C.D.; Beaudet, M.; Messier, C. Shade tolerance, canopy gaps and mechanisms of coexistence of forest trees. *Oikos* **2010**, *119*, 475–484. [CrossRef]

135. Vester, H.F.M. Forest development as a basis for management; tree architecture and tree temperaments. In *Ecology and Management of Tropical Secondary Forest: Science, People, and Policy*; Centro Agronómico Tropical de Investigación y Enseñanza (CATIE): Turrialba, Costa Rica, 1998; pp. 35–48.

136. Bazzaz, F.A. Regeneration of tropical forests: Physiological responses of pioneer and secondary species. In *Rainforest Regeneration and Management*; Gomez-Pompa, A., Whitmore, T.C., Hadley, M., Eds.; Man and the Biosphere Series 6; UNESCO: Paris, France, 1991; pp. 91–118.

137. Whitmore, T.C. Tropical rain forest dynamics and its implications for management. In *Rainforest Regeneration and Management*; Gomez-Pompa, A., Whitmore, T.C., Hadley, M., Eds.; Man and the Biosphere Series 6; UNESCO: Paris, France, 1991; pp. 67–89.

138. Van Breugel, M.; Bongers, F.; Martinez-Ramos, M. Species dynamics during early secondary forest succession: Recruitment, mortality and species turnover. *Biotropica* **2007**, *35*, 610–619. [CrossRef]

139. Clark, D.A.; Clark, D.B. Life history diversity of canopy and emergent trees in a neotropical forest. *Ecol. Monogr.* **1992**, *62*, 315–344. [CrossRef]

140. Whitmore, T.C. Canopy gaps and the two major groups of forest trees. *Ecology* **1989**, *70*, 536–538. [CrossRef]

141. Yamamoto, S. Gap regeneration of major tree species in different forest types of Japan. *Vegetatio* **1996**, *127*, 203–213. [CrossRef]

142. Franklin, J. Regeneration and growth of pioneer and shade-tolerant rain forest trees in Tonga. *N. Z. J. Bot.* **2003**, *41*, 669–684. [CrossRef]

Review

Methods for Monitoring Large Terrestrial Animals in the Wild

Alexander Prosekov [1], Alexander Kuznetsov [2], Artem Rada [2] and Svetlana Ivanova [3,4,*]

[1] Laboratory of Biocatalysis, Kemerovo State University, Krasnaya Street 6, 650043 Kemerovo, Russia; a.prosekov@inbox.ru

[2] Computer Engineering Center, Kemerovo State University, Krasnaya Street 6, 650043 Kemerovo, Russia; adkuz@inbox.ru (A.K.); radaartem@mail.ru (A.R.)

[3] Natural Nutraceutical Biotesting Laboratory, Kemerovo State University, Krasnaya Street 6, 650043 Kemerovo, Russia

[4] Department of General Mathematics and Informatics, Kemerovo State University, Krasnaya Street 6, 650043 Kemerovo, Russia

* Correspondence: pavvm2000@mail.ru; Tel.: +7-384-239-6832

Received: 1 June 2020; Accepted: 23 July 2020; Published: 26 July 2020

Abstract: Reliable information about wildlife is absolutely important for making informed management decisions. The issues with the effectiveness of the control and monitoring of both large and small wild animals are relevant to assess and protect the world's biodiversity. Monitoring becomes part of the methods in wildlife ecology for observation, assessment, and forecasting of the human environment. World practice reveals the potential of the joint application of both proven traditional and modern technologies using specialized equipment to organize environmental control and management processes. Monitoring large terrestrial animals require an individual approach due to their low density and larger habitat. Elk/moose are such animals. This work aims to evaluate the methods for monitoring large wild animals, suitable for controlling the number of elk/moose in the framework of nature conservation activities. Using different models allows determining the population size without affecting the animals and without significant financial costs. Although, the accuracy of each model is determined by its postulates implementation and initial conditions that need statistical data. Depending on the geographical, climatic, and economic conditions in each territory, it is possible to use different tools and equipment (e.g., cameras, GPS sensors, and unmanned aerial vehicles), a flexible variation of which will allow reaching the golden mean between the desires and capabilities of researchers.

Keywords: monitoring methods; large wild animals; elk; moose; hunting; unmanned aerial vehicles

---

1. Introduction

It is difficult to overestimate the importance of monitoring animals in the wild as it is part of the environmental control system. It includes monitoring, evaluating, and predicting the state of the human environment. The problems of the control and monitoring effectiveness of both large and small wild animals are relevant all over the world [1]. Today, modern monitoring methods are being implemented. This is supported by the use of new monitoring technologies [2,3].

Wild animals are at risk during their co-existence with humans due to capture, extermination, and anthropological impact on their habitat areas. Understanding the need to regulate the population of different animals, including commercial species, partly solves the problem of preserving the fauna diversity on our planet. Environmental monitoring issues concern the general public (ecologists, biologists, hunters, statesmen, top managers, researchers, and ordinary citizens). Animal control takes into account various natural and man-made factors that affect the state of the environment.

Monitoring helps identify patterns of such changes and ensure proper environmental management. In addition, it makes it possible to establish a correlation and find connections between biology, ecology, and economics, e.g., hunting and eco-tourism.

Ideally, it is desirable to get an absolute estimate of wildlife populations [4–6]. However, in reality, it is difficult to get data close to absolute. This problem is most acute for animals that live in hardly accessible areas and vast territories, such as impenetrable taiga, jungles, and tropical forests.

Monitoring biodiversity on a large scale and with adequate representation requires significant effort and resources, and presents a logistical challenge for researchers [7]. Monitoring schemes for butterflies and birds [8,9] can be classified as successful projects of civil science, with the presentation of a large amount of data on the occurrence of species as well as in terms of the number and distribution of species. For animals, there are no such monitoring schemes; an attempt was made to initially accumulate data for 21 candidates (genetic composition, species population, species characteristics, community composition, ecosystem functioning, and ecosystem structure, www.geobon.org/ebvs) [7].

Traditional wildlife counting methods (Table 1) are widely used throughout the world [10,11]. In the scientific context, different counting methods have been proposed for decades, and there is a constant debate about their effectiveness.

Table 1. Traditional wildlife monitoring methods.

Method	Animals	Sources
Survey and questionnaire	Large and medium-sized animals	[12–15]
Counting by traces of vital activity (counting indirect signs-the number of burrows, claw marks, the number of feces, etc.)	Large and medium-sized mammals	[16–21]
Sampling and marking	All animal species	[22–27]
Winter route tracking	Large, medium, and small animals, birds	[28,29]
The use of traps, pens, and nets	Large and medium-sized mammals	[11,30–33]
Remote tracking using specialized equipment (camera traps, sensor nets, acoustic sensors, and GPS sensors)	All animal, bird, and insect species	[16,34–48]
Aerial survey (counting, photo, and video shooting from aerial devices and systems)	Large animals	[1,49–52]

Researchers face an acute problem of choosing a method of control and observation for all animal species, both small and large. Among other things, it depends on the habitat. Individual monitoring methods have been developed for marine animals [53,54], insects [55–58], and terrestrial animals [59,60]. The size of the studied animals is also important for choosing the monitoring method. This review is limited to considering only large terrestrial animals. Large animals are usually those with the highest quantitative parameters of ontogenesis (weight, length, height, etc.) in their classes. Large forest animals include elephant, rhinoceros, hippo, giraffe, bear, crocodile, tiger, etc.

The habitat of large terrestrial animals is assumed to be land. Russia occupies the first place in terms of the area among the countries (almost half of its territory consists of forests with a rich variety of flora and fauna). Bear (brown and polar), elk, and bison are considered to be the large animals of Russian forests [61].

Elk are widely distributed throughout the world, including the Scandinavian Peninsula, Alaska, the European part of Russia, and Western Siberia (including Kemerovo oblast—Kuzbass) [62]. An important commercial animal, elk is an object of sport hunting, wildlife watching, and a protected species.

Elks inhabit forests, willows on the shores of steppe lakes, and river floodplains in forest-steppe; they also love cool coniferous forests, where there is swampy soil. Elks do not usually actively use their entire territory, but only a part of it, mainly where there are sufficient stocks of the main seasonal feed and good protective conditions throughout the year. Elks correspond to a high degree of settledness; some individuals can adhere to a small area of the territory for a long time. They look for a new place of residence when the amount of food decreases, for example, in the winter with a significant height of snow cover; however, in the spring they return to their original place. A group of elk is quite grouped—in winter they try not to scatter far from each other, but in spring they show more independence. These animals adapt well to changing environmental conditions.

Some methods make it possible to estimate the population size based on relative counts, one of which is distance sampling [63]. However, they do not apply to large animals for various reasons. Populations of rare or elusive large mammals are difficult to control because they are usually secretive, lone, occur at low densities, and have large domestic ranges, which poses significant methodological problems for population estimation [64,65]. The elk population is rather small, so it is important to assess the effectiveness of monitoring in the field of hunting as an environmental protection measure for biological diversity support and endangered species preservation.

The purpose of this work is to assess methods for monitoring large wild animals that are suitable for controlling elk within the Kuzbass nature protection measures. The need for this review is based on the diversity of methods, the disparity of information about them, and the lack of a comprehensive methodology for counting large wild animals and the prospect of combining different survey methods, including the use of unmanned aerial systems in order to obtain the most reliable information.

2. Methods

Undertaken in November 2019 and updated in July 2020, the literature search considered the articles published from 1 January 2015 until the present. Article databases from Scopus, Web of Sciences, and Google Scholar were used for cross-checking. We used a search strategy based on multiple queries. The final ones were selected after several search passes by qualitative evaluation of the number of results obtained and their relevance.

Table 2 lists the queries used and the number of articles identified by each query. The entire bibliography of the included studies was manually checked for compliance with the search subject by title and abstract. Articles whose titles and/or abstracts contained "moose/elk monitoring methods" were passed to the full-text selection stage by default. To narrow the scope of the "Animal monitoring methods" search, both the Scopus and Web of Science databases excluded publications from the fields unrelated to the search topic (such as Medicine, Pharmacology, Immunology and Microbiology, Neuroscience, Engineering, Chemistry, Computer Science, Physics and Astronomy, Social Science, etc.). Due to the number of publications of the query in Google Scholar, sources from the year 2019 were put under review. After excluding intersections across all search databases, 1205 sources remained under review.

Table 2. Strategies for searching the literature sources for the review.

Database	Search Query	Number of Articles	Matching Search Results*
Scopus	Moose monitoring methods [title/abstract/keywords]	19	8
	Elk monitoring methods [title/abstract/keywords]	20	2
	Animal monitoring methods [title/abstract/keywords]	1976	827
WoS	Moose monitoring methods	17	8
	Elk monitoring methods	19	5
	Animal monitoring methods	499	322
Google Scholar	Moose monitoring methods	11,800	468
	Elk monitoring methods	16,300	231

* Even indirect matching was taken into account

3. Results and Discussion

The global experience of wildlife monitoring is represented by the use of various traditional counting methods. These include observation, trapping, experiments, collection, analysis of demographic data, and surveys of reserve employees [12–14,17,25,66].

3.1. Active Observation Methods

The method of trapping using box traps, trap pens, drip nets, independently or in combination with indirect indicators, allows assessing the impact of traps on the population of wild animals to determine their density and identify the risk of disease. This method is a traditional one, and using several methods together can improve its accuracy. However, the quality of its application is directly dependent on the number of qualified professionals engaged; any attempt to save finances leads to errors in monitoring and external negative impact on the studied animal population with all the resulting negative consequences [59]. It is possible to improve the monitoring accuracy by identifying individual species, but in this case, it is necessary to physically impact the animals, which can be invasive, expensive, and difficult from the point of view of logistics [67].

3.2. Passive Observation Methods

The authors of [15] used surveys of hunters in the Northeastern and upper Midwestern regions of the United States (approximately 11,000 hunters annually between 2012 and 2014) to monitor the number of elk (*Alces alces*) and collect information. The study [68] presents the results of testing a smartphone application for interviewing hunters (Alberta, Canada, during 2012–2014) for reporting the number of elk (*Alces alces*). The potential of the developed application Loose Survey and its cost-effectiveness in comparison with more expensive methods of the aerial survey are underlined. However, the human factor should be taken into account as it can significantly reduce the accuracy of this approach.

Observation of winter routes [69,70] from ungulate tracks is one of the oldest large animal monitoring methods, which are used to track changes in an animal's moving path, habitat, winter shelters, breeding grounds, and to determine the population size. The main advantages of this method are long-term usage possibility, low financial costs, and feasibility. While, the list of disadvantages includes the occurrence of errors and unreliability of the data obtained, whose value is not stable, as well as the presence of a human factor. Obermoller et al. used the method of observing the movement dynamics of female species to estimate the number of elk calves, and the data obtained were extended to the population size by applying modeling [71]. Migration appears to be an effective behavioral strategy for extending access to seasonal resources and can be a sustainable strategy of industrial centers or settlements for ungulates experiencing climate change [72]. If the snow cover is not high enough, it is either not possible to collect data on the number of animals or they have significant distortions [73].

The method of monitoring ungulates based on counting the number of fecal pellets is among the traditional ones [74]. It is used both as the main and complementary methods [75–77]. Blåhed et al. optimized the SNP (single nucleotide polymorphism) genotyping of fecal samples from elk (*Alces alces*) for identification of a species and its gender (489 fecal samples) [78]. Together with other traditional methods, genetic methods allow specifying the number of animals and complete information about the sex ratio, settlement, reproduction, and genetic variability. Pfeffer et al. compared the results of the obtained numbers of elk (*Alces alces*) and roe deer (*Capreolus capreolus*) in Northern Sweden after using fecal count and camera traps together with the random encounter model (REM), which can estimate the density without having to recognize individual species [79]. The authors found that, compared to density estimates from camera traps, fecal counts appear to underestimate the population density for roe deer. For elk, the data obtained from the two methods were comparable. In comparison with other methods, the method of counting groups of pellets has the main advantage—cost effectiveness—and can be used in areas that are not impenetrable forests.

3.3. Remote Monitoring

Remote monitoring involves the use of specialized equipment (photo and video equipment, acoustic devices, and sensors), usually providing not only data collection, but also almost real-time data transmission. The most reliable data can be obtained by organizing constant automatic monitoring—continuous automatic tracking of animals in real-time [80]. There are enough advantages of this control method, although it is difficult to implement it in remote wild areas (lack of a steady connection, inability to ensure uninterrupted operation of batteries, maintenance of equipment, and sensors used). Therefore, the scope of application is limited to small hunting farms or reserves located not too far away from industrial centers.

Tracking methods using a wireless sensor network (WSN) can achieve similar goals. Zhang and team [35] used simulation modeling based on the collected data to obtain a more reliable picture of the sika deer population.

There is a relatively new group of visual and acoustic wild animal monitoring methods using automatic recorders. Tracking technologies allow researchers to study the life of nature with huge spatial and temporal resolution. They have been developed relatively recently to assess biodiversity by measuring the acoustic heterogeneity created by animals in natural habitats [36].

Monitoring methods using camera traps and acoustic sensors are considered cost-effective [3]. The list of advantages includes the ability to simultaneously evaluate various indicators. The effectiveness and adequacy of these methods determine the potential for their use in the long term. With the development of technologies to produce widely available hardware and software, the methods will be extensively applied.

Night photography, telemetering, and camera traps in combination with the search and calculation of indirect indicators (traces of vital activities) were simultaneously used to determine the number of elk (*Cervus canadensis*) [81].

Visual observation methods using GPS location and digital visualization (3D) images are among the relatively new monitoring methods, that have already proven their effectiveness for environmental activities. Remote sensing methods (GPS or geolocation) present a comprehensive analysis of various data. Monitoring methods using 2D and 3D cameras create a "presence effect" in the process of detecting an animal. They provide an automated, non-contact, and cost-effective way to study animal behavior. Digital technologies and Big Data allow us to approach wildlife monitoring in a different way considering its versatility. The integrated use of aerial photography methods with the interpretation of indirect signs (the number of tracks) contributes to the overall effectiveness of the monitoring.

The significant cost of visual observation methods using digital technologies does not expand the possibilities of their application [34,35,38].

Smith et al. determined the migration routes and numbers of elk by observing and collecting data from global positioning system (GPS) collars [82]. The authors modeled the seasonal selection of habitat-related resources. Phillips et al. determined the number of elk using a probabilistic model combined with GPS collar detection data [83]. The reliability of the obtained population size estimates was recognized as an advantage. The risks include the use of this model in territories with different characteristics. It is important to consider the factors under which this model is valid. Bergman et al. jointly analyzed the data obtained from field route observations of elk and GPS collars and came to the conclusion about the economic efficiency and benefits of their joint use [84].

3.4. Methods of Aerial Survey

Many problems with the previously described methods can be avoided with the help of an aerial survey. Bristow et al. used several approaches to estimate the number of elk in Arizona: the efficiency of the hybrid model (double-counting and double-observer methods) was compared to the traditional tracking by tagging and recording individual animals during helicopter flights [85]. In terms of economy, the hybrid model is preferable, while in terms of precision, an aerial survey is better. The development of the hybrid model was more expensive than using traditional counting methods.

The development of unmanned aerial vehicles has opened up new opportunities for using them to solve civil tasks. Unmanned aerial vehicles (UAVs) and continuous shooting (photo, video, and IR) are a sustainable and effective group of modern methods of wildlife recording. Monitoring methods using UAVs can determine the number of animals in a given area. This method can be used for one or several types of species.

The authors [51,86–91] used unmanned aerial vehicles with consumer-level digital cameras to count the number of elk. Havens and Sharp used a modified drone "Predator" for this purpose [2]. Patterson et al. used small unmanned aerial systems (UAS) for aerial photography of caribou (reindeer of North America, Labrador, Canada) and recognized the advantages and capabilities of a small electric-powered fixed-wing UAS Brican TD100E [92].

Xu et al. [93] used aerial photography from a quadrocopter to estimate the number of large horned animals, including elk, in combination with a system for processing RGB images. To detect and count large horned individuals, the RGB image obtained by the drone was processed with a modern machine learning algorithm.

The authors underlined the performance and efficiency of this model. Shao et al. [94] used the method of aerial photography using UAVs to count large horned animals, including elk. A system based on convolutional neural networks (CNNs) using aerial photographs was applied. Witczuk et al. [95] used drones with thermal infrared imaging to count ungulates. Dziki-Michalska et al. [96] analyzed data collected by volunteers, hunters, and foresters in combination with aerial photography. Data on the abundance of large animals, including moose, collected from various sources, were used to construct regression models. The authors note that the use of aerial photography increases the accuracy of the modeling, but significant financial costs do not allow its use on an ongoing basis.

UAV aerial photography is very promising with regard to the development of geospatial data collection methods. They are justified in cases where it is necessary to get accurate information about the area at a certain time. The main advantages of UAVs are the following: the ability to automatically detect almost all animals, both in semi-arid savannas and areas with significant vegetation; the accuracy of the results obtained; high-quality images; and fast data processing. Despite its effectiveness, the latter method is not often used as it is quite costly, even though they are cheaper to operate than helicopters and manned fixed-wing aerial vehicles. UAVs require special piloting skills, have limitations on their use over large spatial and temporal scales, including regulations, and depend on weather conditions [51,86–91].

4. Conclusions

Effective monitoring issues are problematic within the framework of foreign and domestic science. In many territories (Alaska, Canada, Scandinavian countries, and Siberia), biologists, researchers, and hunting farm employees traditionally walk around the property and count the tracks of animals (elk, deer, roe, etc.), making an absolute (counting all individuals of the territory) or relative (counting a part of the animals and obtaining an idea of the entire population using the appropriate conversion factors) census [73,97,98]. The size of the territories, their inaccessibility, a lack of funding for hunting farms and natural territories, as well as adverse weather conditions impede the spread of modern wildlife control and monitoring methods. World practice confirms the need to modernize monitoring technologies and the readiness to use new methods, if they take into account most of the disadvantages of the traditional methods (full snow cover, low accuracy, and non-dependence on weather conditions and vegetation cover), if there is technical support or instructions, and in case they do not exceed the cost of existing methods.

It has already been proven effective to use several methods simultaneously with specialized equipment, such as photo traps, video-, or IR cameras. If the observation area is accessible and limited, and the number of animals is insignificant, to obtain reliable information about the population size, it is sufficient to mark them, as described in [99], with GPS collars [42]. If the terrain conditions allow, a stationary camera with a GPS sensor will allow getting almost complete information at a fairly low cost. However, there are areas with vegetation that does not allow free movement, while there is no other way to monitor elk or other large animals with UAVs. It is known that the effectiveness of animal detection by UAVs significantly increases with the use of infrared detectors in low-altitude flights [100]. The use of these technologies to detect and enumerate groups of wild animals in wildlife monitoring is superior to conventional remote sensing technologies [101], but improvements in detection and identification technology are needed to exceed the accuracy of traditional aerial photography of animals [102,103].

Author Contributions: Conceptualization, A.P., A.K. and A.R.; methodology, A.K.; analyzed and interpreted the data, S.I.; formal analysis, A.R.; resources, A.K.; data curation, A.K. and S.I.; writing—review and editing, S.I. and A.P.; project administration, A.P. All authors have read and agreed to the published version of the manuscript.

Funding: This research was financial supported MINISTRY OF SCIENCE AND HIGHER EDUCATION OF THE RUSSIAN FEDERATION, project number 075-15-2020-540 (unique project identifier RFMEFI62120X0037), and partially financial supported by theCOUNCIL ON GRANTS OF THE PRESIDENT OF THE RUSSIAN FEDERATION,grant number[SS-2694.2020.4].

Conflicts of Interest: The authors declare no conflict of interest.

References

1. Haalck, L.; Mangan, M.; Webb, B.; Risse, B. Towards image-based animal tracking in natural environments using a freely moving camera. *J. Neurosci. Methods* **2020**, *330*, 108–455. [CrossRef] [PubMed]
2. Havens, K.J.; Sharp, E.J. *Thermal Imaging Techniques to Survey and Monitor. Animals in the Wild*; Elsevier Inc.: Amsterdam, the Netherlands; Academic Press: New York, NY, USA, 2016. [CrossRef]
3. Buxton, R.T.; Lendrum, P.E.; Crooks, K.R.; Wittemyer, G. Pairing camera traps and acoustic recorders to monitor the ecological impact of human disturbance. *Glob. Ecol. Conserv.* **2018**, *16*, e00493. [CrossRef]
4. Hammond, T.T.; Springthorpe, D.; Walsh, R.E.; Berg-Kirkpatrick, T. Using accelerometers to remotely and automatically characterize behavior in small animals. *J. Exp. Biol.* **2016**, *219*, 1618–1624. [CrossRef] [PubMed]
5. Merrick, M.J.; Koprowski, J.L. Should we consider individual behavior differences in applied wildlife conservation studies? *Biol. Conserv.* **2017**, *209*, 34–44. [CrossRef]
6. Said, M.Y.; Ogutu, J.O.; Kifugo, S.C.; Maqui, O.; Reid, R.S.; De Leeuwa, J. Effects of extreme land fragmentation on wildlife and livestock population abundance and distribution. *J. Nat. Conserv.* **2016**, *34*, 151–164. [CrossRef]
7. Cretois, B.; Linnell, J.D.C.; Grainger, M.; Nilsen, E.B.; Røa, J.K. Hunters as citizen scientists: Contributions to biodiversity monitoring in Europe. *Glob. Ecol. Conserv.* **2020**, *23*, e01077. [CrossRef]

8. Cima, V.; BenoîtFontaine, B.; IsabelleWitté, I.; Dupont, P.; Jeanmougin, M.; Touroult, J. A test of six simple indices to display the phenology of butterflies using a large multi-source database. *Ecol. Indic.* **2020**, *110*, 105885. [CrossRef]

9. Neate-Clegg, M.H.C.; Hornsa, J.J.; Adler, F.R.; Aytekin, M.Ç.K.; Şekercioğlu, Ç.H. Monitoring the world's bird populations with community science data. *Biol. Conserv.* **2020**, *248*, 108653. [CrossRef]

10. Bobek, B.; Merta, D.; Furtek, J. Winter food and cover refuges of large ungulates in lowland forests of south-western Poland. *For. Ecol. Manag.* **2016**, *359*, 247–255. [CrossRef]

11. Valente, A.M.; Binantel, H.; Villanua, D.; Acevedo, P. Evaluation of Methods to Monitor Wild Mammals on Mediterranean Farmland. *Mamm. Biol.* **2018**, *91*, 23–29. [CrossRef]

12. Kubo, T.; Mieno, T.; Kuriyama, K. Wildlife viewing: The impact of money-back guarantees. *Tour. Manag.* **2018**, *70*, 49–55. [CrossRef]

13. Störmer, N.; Weaver, L.C.; Stuart-Hill, G.; Diggle, R.W.; Naidoo, R. Investigating the effects of community-based conservation on attitudes towards wildlife in Namibia. *Biol. Conserv.* **2019**, *233*, 193–200. [CrossRef]

14. Willcox, D.; Nash, H.C.; Trageser, S.; Kim, H.J.; Hywood, L.; Connelly, E.; Ichu Ichu, G.; Kambale, N.; Mousset, M.; Ingram, L.S.; et al. Evaluating methods for detecting and monitoring pangolin populations (Pholidota: Manidae). *Glob. Ecol. Conserv.* **2019**, *17*, e00539. [CrossRef]

15. Crum, N.J.; Fuller, A.K.; Sutherland, C.S.; Cooch, E.G.; Hurst, J. Estimating occupancy probability of moose using hunter survey data. *Wildl. Manag.* **2017**, *3*, 521–534. [CrossRef]

16. Khwaja, H.; Buchan, C.; Wearn, O.R.; Bahaa-el-din, L.; Bantlin, D.; Bernard, H.; Bitariho, R.; Bohm, T.; Borah, J.; Brodie, J.; et al. Pangolins in global camera trap data: Implications for ecological monitoring. *Glob. Ecol. Conserv.* **2019**, *20*, 1–14. [CrossRef]

17. Cromsigt, J.P.G.M.; Van Rensburg, S.J.; Etienne, R.S.; Olff, H. Monitoring large herbivore diversity at different scales: Comparing direct and indirect methods. *Biodivers. Conserv.* **2009**, *18*, 1219–1231. [CrossRef]

18. Eggert, L.S.; Eggert, J.A.; Woodruff, D.S. Estimating population sizes for elusive animals: The forest elephants of Kakum National Park, Ghana. *Mol. Ecol.* **2003**, *12*, 1389–1402. [CrossRef]

19. Lioy, S.; Braghiroli, S.; Dematteis, A.; Meneguz, P.G.; Tizzani, P. Faecal pellet count method: Some evaluations of dropping detectability for *Capreolus capreolus* Linnaeus, 1758 (Mammalia: Cervidae), Cervus elaphus Linnaeus, 1758 (Mammalia: Cervidae) and Lepus europaeus Pallas, 1778 (Mammalia: Leporidae). *Ital. J. Zool.* **2015**, *82*, 231–237. [CrossRef]

20. Plhal, R.; Kamler, J.; Homolka, M.; Drimaj, J. An assessment of the applicability of dung count to estimate the Wild boar population density in a forest environment. *J. For. Sci.* **2014**, *60*, 174–180. [CrossRef]

21. De Assis Morais, T.; Da Rosa, C.A.; Viana-Junior, A.B.; Santos, A.P.; Passamani, M.; De Azevedo, C.S. The influence of population-control methods and seasonality on the activity pattern of wild boars (*Sus scrofa*) in high-altitude forests. *Mamm. Biol.* **2020**, *100*, 101–106. [CrossRef]

22. Field, K.A.; Paquet, P.C.; Artelle, K.; Proulx, G.; Brook, R.K.; Darimont, C.T. Correction: Publication reform to safeguard wildlife from researcher harm. *PLoS Biol.* **2019**, *18*, e3000752. [CrossRef] [PubMed]

23. Walker, K.A.; Mellish, J.A.E.; Weary, D.M. Behavioural responses of juvenile Steller sea lions to hot-iron branding. *Appl. Anim. Behav. Sci.* **2010**, *122*, 58–62. [CrossRef]

24. Ferreira, V.H.B.; Silva, C.P.C.D.; Fonseca, E.D.P.; Chagas, A.C.C.S.D.; Pinheiro, L.G.M.; Almeida, R.N.; De Sousa, M.B.C.; Silva, H.P.A.D.; Galvão-Coelho, N.L.; Ferreira, R.G. Hormonal correlates of behavioural profiles and coping strategies in captive capuchin monkeys (Sapajus libidinosus). *Appl. Anim. Behav. Sci.* **2018**, *207*, 108–115. [CrossRef]

25. Zemanova, M.A. Towards more compassionate wildlife research through the 3Rs principles: Moving from invasive to non-invasive methods. *Wildl. Biol.* **2020**. [CrossRef]

26. Cordier, T.; Alonso-Sáez, L.; Apothéloz-Perret-Gentil, L.; Aylagas, E.; Bohan, D.A.; Bouchez, A.; Chariton, A.; Creer, S.; Frühe, L.; Keck, F.; et al. Ecosystems monitoring powered by environmental genomics: A review of current strategies with an implementation roadmap. *Mol. Ecol.* **2020**. [CrossRef]

27. Shury, T.K.; Pybus, M.J.; Nation, N.; Cool, N.L.; Rettie, W.J. Fascioloides magna in Moose (*Alces alces*) From Elk Island National Park, Alberta. *Vet. Pathol.* **2019**, *56*, 476–485. [CrossRef] [PubMed]

28. Sieber, A.; Uvarov, N.V.; Baskin, L.M.; Radeloff, V.C.; Bateman, B.L.; Pankov, A.B.; Kuemmerle, T. Post-Soviet land-use change effects on large mammals' habitat in European Russia. *Biol. Conserv.* **2015**, *191*, 567–576. [CrossRef]

29. Lee, J.-B.; Kim, Y.-K.; Bae, Y.-S. A study of methods for monitoring wild mammals in Unmunsan, Korea. *J. Asia-Pac. Biodivers.* **2019**, *12*, 541–544. [CrossRef]

30. Davis, A.J.; Leland, B.; Bodenchuk, M.; Vercauteren, K.C.; Pepin, K.M. Estimating population density for disease risk assessment: The importance of understanding the area of influence of traps using wild pigs as an example. *Prev. Vet. Med.* **2017**, *14*, 33–37. [CrossRef]

31. Sugimoto, T.; Aramilev, V.V.; Nagata, J.; McCullough, D.R. Winter food habits of sympatric carnivores, Amur tigers and Far Eastern leopards, in the Russian Far East. *Mamm. Biol.* **2016**, *81*, 214–218. [CrossRef]

32. Zaumyslova, O.Y.; Bondarchuk, S.N. The Use of Camera Traps for Monitoring the Population of Long-Tailed Gorals. *Achiev. Life Sci.* **2015**, *9*, 15–21. [CrossRef]

33. Markov, N.; Pankova, N.; Morelle, K. Where winter rules: Modeling wild boar distribution in its north-eastern range. *Sci. Total Environ.* **2019**, *68*, 1055–1064. [CrossRef] [PubMed]

34. Badescu, A.-M.; Cotofana, L. A wireless sensor network to monitor and protect tigers in the wild. *Ecol. Indic.* **2015**, *57*, 447–451. [CrossRef]

35. Zhang, L.; Ameca, E.I.; Jiang, Z. Viability analysis of the wild sika deer (*Cervus nippon*) population in China: Threats of habitat loss and effectiveness of management interventions. *J. Nat. Conserv.* **2018**, *43*, 117–125. [CrossRef]

36. Papin, M.; Aznar, M.; Germain, E.; Guérold, F.; Pichenot, J. Using acoustic indices to estimate wolf pack size. *Ecol. Indic.* **2019**, *103*, 202–211. [CrossRef]

37. Kalan, A.K.; Mundry, R.; Wagner, O.J.J.; Heinicke, S.; Boesch, C.; Kühl, H.S. Towards the automated detection and occupancy estimation of primates using passive acoustic monitoring. *Ecol. Indic.* **2015**, *54*, 217–226. [CrossRef]

38. Sugai, L.S.M.; Llusia, D. Bioacoustic time capsules: Using acoustic monitoring to document biodiversity. *Ecol. Indic.* **2019**, *99*, 149–152. [CrossRef]

39. Enari, H.; Enari, H.S.; Okuda, K.; Maruyama, T.; Okuda, K. An evaluation of the efficiency of passive acoustic monitoring in detecting deer and primates in comparison with camera traps. *Ecol. Indic.* **2018**, *98*, 753–762. [CrossRef]

40. Hertel, A.G.; Leclerc, M.; Warren, D.; Pelletier, F.; Zedrosser, A.; Mueller, T. Don't poke the bear: Using tracking data to quantify behavioural syndromes in elusive wildlife. *Anim. Behav.* **2019**, *147*, 91–104. [CrossRef]

41. Nicholson, K.L.; Warren, M.J.; Rostan, C.R.; Mansson, J.; Paragi, T.F.; Sand, H. Using fine-scale movement patterns to infer ungulate parturition. *Ecol. Indic.* **2019**, *101*, 22–30. [CrossRef]

42. Boughton, R.K.; Allen, B.L.; Tillman, E.A.; Wisely, S.M.; Engeman, R.M. Road hogs: Implications from GPS collared feral swine in pastureland habitat on the general utility of road-based observation techniques for assessing abundance. *Ecol. Indic.* **2019**, *99*, 171–177. [CrossRef]

43. Nicheporchuk, V.; Gryazin, I.; Favorskaya, M.N. Framework for Intelligent Wildlife Monitoring. In *International Conference on Intelligent Decision Technologies IDT 2020: Intelligent Decision Technologies*; Springer: Singapore, 2020; Volume 193, pp. 167–177. [CrossRef]

44. Gilbert, N.A.; Clare, J.D.J.; Stenglein, J.L.; Zuckerberg, B. Abundance estimation of unmarked animals based on camera-trap data. *Conserv. Biol.* **2020**. [CrossRef] [PubMed]

45. Smith, J.A.; Suraci, J.P.; Hunter, J.S.; Gaynor, K.M.; Keller, C.B.; Palmer, M.S.; Atkins, J.L.; Castañeda, I.; Cherry, M.J.; Garvey, P.M.; et al. Zooming in on mechanistic predator–prey ecology: Integrating camera traps with experimental methods to reveal the drivers of ecological interactions. *J. Anim. Ecol.* **2020**. [CrossRef] [PubMed]

46. Davies, C.; Wright, W.; Hogan, F.E.; Davies, H. Detectability and activity patterns of sambar deer (Rusa unicolor) in Baw Baw National Park, Victoria. *Aust. Mammal.* **2020**. [CrossRef]

47. Atwood, M.P.; Kie, J.G.; Millspaugh, J.J.; Matocq, M.D.; Bowyer, R.T. Condition of mule deer during winter: Stress and spatial overlap with North American elk. *Mammal Res.* **2020**, *65*, 349–358. [CrossRef]

48. Milleret, C.; Ordiz, A.; Chapron, G.; Andreassen, H.P.; Kindberg, J.; Månsson, J.; Tallian, A.; Wabakken, P.; Wikenros, C.; Zimmermann, B.; et al. Habitat segregation between brown bears and gray wolves in a human-dominated landscape. *Ecol. Evol.* **2018**, *8*, 11450–11466. [CrossRef]

49. Fang, Y.; Du, S.; Abdoola, R.; Djouani, K.; Richards, C. Motion Based Animal Detection in Aerial Videos. *Procedia Comput. Sci.* **2016**, *92*, 13–17. [CrossRef]

50. Prugh, L.R.; Sivy, K.J.; Mahoney, P.J.; Ganz, T.R.; Ditmer, M.A.; de Kerk, M.; Gilbert, S.L.; Montgomery, R.A. Designing studies of predation risk for improved inference in carnivore-ungulate systems. *Biol. Conserv.* **2019**, *232*, 194–207. [CrossRef]

51. Ezat, M.A.; Fritsch, C.J.; Downs, C.T. Use of an unmanned aerial vehicle (drone) to survey Nile crocodile populations: A case study at Lake Nyamithi, Ndumo game reserve, South Africa. *Biol. Conserv.* **2018**, *223*, 76–81. [CrossRef]

52. Lethbridge, M.; Stead, M.; Wells, C. Estimating kangaroo density by aerial survey: A comparison of thermal cameras with human observers. *Wildl. Res.* **2019**, *46*, 639–648. [CrossRef]

53. Aguzzi, J.; Iveša, N.; Gelli, M.C.; Costa, C.; Gavrilovic, A.; Cukrov, N.; Cukrov, M.; Cukrov, N.; Omanovic, D.; Štifanić, M.; et al. Ecological video monitoring of Marine Protected Areas by underwater cabled surveillance cameras. *Mar. Policy* **2020**, *119*, 104052. [CrossRef]

54. Verfuss, U.K.; Aniceto, A.S.; Harris, D.V.; Gillespie, D.; Fielding, S.; Jiménez, G.; Johnston, P.; Sinclair, R.R.; Sivertsen, A.; Solbø, S.A.; et al. A review of unmanned vehicles for the detection and monitoring of marine fauna. *Mar. Pollut. Bull.* **2019**, *140*, 17–29. [CrossRef] [PubMed]

55. Broughton, S.; Harrison, J. Evaluation of monitoring methods for thrips and the effect of trap colour and semiochemicals on sticky trap capture of thrips (Thysanoptera) and beneficialinsects (Syrphidae, Hemerobiidae) in deciduous fruit trees in Western Australia. *Crop Prot.* **2012**, *42*, 156–163. [CrossRef]

56. Schoeny, A.; Gognalons, P. Data on winged insect dynamics in melon crops in southeastern France. *Data Brief* **2020**, *29*, 105132. [CrossRef]

57. Van Swaay, C.; Regan, E.; Ling, M.; Bozhinovska, E.; Fernandez, M.; Marini-Filho, O.J.; Huertas, B.; Phon, C.K.; Korösi, A.; Meerman, J.; et al. *Guidelines for Standardised Global Butterfly Monitoring. GEO BON Technical Series 1*; Group on Earth Observations Biodiversity Observation Network: Leipzig, Germany, 2015.

58. Zhang, C.; Harpke, A.; Kühn, E.; Páramo, F.; Setteleb, J.; Stefanescu, C.; Wiemers, M.; Zhang, Y.; Schweiger, O. Applicability of butterfly transect counts to estimate species richness in different parts of the palaearctic region. *Ecol. Indic.* **2018**, *95*, 735–740. [CrossRef]

59. Hoffmann, A.; Decher, J.; Rovero, F.; Schaer, J.; Voigt, C.; Wibbelt, G. Field methods and techniques for monitoring mammals. In *Manual on Field Recording Techniques and Protocols for All Taxa Biodiversity Inventories and Monitoring*; Degreef, J., Häuser, C., Mohje, J.C., Samyn, Y., Spiegel, V.D., Eds.; Abc Taxa: Brussels, Belgium, 2010; pp. 482–529.

60. Leoni, J.; Tanelli, M.; Strada, S.C.; Berger-Wolf, T. Ethogram-based automatic wild animal monitoring through inertial sensors and GPS data. *Ecol. Inform.* **2020**, *59*, 101112. [CrossRef]

61. Skalon, N.V.; Skalon, T.A. *Zveri Sibiri [Animals of Siberia]*; SPPE «Kuzbass»: Kemerovo, Russia, 2008. (In Russian)

62. Skalon, N.; Stepanov, P.; Prosekov, A. Features of seasonal migrations and wintering of epy elks (*Alces alces*) in the Kuznetsk-Salair mountain region. In Proceedings of the IOP Conference Series: Earth and Environmental Science: 012020International Conference on Sustainable Development of Cross-Border Regions (SDCBR 2019), Barnaul, Russia, 19–20 April 2019; Volume 395, p. 156286. [CrossRef]

63. Buckland, S.T.; Anderson, D.R.; Burnham, K.P.; Laake, J.L.; Borchers, D.L.; Thomas, L. *Introduction to Distance Sampling: Estimating Abundance of Biological Populations*; Oxford University Press: Oxford, UK, 2001.

64. Ripple, W.J.; Estes, J.A.; Beschta, R.L.; Wilmers, C.C.; Ritchie, E.G.; Hebblewhite, M.; Berger, J.; Elmhagen, B.; Letnic, M.; Nelson, M.P.; et al. Status and ecological effects of the world's largest carnivores. *Science* **2014**, *343*, 1241484. [CrossRef]

65. Popescu, V.D.; Artelle, K.A.; Pop, M.I.; Manolache, S.; Rozylowicz, L. Assessing biological realism of wildlife population estimates in data-poor systems. *J. Appl. Ecol.* **2016**, *53*, 1248–1259. [CrossRef]

66. Burton, A.C.; Neilson, E.; Moreira, D.; Ladle, A.; Steenweg, R.; Fisher, J.T.; Bayne, E.; Boutin, S. REVIEW: Wildlife camera trapping: A review and recommendations for linking surveys to ecological processes. *J. Appl. Ecol.* **2015**, *52*, 675–685. [CrossRef]

67. Chandler, R.B.; Andrew Royle, J. Spatially explicit models for inference about density in unmarked or partially marked populations. *Ann. Appl. Stat.* **2013**, *7*, 936–954. [CrossRef]

68. Boyce, M.S.; Corrigan, R. Moose survey app for population monitoring. *Wildl. Soc. Bull.* **2017**, *41*, 125–128. [CrossRef]

69. Pacyna, A.D.; Frankowski, M.; Kozioł, K.; Węgrzyn, M.H.; Wietrzyk-Pełka, P.; Lehmann-Konera, S.; Polkowska, Ż. Evaluation of the use of reindeer droppings for monitoring essential and non-essential elements in the polar terrestrial environment. *Sci. Total Environ.* **2019**, *658*, 1209–1218. [CrossRef] [PubMed]

70. Arsenault, A.; Rodgers, A.R.; Whaley, K. Demographic Status of Moose populations in the boreal plain Ecozone of Canada. *Alces A J. Devoted Biol. Manag. Moose* **2020**, *55*, 43–60, ISSN: 2293-6629.

71. Obermoller, T.R.; Delgiudice, G.D.; Severud, W.J. Maternal Behavior Indicates Survival and Cause-Specific Mortality of Moose Calves. *J. Wildl. Manag.* **2019**, *83*, 790–800. [CrossRef]

72. Rickbeil, G.J.M.; Merkle, J.A.; Anderson, G.; Atwood, M.P.; Beckmann, J.P.; Cole, E.K.; Courtemanch, A.B.; Dewey, S.; Gustine, D.D.; Kauffman, M.J.; et al. Plasticity in elk migration timing is a response to changing environmental conditions. *Glob. Chang. Biol.* **2019**, *25*, 2368–2381. [CrossRef]

73. Kellie, K.; Colson, K.; Reynolds, J. Challenges to Monitoring Moose in Alaska. *Alsk. Dep. Fish Game. Wildl. Manag. Rep.* **2019**. [CrossRef]

74. Alves, J.; Alves da Silva, A.; Soares, A.M.V.M.; Fonseca, C. Pellet group count methods to estimate red deer densities: Precision, potential accuracy and efficiency. *Mamm. Biol.* **2013**, *78*, 134–141. [CrossRef]

75. Forsyth, D.M.; Barker, R.J.; Morriss, G.; Scroggie, M.P. Modeling the Relationship between Fecal Pellet Indices and Deer Density. *J. Wildl. Manag.* **2007**, *71*, 964–970. [CrossRef]

76. Månsson, J.; Andrén, H.; Sand, H. Can pellet counts be used to accurately describe winter habitat selection by moose Alces alces? *Eur. J. Wildl. Res.* **2011**, *57*, 1017–1023. [CrossRef]

77. Jung, T.S.; Kukka, P.M. Influence of habitat type on the decay and disappearance of elk *Cervus canadensis* pellets in boreal forest of northwestern Canada. *Wildl. Biol.* **2016**, *22*, 160–166. [CrossRef]

78. Blåhed, I.M.; Ericsson, G.; Spong, G. Noninvasive population assessment of moose (*Alces alces*) by SNP genotyping of fecal pellets. *Eur. J. Wildl. Res.* **2019**, *65*, 96. [CrossRef]

79. Pfeffer, S.E.; Spitzer, R.; Allen, A.M.; Hofmeester, T.R.; Ericsson, G.; Widemo, F.; Singh, N.J.; Cromsigt, J.P.G.M. Pictures or pellets? Comparing camera trapping and dung counts as methods for estimating population densities of ungulates. *Remote Sens. Ecol. Conserv.* **2018**, *4*, 173–183. [CrossRef]

80. Busse, M.; Schwerdtner, W.; Siebert, R.; Doernberg, A.; Kuntosch, A.; König, B.; Bokelman, W. Analysis of animal monitoring technologies in Germany from an innovation system perspective. *Agric. Syst.* **2015**, *138*, 55–65. [CrossRef]

81. Found, R.; Clair, C.C.S. Behavioural syndromes predict loss of migration in wild elk. *Anim. Behav.* **2016**, *115*, 35–46. [CrossRef]

82. Smith, T.N.; Rota, C.T.; Keller, B.J.; Chitwood, M.C.; Bonnot, T.W.; Hansen, L.P.; Millspaugh, J.J. Resource selection of a recently translocated elk population in Missouri. *J. Wildl. Manag.* **2019**, *83*, 365–378. [CrossRef]

83. Phillips, E.C.; Lehman, C.P.; Klaver, R.W.; Jarding, A.R.; Rupp, S.P.; Jenks, J.A.; Jacques, C.N. Evaluation of an Elk Detection Probability Model in the Black Hills, South Dakota. *West. North Am. Nat.* **2020**, *79*, 551–565. [CrossRef]

84. Bergman, E.J.; Hayes, F.P.; Lukacs, P.M.; Bishop, C.J. Moose calf detection probabilities: Quantification and evaluation of a ground-based survey technique. *Wildl. Biol.* **2020**, *2*. [CrossRef]

85. Bristow, K.D.; Clement, M.J.; Crabb, M.L.; Harding, L.E.; Rubin, E.S. Comparison of aerial survey methods for elk in Arizona. *Wildl. Soc. Bull.* **2019**, *43*, 77–92. [CrossRef]

86. Andreozzi, H.A.; Pekins, P.J.; Kantar, L.E. Using Aerial Survey Observations to Identify Winter Habitat Use of Moose in Northern Maine. *Alces A J. Devoted Biol. Manag. Moose* **2016**, *52*, 41–53, ISSN: 0835-5851.

87. Chrétien, L.P.; Théau, J.; Ménard, P. Visible and thermal infrared remote sensing for the detection of white-tailed deer using an unmanned aerial system. *Wildl. Soc. Bull.* **2016**, *40*, 181–191. [CrossRef]

88. Rey, N.; Volpi, M.; Joost, S.; Tuia, D. Detecting animals in African Savanna with UAVs and the crowds. *Remote Sens. Environ.* **2017**, *200*, 341–351. [CrossRef]

89. Kellenberger, B.; Marcos, D.; Tuia, D. Detecting mammals in UAV images: Best practices to address a substantially imbalanced dataset with deep learning. *Remote Sens. Environ.* **2018**, *216*, 139–153. [CrossRef]

90. Rivas, A.; Chamoso, P.; González-Briones, A.; Corchado, J.M. Detection of cattle using drones and convolutional neural networks. *Sensors* **2018**, *18*, 2048. [CrossRef] [PubMed]

91. Barbedo, J.G.A.; Koenigkan, L.V.; Santos, T.T.; Santos, P.M. A study on the detection of cattle in UAV images using deep learning. *Sensors* **2019**, *19*, 5436. [CrossRef] [PubMed]

92. Patterson, C.; Koski, W.; Pace, P.; McLuckie, B.; Bird, D.M. Evaluation of an unmanned aircraft system for detecting surrogate caribou targets in Labrador. *J. Unmanned Veh. Syst.* **2016**, *4*, 53–69. [CrossRef]

93. Xu, B.; Wang, W.; Falzon, G.; Kwan, P.; Guo, L.; Chen, G.; Tait, A.; Schneider, D. Automated cattle counting using Mask R-CNN in quadcopter vision system. *Comput. Electron. Agric.* **2020**, *171*, 105–300. [CrossRef]

94. Shao, W.; Kawakami, R.; Yoshihashi, R.; You, S.; Kawase, H.; Naemura, T. Cattle detection and counting in UAV images based on convolutional neural networks. *Int. J. Remote Sens.* **2020**, *41*, 31–52. [CrossRef]

95. Witczuk, J.; Pagacz, S.; Zmarz, A.; Cypel, M. Exploring the feasibility of unmanned aerial vehicles and thermal imaging for ungulate surveys in forests-preliminary results. *Int. J. Remote Sens.* **2018**, *39*, 5504–5521. [CrossRef]

96. Dziki-Michalska, K.; Tajchman, K.; Budzyńska, M. Increase in the moose (*Alces alces* L. 1758) population size in Poland: Causes and consequences. *Ann. Wars. Univ. Life Sci. -Sggw. Anim. Sci.* **2019**, *58*, 203–214. [CrossRef]

97. Lavsund, S.; Nygrén, T.; Solberg, E.J. Status of Moose Populations and Challenges to Moose Management in Fennoscandia. *Alces A J. Devoted Biol. Manag. Moose* **2003**, *39*, 109–130.

98. Myslenkov, A.I.; Miquelle, D.G. Comparison of Methods for Counting Hoofed Animal Density in Sikhote-Alin. *Achiev. Life Sci.* **2015**, *9*, 1–8. [CrossRef]

99. Paragi, T.F.; Kellie, K.A.; Peirce, J.M.; Warren, M.J. Movements and sightability of moose in Game Management Unit 21E. In *Final Wildlife Research Report 2017 ADF&G/DWC/WRR-2017-2*; Alaska Department of Fish and Game: Juneau, AK, USA, 2017.

100. Crusiol, L.G.T.; Nanni, M.R.; Furlanetto, R.H.; Cezar, E.; Silva, G.F.C. Reflectance calibration of UAV-based visible and near-infrared digital images acquired under variant altitude and illumination conditions. *Remote Sens. Appl. Soc. Environ.* **2020**, *18*, 100312. [CrossRef]

101. Alladi, T.; Chamola, V.; Sahu, N.; Guizani, M. Applications of blockchain in unmanned aerial vehicles: A review. *Veh. Commun.* **2020**, *23*, 100249. [CrossRef]

102. Gentle, M.; Finch, N.; Speed, J.; Pople, A. A comparison of unmanned aerial vehicles (drones) and manned helicopters for monitoring macropod populations. *Wildl. Res.* **2018**, *45*, 586–594. [CrossRef]

103. Harris, C.M.; Herat, H.; Hertel, F. Environmental guidelines for operation of Remotely Piloted Aircraft Systems (RPAS): Experience from Antarctica. *Biol. Conserv.* **2019**, *236*, 521–531. [CrossRef]

Article

Genetic Diversity and Range Dynamics of *Helleborus odorus* subsp. *cyclophyllus* under Different Climate Change Scenarios

Georgia Fassou [1,†], Konstantinos Kougioumoutzis [1,2,†], Gregoris Iatrou [1], Panayiotis Trigas [3] and Vasileios Papasotiropoulos [4,*]

[1] Division of Plant Biology, Laboratory of Botany, Department of Biology, University of Patras, 26504 Patras, Greece; fassoug@upatras.gr (G.F.); kkougiou@biol.uoa.gr (K.K.); iatrou@upatras.gr (G.I.)

[2] Department of Ecology and Systematics, Faculty of Biology, National and Kapodistrian University of Athens, Panepistimiopolis, 15701 Athens, Greece

[3] Laboratory of Systematic Botany, Department of Crop Science, Agricultural University of Athens, Iera Odos 75, 11855 Athens, Greece; trigas@aua.gr

[4] Laboratory of Agricultural Genetics and Breeding, Department of Agriculture, University of Patras, Theodoropoulou Str, 27200 Amaliada, Greece

* Correspondence: vpapasot@upatras.gr

† These authors contributed equally to this work.

Received: 9 May 2020; Accepted: 27 May 2020; Published: 1 June 2020

Abstract: *Research Highlights*: The effects of climate change on habitat loss, range shift and/or genetic impoverishment of mid-elevation plants has received less attention compared to alpine species. Moreover, genetic diversity patterns of mountain forest herbaceous species have scarcely been explored in the Balkans. In this context, our study is the first that aims to examine *Helleborus odorus* subsp. *cyclophyllus*, a medicinal plant endemic to the southern part of the Balkan Peninsula. *Background and Objectives*: We compare its genetic diversity and structure along the continuous mountain range of western Greece with the topographically less structured mountains of eastern Greece, and predict its present and future habitat suitability, using several environmental variables. *Materials and Methods:* Inter Simple Sequence Repeat (ISSR) markers were used to genotype 80 individuals from 8 populations, covering almost the species' entire distribution range in Greece. We investigated the factors shaping its genetic composition and driving its current and future distribution. *Results*: High gene diversity (0.2239–0.3319), moderate population differentiation (0.0317–0.3316) and increased gene flow (Nm = 1.3098) was detected. According to any GCM/RCP/climate database combination, *Helleborus odorus* subsp. *cyclophyllus* is projected to lose a significant portion of its current distribution by 2070 and follow a trend towards genetic homogenization. *Conclusions*: Populations exhibit in terms of genetic structure a west–east genetic split, which becomes more evident southwards. This is mainly due to geographic/topographic factors and their interplay with Quaternary climatic oscillations, and to environmental constraints, which may have a negative impact on the species' future distribution and genetic composition. Pindos mountain range seems to buffer climate change effects and will probably continue to host several populations. On the other hand, peripheral populations have lower genetic diversity compared to central populations, but still hold significant evolutionary potential due to the private alleles they maintain.

Keywords: climate alterations; mid-elevation plants; range distribution; genetic variability; molecular markers

---

1. Introduction

Mountains are topographically complex regions undergoing substantial changes over geologically short time scales [1]. Topographic complexity and elevation per se create a large variety of habitats and promote speciation processes, resulting in the exceptionally high biodiversity of mountain systems. Moreover, mountains often harbour aggregations of small-ranged species forming highly diverse endemism centres [1–3]. Mediterranean mountains, a globally important biodiversity hotspot for plant species richness and endemism, are expected to experience a severe species loss and turnover in the near future due to climate warming [4]. Understanding the fluctuating impacts of mountains on the distribution, diversification and ecological adaptation of individual species is crucial for managing mountain biodiversity in the face of the impending climate change.

The Balkan Peninsula is a highly important region for European and Mediterranean plant diversity and endemism [5]. Topographic heterogeneity, climatic diversity and long-term environmental stability have been proposed as the underlying factors influencing the exceptionally high plant diversity of the Balkans [6,7]. Divergence in the multiple Pleistocene refugia of the Balkan Mountains has been highlighted as a key factor generating intraspecific plant diversity [8–10]. Despite the importance of Balkan plant diversity for plant conservation in Europe and the Mediterranean, the patterns of intraspecific variability across the Balkan Mountains remain poorly understood. This knowledge gap becomes even more crucial when considering the large number of socio-economically important plant species distributed in the Balkans, like crop wild relatives, ornamental and medicinal plants.

The effects of the current climate warming are anticipated to have the strongest deleterious effects for the cold adapted plants distributed on mountain summits [1,11]. In this respect, the effects of climate change on habitat loss, range shift and/or genetic impoverishment of mid-elevation plants has received much less attention compared to the alpine species. Moreover, the influence of topographical mountain setting in migration corridors of mountain plants, and its fingerprint to species genetic variability has been largely overlooked. Possibly, the north-south directed Pindos mountain range that dominates western Greece has acted as a corridor for mountain species migration during Quaternary climatic oscillations, leading to increased population genetic homogenization. Mountain plants distributed in the more isolated mountain massifs of the eastern Greek mainland facing the Aegean archipelago, especially in the south, most likely responded with elevational range shifts in small geographical scales, resulting in stronger genetic differentiation of eastern-south-eastern populations compared to western ones. Current climate warming could also strongly affect habitat suitability for mountain species distributed in small isolated mountains with limited or no chance for upslope range shifts. In many cases, these changes will probably reduce genetic diversity in populations and species, in extreme situations to the point where genetic impoverishment will contribute to the diminishment of population viability [12].

Helleborus L. (Ranunculaceae) is a small genus comprising 22 species, distributed in Europe and Asia [13], with the Balkan Peninsula being the genus' diversity centre. *Helleborus odorus* Willd. is an outcrossing, insect-pollinated, Balkan species also reaching northern Italy and Hungary, with three subspecies; among them, subsp. *cyclophyllus* (A. Braun) Maire & Petitm. is endemic to the southern Balkan Peninsula [14] and mainly distributed in Greece. The Northern Peloponnese constitutes the southern edge of this subspecies' range, which is locally common in Greece in mid-elevation woodlands, scrubs, and meadows.

Helleborus spp. have been recognized as important herbs for their therapeutic use since antiquity, e.g., as a cure for mania and depression or as a painkiller and emetic [15]. Recently, some hellebore compounds, like hellebrin and hellebrigenin seem promising remedies for severe diseases such as cancer, ulcer, and diabetes [16,17].

The global increase on the demand of medicinal plant species has led to over-exploitation of natural populations. Understanding the patterns and distribution of genetic variation within and among populations is essential to ensure efficient conservation management and sustainable use of medicinal plants' genetic resources. Moreover, maintenance of sufficient genetic diversity is an important consideration in species conservation, because genetic diversity is required for populations to evolve in response to environmental changes [18]. When the amount of diversity is reduced, inbreeding depression and higher homozygosity appear, undermining the adaptive potential of populations [19]. Genetic characterization is essential to understand the relationships and assess diversity within and among species. Molecular marker technologies, which rely on DNA analysis provide powerful tools to examine biodiversity at different levels. One of the biggest advantages is that they are unaffected by the environment and the developmental stages of the plants and free of pleiotropic or epistatic effects. Among them, Inter Simple Sequence Repeats (ISSRs) is a dominant marker system, not influenced by the presence of organellar DNA, often used in plant conservation genetics and diversity studies, due to high reproducibility, low cost and universality they exhibit [20,21].

Herein, we examine the genetic structure and variability of *Helleborus odorus* subsp. *cyclophyllus* populations across the Greek mountains, using ISSR markers, and assess climate change impacts on its current and future distribution and genetic composition. By doing so, we aim to contribute to the better understanding of the processes shaping mountain biodiversity patterns and to provide new insights into the range dynamics of mid-elevation mountain species influenced by global warming.

2. Materials and Methods

2.1. Plant Material

Individuals from eight populations of *Helleborus odorus* subsp. *cyclophyllus*, covering almost the entire distribution of the taxon in Greece, were used in the present study (Table 1). The populations consisted of many individuals, occupying large areas, which makes population size estimation difficult. All samples were collected from late February until early March 2016. Fresh and healthy leaves from 10 individuals of each population were collected and dried in silica gel.

Table 1. Abbreviations used for the analyses, locality, and collection information of the *Helleborus odorus* subsp. *cyclophyllus* populations under study.

Population	Population ID	Latitude	Longitude	Altitude	Part of Greece	Habitat
Erymanthos	ERY	37.9504	21.7669	950 m	West	Forest openings with *Abies cephalonica*
Velouchi	VEL	38.9285	21.8683	1400 m	West	Shrubs
Parnassos	PAR	38.5397	22.7038	700 m	East	Forest openings with *Abies cephalonica*
Vourgareli	VOU	39.3684	21.1919	750 m	West	Forest edges with *Fagus sylvatica* & *Abies borisii-regis*
Fragma Mesachoras	FRA	39.4683	21.3199	850 m	West	Forest edges with *Fagus sylvatica* & *Abies borisii-regis*
Naousa	NAOU	40.6548	22.0024	1050 m	East	Forest edges with *Fagus*
Olympos	OLY	40.1178	22.4881	400 m	East	Forest openings with *Quercus ilex* & *Q. rotundifolia*
Dirfi	DIR	38.6112	23.8602	1100 m	East	Forest openings with *Abies cephalonica*

2.2. Genetic Analysis

2.2.1. DNA Extraction

Total genomic DNA was isolated from individual dried leaves (100 mg) using the DNeasy Plant Mini Kit (Qiagen). Quality assessment and concentration of the purified DNAs were measured using a NanoDrop ND-1000 spectrophotometer (Thermo Fisher Scientific, Waltham, MA, USA), while DNA integrity was evaluated by agarose gel (0.8%) electrophoresis (Sigma).

2.2.2. ISSR Genotyping

Initially, we performed preliminary screening of several ISSR markers from the University of British Columbia UBC primer set # 9 (Vancouver, Canada), to select those exhibiting higher reproducibility and polymorphism. Based on this, ten ISSR primers were chosen, yielding the maximum numbers of reliable and reproducible bands for population genotyping (Table 2).

Table 2. Primers used for ISSR analysis; primers' sequences; annealing temperatures; number and range of bands scored for each primer used.

Primer ID	Primer Sequence (5′–3′)	Tm (°C)	No of Bands Scored	Range (bp)
UBC-807	$(AG)_8T$	52	19	420–2000
UBC-811	$(GA)_8T$	52	22	430–1980
UBC-819	$(GT)_8A$	52	9	650–1190
UBC-821	$(GT)_8T$	48.6	7	950–1780
UBC-822	$(CT)_8A$	45.7	13	680–2000
UBC-826	$(AC)_8C$	52	14	670–1800
UBC-840	$(GA)_8YT$	45.7	17	470–1500
UBC-845	$(CT)_8RG$	45.7	10	1000–2400
UBC-869	$(GTT)_6$	51	9	1100–2470
UBC-880	$(GGAG)4$	51	14	470–1560

The PCR amplification reactions were performed within a 20 μL volume, containing 20 ng genomic DNA template, 1X PCR buffer, 1.6 μL of $MgCl_2$ (25 mM), 1.2 μL dNTPs (100 mM each), 10 pmol from each primer, 1 U thermostable Taq DNA Polymerase (Kapa Biosystems). Reactions were carried out in MyCycler thermal cycler (Bio-Rad), under the following conditions: an initial denaturation step at 95 °C for 4 min followed by 35 cycles of denaturation at 94 °C for 45 s, annealing at 45.7–52 °C (depending on the primer set) for 45 s, extension at 72 °C for 2 min. The program was ended with a final extension step at 72 °C for 5 min. The PCR amplified fragments were separated by agarose (2%) gel electrophoresis and visualized with a GelDocEZ imaging system (Bio-Rad).

Fragments between 420 and 2470 bp, consistently amplified, were scored manually to form a binary matrix, where the presence of a band was scored as 1 and the absence as 0, assuming that each one of them corresponds to one locus with two alleles.

2.2.3. Statistical Analysis

For the estimation of the genetic diversity parameters, POPGENE version 1.31 [22], AFLPsurv version 1.0 [23] and GENALEX 6.5 [24,25] were employed. These parameters included the percentage of polymorphic bands (PPB, an index equivalent to polymorphic loci), calculated by dividing the number of polymorphic bands at the population and species levels by the total number of bands scored, Nei's gene diversity (Hj) [26] and Shannon's diversity index (I). Because there was no prior information regarding the level of inbreeding, we ran AFLPsurv 1.0 with different Fis values (0 assuming Hardy-Weinberg equilibrium to 1 complete lack of heterozygotes).

The estimation of gene flow (Nm), calculated as the number of migrants entering a population in each generation, was made by using POPGENE version 1.31 [22].

Pairwise Fst between the *Helleborus odorus* subsp. *cyclophyllus* populations was calculated using AFLPsurv 1.0. To depict relatedness at individual level we constructed a UPGMA polar dendrogram based on Jaccard's coefficient of similarity using FREETREE software [27] visualized with FigTree version 1.4.3 [28]. Two samples of *Ranunculus ficaria* collected from North Peloponnesus were used as outgroup.

Analysis of Molecular Variance (AMOVA) and Principal Coordinates Analysis (PCoA) were performed with Genalex 6.5. For PCoA, the method for dominant data by Huff et al. [29] was followed. The test of significance for the AMOVA was carried out on 9999 permutations of the data.

We generated a Moran's spatial correlogram (9999 permutations) with 'fossil' 0.3.7 [30] and 'vegan' 2.5.2 [31] packages, to check if there was any correlation between populations' genetic and geographic distances.

To infer population structure and assign individuals to populations, we used STRUCTURE version 2.3.1 [32]. We used the admixture model, with correlated allele frequencies and the options RECESSIVE ALLELES and LOCPRIOR activated. No prior knowledge of the populations was included in the analysed dataset. To determine the optimal number of groups (K), we ran STRUCTURE with K varying from 1 to 8 (equal to the number of populations), with ten runs for each K value using a burn-in period of 100,000 interactions followed by 50,000 additional Markov Chain Monte Carlo (MCMC) interactions. The ΔK method [33] was used to identify the most likely number of clusters (K) using STRUCTURE HARVESTER 0.6.94 [34]. Each accession was assigned to its corresponding group based on maximum membership probability, as indicated by Remington et al. [35].

2.3. Species Occurrence Data

Helleborus odorus subsp. *cyclophyllus* occurrence data (167 occurrences) were obtained from the Global Biodiversity Information Facility database through functions from the 'rgbif' 1.4.0 [36] package. Data cleaning and organizing procedure follows Robertson et al. (see Supplementary) [37].

2.4. Environmental Data

Current and future climatic data were obtained from the WorldClim database [38] and the CHELSA database [39,40] at a 30 s resolution. We used two climate databases to assess the bioclimatic consistency and congruence of model predictions [41], a crucial source of uncertainty in SDMs [42].

An initial set of 50 predictors was constructed (Supplementary), including:

Sixteen climatic variables based on the 19 bioclimatic variables from WorldClim for current and future climate conditions.

a. Three Global Circulation Models (GCMs) that are rendered more suitable and realistic for the study area's future climate.
b. Four different IPCC scenarios from the Representative Concentration Pathways (RCP) family.
c. Seven soil variables providing predicted values for the surface soil layer at varying depths.
d. Elevation data from the CGIAR-CSI data-portal [43] were aggregated and resampled to match the resolution of the other environmental variables. Additional topographical variables (slope, aspect, heat load index, topographic position index and terrain ruggedness index) were computed based on elevation data using functions from the 'raster' 2.6.7 package [44] and 'spatialEco' 1.2–0 package [45].
e. Geological data from the Geological Map of Greece [46].

From this set of predictors, only fifteen and seventeen variables (depending on the extent of the distributional area) were not highly correlated (Spearman rank correlation < 0.7 and VIF < 5 [47]) (see Supplementary).

2.5. Species Distribution Models

2.5.1. Model Parameterization and Evaluation

The realized climatic niche of *Helleborus odorus* subsp. *cyclophyllus* was modelled by combining the available occurrence data to current environmental predictors in an ensemble modelling scheme (see Supplementary) [48].

2.5.2. Model Projections

Calibrated models with a TSS score > 0.8 (to avoid poorly calibrated ones) were used to project the suitable area for *Helleborus odorus* subsp. *cyclophyllus* taxon under current and future conditions through an ensemble forecast approach (see Supplementary) [48].

2.5.3. Area Range Change

To assess whether *Helleborus odorus* subsp. *cyclophyllus* will experience range contraction or expansion under future conditions, we used the 'biomod' 3.3.7 package [49]. The taxon was not assumed to have unlimited dispersal capability, since this assumption could be overoptimistic.

2.6. Bioclimatic Congruence and Consistency

We followed the framework of Morales-Barbero and Álvarez [41] in order to construct the bioclimatic congruence and consistency maps for *Helleborus odorus* subsp. *cyclophyllus* for every time period that was available in both climate databases.

2.7. Generalized Dissimilarity Modelling

We used Generalized Dissimilarity Modelling (GDM–) [50] in the framework laid out by Fitzpatrick and Keller [51] to investigate the spatial and environmental drivers of genetic beta diversity, as well as to explore the potential variation in future genetic diversity patterns due to climate change. We used the same environmental variables as in the SDM analyses, with the significance of all variables assessed through a Monte Carlo permutation test (see Supplementary). We followed Fitzpatrick and Keller [51] to visualise the spatial patterns of genetic variation (see Supplementary). Finally, we projected our GDM based on current environmental conditions to every GCM/RCP combination we included in our study, so as to predict the areas where the relationship between genetic composition and future environmental conditions will experience the greatest change ([51]—'genetic offset'; see Supplementary).

All GDM analyses were performed with the 'gdm' 1.3.7 [52] R package in the R 3.5.3 [53].

3. Results

3.1. Molecular Analysis

After preliminary screening, ten ISSR primers were eventually selected for genotyping. In total 134 bands were scored, with their size ranging from 420 bp to 2470 bp. Among them, primer UBC–811 had the highest number of bands (22), while UBC–821 the lowest (7) (Table 2).

The highest number of bands (115) was observed in Velouchi population, while the lowest (84) in Olympos. Moreover, the highest number of private bands (26) were revealed in Olympos, while the lowest (9) in Parnassos. The percentage of polymorphic bands (PPB) per population ranged from 43.28% to 76.87% (Olympos and Velouchi, respectively).

Pairwise Fst values ranged from 0.0317 between Vourgareli and Fragma Mesachoras to 0.3316 between Olympos and Dirfi, while the average Fst value among the eight populations was 0.1505 (Table S1).

There were no significant differences detected for Nei's gene diversity (Hj) values regardless populations were in Hardy–Weinberg equilibrium or not. Assuming Hardy–Weinberg equilibrium, Hj ranged from 0.2239 ± 0.016 (Olympos) to 0.3319 ± 0.013 (Velouchi), which was the richest in terms of genetic diversity. The average gene diversity within the populations (Hw) was 0.2898 ± 0.013, while the total gene diversity for the taxon (Ht) was 0.3410. Assuming no Hardy–Weinberg equilibrium, the population of Velouchi had the highest Hj value (0.3490 ± 0.0158), while the population from Olympos the lowest (0.2010 ± 0.0183). The average gene diversity within the populations of *H. odorus* subsp. *cyclophyllus* (Hw) was 0.2901 ± 0.019, while the total gene diversity for the taxon (Ht) was 0.3763.

Shannon's information index (*I*) ranged from 0.2322 ± 0.025 (Olympos) to 0.3951 ± 0.022 (Velouchi). Finally, gene flow (Nm) among populations was 1.3098 (Table 3).

Table 3. Genetic diversity as detected by ISSR markers; N = numbers of individuals surveyed from each population; No. B = number of scored bands PPB = Percentage of Polymorphic Bands; PrB = private bands; *I* = Shannon's Information index; *Hj* = Nei's gene diversity after Lynch and Milligan (1994) assuming *Hw* equilibrium (*Hj*$_{-HWE}$) or complete shortage of heterozygotes (*Hj*$_{-non\ HWE}$); Nm = estimate of gene flow among populations.

Population ID	N	No.B	PPB (%)	PrB	*I*	*Hj*$_{-HWE}$	*Hj*$_{-non\ HWE}$	Nm
ERY	10	87	55.22	13	0.3167 ± 0.026	0.2719 ± 0.017	0.2317 ± 0.018	
VEL	10	115	76.87	12	0.3951 ± 0.022	0.3319 ± 0.013	0.3490 ± 0.015	
PAR	10	98	66.42	9	0.3368 ± 0.024	0.2781 ± 0.015	0.2910 ± 0.017	
VOU	10	112	75.37	11	0.3843 ± 0.023	0.3208 ± 0.014	0.3395 ± 0.016	
FRA	10	111	75.37	10	0.3788 ± 0.023	0.3131 ± 0.014	0.3274 ± 0.016	
NAOU	10	108	71.64	12	0.3745 ± 0.024	0.3171 ± 0.015	0.3214 ± 0.017	
OLY	10	84	43.28	26	0.2322 ± 0.025	0.2239 ± 0.016	0.2010 ± 0.018	
DIR	10	94	58.96	15	0.2983 ± 0.024	0.2613 ± 0.016	0.2592 ± 0.018	
Mean					0.3396 ± 0.009	0.2898 ± 0.013	0.2901 ± 0.019	1.3098

The analysis of molecular variance (AMOVA) showed that most of the genetic diversity was attributable to differences among individuals within populations (74%), while significantly less was observed among populations (26%).

Non-significant correlation between the geographic and the genetic distance values was observed (r = 0.24; *p* > 0.05) for the populations under study.

The PCoA analysis among individuals (Figure 1), showed that 60.22% of total variation is explained by the first three axis (26.29%, 18.79% and 15.14%, respectively). The populations from Olympos and Dirfi, which are separated by the first coordinate are the most homogeneous and divergent ones with less intrapopulation variability. The populations from Parnassos and Naousa are separated by the second coordinate, although in some limited cases individuals from these population seem to interact. The remaining populations from Vourgareli, Velouchi, Erymanthos and Fragma Mesachoras seem to interact with each other forming a mix.

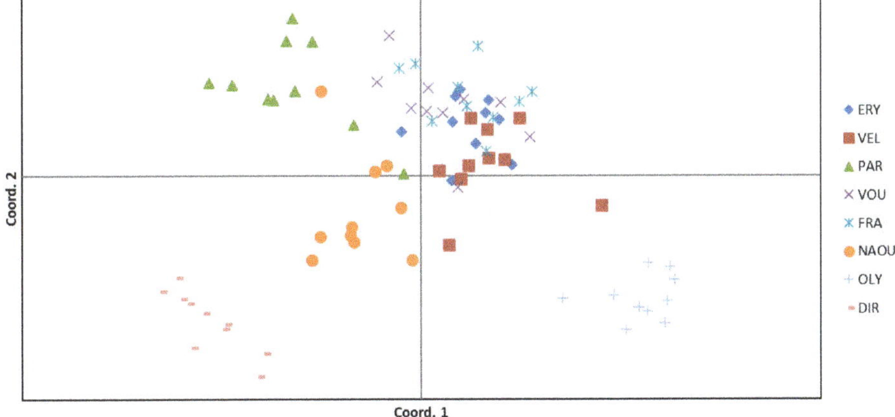

Figure 1. A two-dimensional plot of the principal coordinates analysis (PCoA) of ISSR data showing the clustering of the populations.

The ΔK statistic in the STRUCTURE analysis presented a maximum peak at K = 3 (Figure S1), suggesting that three different genetic clusters exist. The individuals from Olympos and Dirfi were categorised into their own unique clusters, while most of the individuals from Parnassos formed a third one. The individuals from the remaining populations were assigned to Parnassos' and Olympos' genetic clusters, except those from Naousa which were assigned to all three existing clusters (Figure 2).

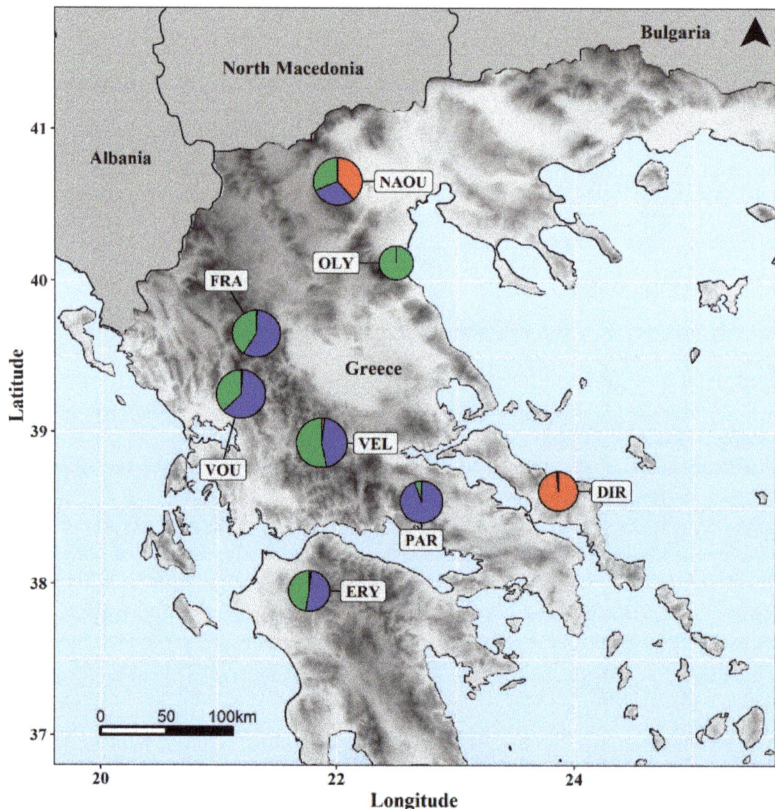

Figure 2. Genetic diversity and structure of eight populations of *Helleborus odorus* subsp. *cyclophyllus* from Greece. Genetic structure derived from Bayesian analysis using STRUCTURE at K = 3. The size of the dots is directly proportional to the depicted values of Nei's gene diversity (*Hj*) for each population.

The polar UPGMA dendrogram obtained using Jaccard's coefficient of similarity shows that in most cases all the individuals were clustered in their respective populations. Individuals from Vourgareli and Fragma Mesachoras populations are grouped together forming a big cluster, while one individual from Velouchi was positioned separately next to those from Olympos population. The most distinct clade is the one formed by the individuals from Dirfi, followed by that formed by those from Parnassos, while the populations from Olympos, Erymanthos and Velouchi belong to a wider clade (Figure 3).

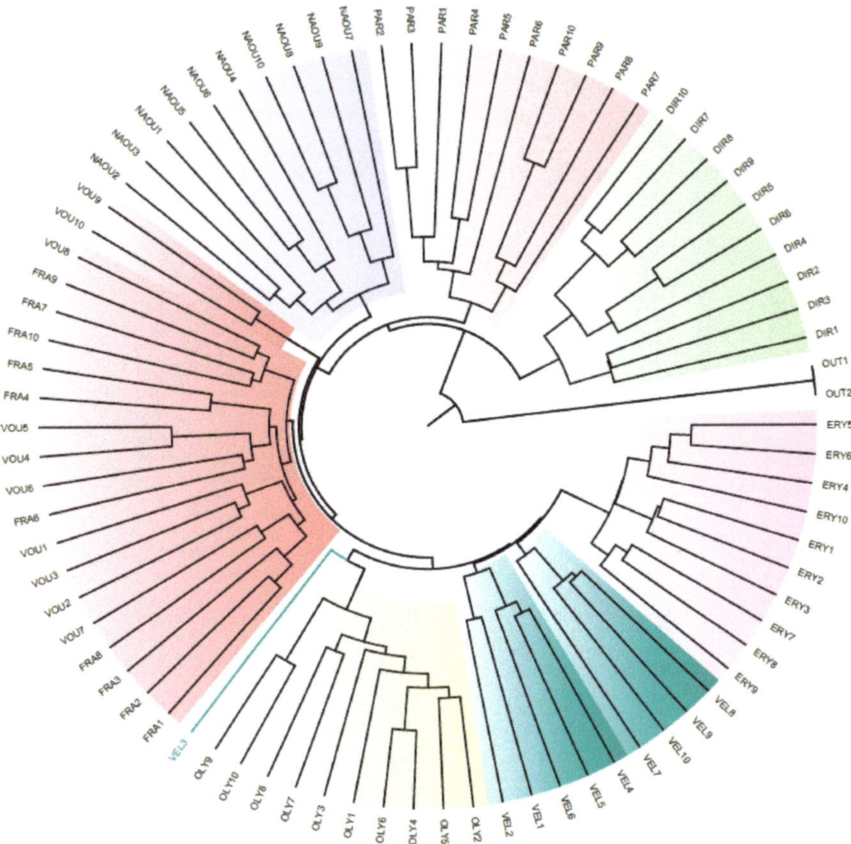

Figure 3. UPGMA dendrogram based on Jaccard's coefficient illustrating the genetic similarities of 80 *Helleborus odorus* subsp. *cyclophyllus* individuals from different populations.

The topology obtained is in concordance with the grouping deduced by the 1st coordinate of PCoA and follows the pattern of diversity clustering revealed by the STRUCTURE analysis.

3.2. Species Distribution Modelling

The ensemble of small models (ESM) framework predictions were very good, with sufficient predictive power (Figure S2). Overall, the model based on the geographical thinning procedure, the alpha-hull method and the CHELSA database had the best predictive accuracy (Figure S2). Only intra-climate database variation was statistically insignificant (KWA: H = 2.18, d.f. = 1, $p = 13.97$; Figure S2); all other uncertainty sources varied significantly (KWA: $p < 0.01$ for all cases; Figure S2). Mean temperature and potential evapotranspiration of the wettest quarter (MAT and PET_{WQ}, respectively) had the highest contribution among the response variables for both thinning procedures for CHELSA and WorldClim, respectively. The resulting habitat suitability maps (Figures S3 and S4) had high bioclimatic consistency for every combination of the thinning procedures and distribution areas (Figures S5 and S6).

They were converted into binary maps and then compared to the binary maps obtained for each Global Circulation Model (GCM), Representative Concentration Pathway (RCP) scenario, time period, thinning procedure, and climate database. Since the trends for the future potential distribution of

Helleborus odorus subsp. *cyclophyllus* were largely identical across all sources of uncertainty, we present only the area range change for the WorldClim database and the CCSM4 GCM and the RCP 2.6 scenario.

Our results indicate that by 2070 *H. odorus* subsp. *cyclophyllus* is projected to lose a significant portion of its current distribution under any GCM/RCP/climate database combination, with varying magnitude across GCMs, RCPs and climate databases (3.9%–100.0%; Figure 4 and Figures S7–S17), but the median range contraction is 29.2%. The median range contraction differs between the two climate databases (KWA: H = 36.07, d.f. = 1, $p < 0.001$) and it was significantly higher in the CHELSA database.

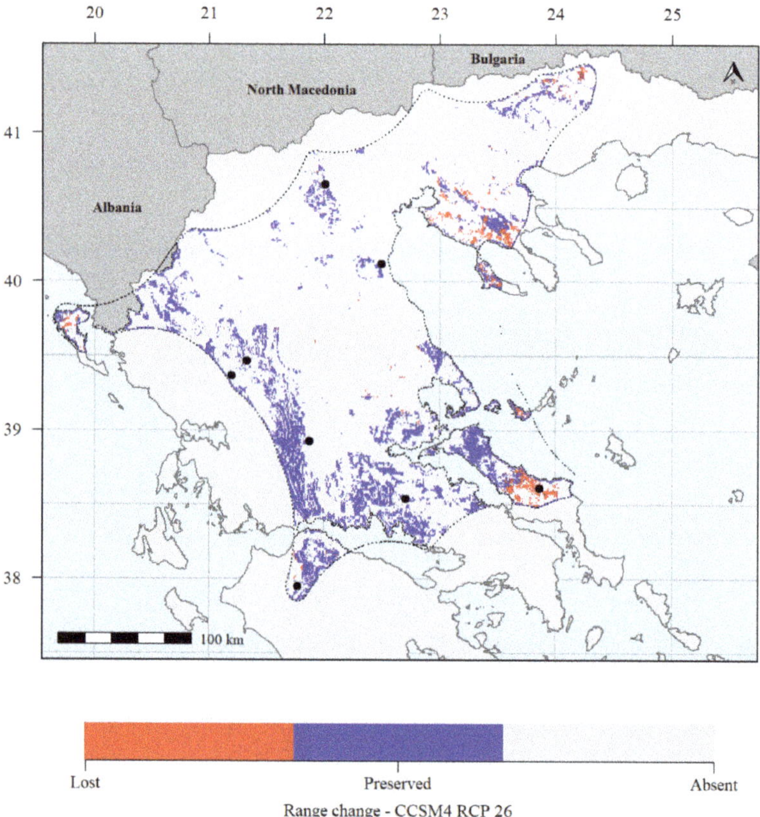

Figure 4. Predicted potential distribution map for 2070 and the CCSM4 GCM and the RCP 2.6 scenario. Red grid cells: the species currently occupies those areas but will not in the future. Blue grid cells: the species currently occupies those areas and will continue to occupy them in the future. Grey grid cells: the species does not currently occupy those areas and it will not occupy them in the future. The dotted line indicates the distribution area of *Helleborus odorus* subsp. *cyclophyllus* based on the alpha-hull method. Climate data refer to the WorldClim database.

3.3. Generalized Dissimilarity Modelling

GDM helped disentangle the relative contribution of geographic and environmental drivers of genetic composition across time and space. Our model explained 81.5% of deviance in genetic composition. The most important gradient for determining genetic diversity turnover was mean diurnal range (MDR), followed by geographical distance (Figure S18). The fitted functions describing the turnover rate and magnitude along each gradient were nonlinear, with turnover rate varying with position along gradients (Figure S18). The MDR I-spline indicates rapid and abrupt changes in genetic variation for values above a certain threshold (Figure S18).

Rapid turnover is predicted in the area between the mountain ranges crossing Greece in an NW–SE axis, as well as in Evvia and NE Greece (Figure S19). Our analyses highlighted five different geographical clusters (Figures S19–S20) and this situation is predicted to change drastically under any GCM/RCP (Figures S21–S44), showing a trend towards genetic homogenization. Based on the V-measure index, the highest and the lowest similarity is observed for the BCC RCP 2.6 and the HadGEM2 RCP 8.5, respectively (Figure S45; median V-measure index: 0.61).

Under any GCM/RCP, the area in northern Greece and the Pindos mountain range is predicted to exhibit greater genetic turnover than any other region (Figure 5). These areas harbour populations for which the genetic offset is predicted to be greatest under climate change (Figure 5). Low genetic offset is predicted in valleys and low-elevation areas in central Greece, as well as in the southern edge of the species' range (Figure 5).

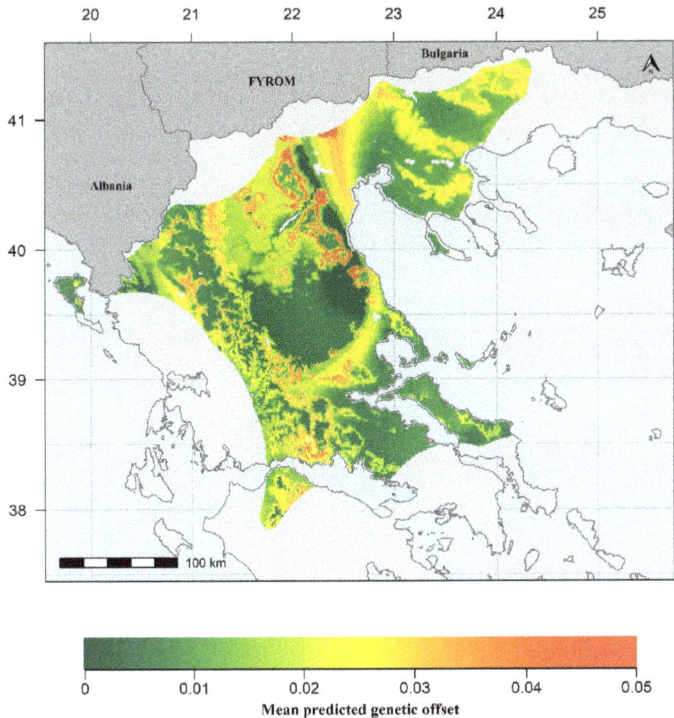

Figure 5. Mean predicted genetic offset from the Generalized Dissimilarity Modelling analyses based on the four Representative Concentration Pathways and the three Global Circulation Models included in our study from the WorldClim climate database.

4. Discussion

4.1. Genetic Diversity and Structure

The relationship between species richness and elevation is predominantly hump-shaped across the mountains of the globe [54], rendering mid-elevation forest plant species as the backbone of mountain plant diversity. Despite their importance in building biodiversity across different spatial scales, mid-elevation plants have received minor attention in relation to the factors shaping intraspecific variability and range dynamics, compared to high elevation species, especially in the face of climate change.

Intraspecific genetic variation is the most fundamental level of biodiversity, and provides the basis for any evolutionary change, being crucial for maintaining the ability of species to adapt to new environmental conditions [12]. Climatic alterations impact intraspecific genetic diversity by two major processes, either by inducing plants to contract or expand their ranges and/or by promoting changes through adaptation to the local selection regime [55,56].

In this study, genetic diversity, and differentiation among *Helleborus odorus* subsp. *cyclophyllus*, populations from various mountainous locations in Greece, were analysed using ISSR markers. A total of 134 genetic loci were genotyped, which according to Nelson and Anderson [57] are adequate for dominant markers to yield acceptable results. Within population diversity in terms of percentage of polymorphic bands, Shannon's information index (I) and Nei's gene diversity (Hj) was higher in the north and northwest populations, except Olympos, becoming lower in the more isolated populations of east and southern Greece. The average gene diversity within populations was high ($Hw = 0.29$), similar to the values reported by Nybom [58] for plants with narrow distribution range ($H_{pop} = 0.28$), and much higher than those reported for endemic plants ($H_{pop} = 0.20$) using dominant markers.

The number of private bands was higher in the more isolated population of east and southern Greece, being remarkably high (26) in the population of Olympos. The Analysis of Molecular Variance (AMOVA) showed that the most of genetic variation was attributable to differences among individuals within populations (74%), and less among populations (26%). Genetic differentiation was moderate (Fst = 0.1505) and in line with AMOVA. Hogbin and Peakall [59] summarized the results of RAPD analysis (dominant markers like ISSRs) in several species, and identified a similar pattern according to which genetic variation of outcrossed species as *Helleborus odorus* subsp. *cyclophyllus*, is mainly distributed within populations, while diversity among populations accounted for less than 27%, an observation that is also in agreement with Nybom [58]. The high value of gene flow (Nm = 1.3098) largely explains the moderate inter-population genetic differentiation observed. Interestingly, gene flow was much higher among western (FRA, VOU, VEL, ERY) populations ($Nm_W = 2.408$) and relatively limited among eastern (NAOU, OLY, PAR, DIR) ones ($Nm_E = 1.062$). Our analyses based on the spatial distribution of the species' genetic diversity identified five different geographical clusters (Figure S19). This bioregionalization might explain the restricted gene flow observed among the eastern populations of *H. odorus* subsp. *cyclophyllus* compared to the western ones.

The reduced genetic diversity observed for Olympos and Dirfi, could be the result of genetic drift. Climatic oscillations and resulting migrations along the altitudinal gradient, may have caused fluctuations in the size of populations promoting the action of genetic drift. Additionally, range expansions due to a series of repeated founder events and allele surfing may lead to a high increase in the frequency of specific alleles [60–62]. We cannot, however, rule out the possibility that under climate change scenarios adaptive micro-evolutionary processes (e.g., selection of local genotypes better adapted to local environmental conditions) tend to reduce variation on specific loci upon which natural selection acts, including also regions of the genome that are hitchhiking [56].

The present distribution of *Helleborus odorus* subsp. *cyclophyllus* is confined to fragmented mountainous areas, where most of the studied populations are separated by deep valleys and plains, and by the sea in the cases of Erymanthos and Dirfi, unfavourable for species growth. The moderate among populations genetic differentiation indicates a more continuous distribution in the recent past, at least for a part of its distribution range. During Quaternary glacial episodes numerous European mountain plant species managed to survive by shifting their ranges downwards in elevation and/or southwards in latitude, followed by the opposite elevational and latitudinal range shifts during interglacial periods [9,62–64]. It is reasonable to assume that during cold stages *H. odorus* subsp. *cyclophyllus* was distributed in lower elevation than today, allowing in some cases contacts of genetically divergent populations and inter-population hybridization.

Phylogeographic studies have provided evidence that postglacial secondary contact has also led to climate-driven hybridization events in several species increasing local genetic diversity [56]. Connectivity among populations might have been maximal during the glaciation periods and interrupted in the post-glacial period with remigration to higher altitudes, resulting in significant and moderate levels of genetic differentiation among populations. In these cases, admixed populations (e.g., Naousa in our study) are expected to occur, combining different gene pools. All populations of *Helleborus odorus* subsp. *cyclophyllus* distributed along the mountains of western Greece combine the same two gene pools, represented in their pure forms in Olympos and Parnassos as deduced by the STRUCTURE analysis. The close connection of the Erymanthos population to the populations of the Pindos mountain range confirms previous evidence that the Corinthian Gulf that separates mainland Greece from the Peloponnese is only a weak biogeographical barrier [65]. The Dirfi population was most probably isolated for the longest time. It shares a common ancestral population only with the distant population of Naousa. The latter population is the only that combines all detected gene pools, indicating the northern origin of the studied subspecies.

Helleborus odorus is a mainly Balkan species with a distribution centre at the Dinaric Alps, and subsp. *cyclophyllus* occupies the southern range limits of the species. The Greek populations of *H. odorus* subsp. *cyclophyllus* exhibit a west–east genetic split in terms of genetic structure, supported by the ISSR data, which becomes more evident southwards, indicating more effective genetic barriers separating western from eastern populations at the southern range limits of the subspecies. The Pindos mountain range, which runs through western mainland Greece southwards to the Peloponnese, is separated from the mountains of eastern Greece by low mountainous areas in the north, replaced by extensive plains to the south. It is possible that the low mountainous areas of northern Greece have allowed dispersal of mid-elevation plants, like *H. odorus* subsp. *cyclophyllus*, during Pleistocene glacial periods, while the plains of southern Greece have acted as more effective barriers, triggering genetic divergence of the southeast populations. The marked climatic differences between west and east Greece, especially regarding annual precipitation [66], could further promote divergent selection.

4.2. Climate Change Vulnerability

Contemporary climate change is influencing species distribution and population structure, with important consequences for patterns of genetic diversity and species' evolutionary potential [67]. The climatic factors (MAT, PET_{WQ} and MDR) emerged as the most significant variables in shaping the species' distribution and its genetic diversity turnover. Our analyses were robust and bioclimatically consistent (Figures S2–S6), for every source of uncertainty. We were able to discern the physiological threshold above which the species' local adaptation and genetic make-up seems to change drastically (Figure S18) and is responsible for the differences in within and among population genetic diversity observed in the northern and southern populations of *Helleborus odorus* subsp. *cyclophyllus*. It is very likely that the species is going to experience a moderate to severe decline in its distribution under any GCM/RCP/climate database combination (Figure 4 and Figures S7–S17). These changes will probably be more pronounced in the species' southern and NE range edge in Greece (Figure 4 and Figures S7–S17). The Pindos mountain range seems to buffer the climate change effects and will probably continue to host several populations of *H. odorus* subsp. *cyclophyllus*.

The species' local adaptation seems to be driven by MDR and as such, areas with high MDR differences constitute steep barriers for gene flow (e.g., the vast plains in Central Greece–Figures S19–S44). Analysis based on the spatial distribution of the species' genetic diversity identified a strong bioregionalization comprised of five different geographical clusters (Figure S19), which might be accounted for the restricted levels of gene flow. However, we predict a trend towards genetic homogenization (Figures S21–S44) under any GCM/RCP combination, with highest similarity between the current and the future scheme observed in the BCC 2.6 GCM/RCP combination based on the V-measure index (Figure S45).

Overall, the highest similarity is anticipated for the RCP 2.6 group of climate scenarios and the lowest for the RCP 8.5 group (they represent the best–and worst–case climate scenarios, respectively). Our results indicate that the area in northern Greece is predicted to exhibit greater genetic turnover than any other region (Figure 5). These areas harbour populations for which the genetic offset is predicted to be greatest under climate change (Figure 5). Low genetic offset is predicted in the southern edge of the species' range and the Pindos mountain range (Figure 5).

According to the SDM predictions, the species is probably going to experience a reduction in both of its edges in Greece, yet the genetic offset seems to be more pronounced in its NE Greek edge, where the Olympos and Naousa populations are predicted to experience the highest genetic diversity decline (Figure 5). Thus, the species' adaptive genetic variability seems to be more resilient in the species' core distribution in Greece (the Pindos mountains) and the southern limits of its overall distribution. This aligns with the fact that leading edge populations provide the majority of surviving lineages and persisting alleles, while trailing edge lineages and alleles are probably facing higher extinction risk [56,68]. These areas may harbour populations that through adaptive micro-evolutionary processes could enhance the species' ability to cope with climate change, as well as reveal locations expected to be highly diverse in genetic composition, and thus, if sampled, help ensure its genetic diversity and its long-term survival, due to their ecological plasticity and/or adaptation [55,69,70]. The populations in the NE edge of the species' range should also be prioritized in terms of conservation concern, since they exhibit both high genetic diversity and extinction risk.

Moreover, the species' peripheral populations are characterised by lower genetic diversity and higher genetic differentiation compared to its central populations, as expected ('abundant-centre hypothesis') [71,72], but they seem to contain a large number of PrBs (especially the Olympos population), and thus they hold significant future evolutionary potential.

5. Conclusions

Our work aims to assess genetic diversity and population differentiation in *Helleborus odorus* subsp. *cyclophyllus*, and to investigate the influence of climate change on the distribution range and habitat suitability for this species, as a case study for mid-elevation plants in the Balkan Peninsula. The Greek populations of *H. odorus* subsp. *cyclophyllus* exhibit significant gene flow, which largely explains the moderate inter-population genetic differentiation observed, while most of the genetic variance is attributed to individuals within populations. Genetic structuring revealed the existence of three genetic clusters. The populations based on genetic structuring exhibit a split in the W-E axis, which becomes more evident in the southern part of the species' distribution.

This differentiation is due to environmental constraints, which are projected to have a negative impact on the species' future distribution and genetic composition. Apart from contemporary drivers, historical processes such as the Pleistocene climate oscillations might have contributed to the genetic structure of the species' populations in Greece, since it is reasonable to assume that the low mountainous areas separating high mountains of northern Greece have allowed dispersal of mid-elevation plants, while the plains of southern Greece have acted as more effective barriers, triggering genetic divergence of the southeast populations. These leading-edge populations, along with the ones present in the species' core distribution in Greece (the Pindos mountains), seem to be more resilient against environmental/climate change in terms of their adaptive genetic variability.

Helleborus odorus subsp. *cyclophyllus* is a typical mid-elevation forest plant species, a group that has caused less attraction and remained away from the research spotlight regarding the climate-change agenda. In this respect it might be prudent and cost-effective to focus any conservation initiatives regarding this type of species to their leading-edge populations, as they might be highly diverse in genetic composition and better-suited to cope with the effects of climate change.

Supplementary Materials: The following are available online at http://www.mdpi.com/1999-4907/11/6/620/s1.

Author Contributions: Conceptualization, V.P., P.T., and G.I.; Investigation, G.F. and K.K.; Methodology and formal analysis, G.F., V.P, and K.K.; Resources, G.F., G.I., and K.K.; Supervision, V.P., and P.T.; Writing—Original Draft Preparation, G.F., K.K., P.T., and V.P.; Writing—Review and Editing, G.F., K.K., G.I., P.T., and V.P.; Visualization, K.K., G.F., and V.P. All authors have read and agreed to the published version of the manuscript.

Funding: "PlantUP" (MIS 5002803) which is implemented under the Action "Reinforcement of the Research and Innovation Infrastructure", funded by the Operational Programme "Competitiveness, Entrepreneurship and Innovation" (NSRF 2014–2020) and co-financed by Greece and the European Union (European Regional Development Fund).

Conflicts of Interest: The authors declare no conflict of interest.

References

1. Rahbek, C.; Borregaard, M.K.; Antonelli, A.; Colwell, R.K.; Holt, B.G.; Nogues-Bravo, D.; Rasmussen, C.M.; Richardson, K.; Rosing, M.T.; Whittaker, R.J.; et al. Building mountain biodiversity: Geological and evolutionary processes. *Science* **2019**, *365*, 1114–1119. [CrossRef] [PubMed]

2. Rahbek, C.; Gotelli, N.J.; Colwell, R.K.; Entsminger, G.L.; Rangel, T.F.; Graves, G.R. Predicting continental-scale patterns of bird species richness with spatially explicit models. *Proc. R. Soc. B: Boil. Sci.* **2006**, *274*, 165–174. [CrossRef] [PubMed]

3. Steinbauer, M.; Field, R.; Grytnes, J.A.; Trigas, P.; Ah-Peng, C.; Attorre, F.; Birks, H.J.B.; Borges, P.A.V.; Cardoso, P.; Chou, C.H.; et al. Topography-driven isolation, speciation and a global increase of endemism with elevation. *Glob. Ecol. Biogeogr.* **2016**, *25*, 1097–1107. [CrossRef]

4. Thuiller, W.; Lavorel, S.; Araujo, M.B.; Sykes, M.T.; Prentice, I.C. Climate change threats to plant diversity in Europe. *Proc. Natl. Acad. Sci. USA* **2005**, *102*, 8245–8250. [CrossRef]

5. Kier, G.; Kreft, H.; Lee, T.M.; Jetz, W.; Ibisch, P.L.; Nowicki, C.; Mutke, J.; Barthlott, W. A global assessment of endemism and species richness across island and mainland regions. *Proc. Natl. Acad. Sci. USA* **2009**, *106*, 9322–9327. [CrossRef]

6. Kryštufek, B.; Reed, J.M. Pattern and Process in Balkan Biodiversity—An Overview. In *Balkan Biodiversity*; Springer: Dordrecht, The Netherlands, 2004; pp. 1–8. [CrossRef]

7. Tzedakis, P.C. The Balkans as Prime Glacial Refugial Territory of European Temperate Trees. In *Balkan Biodiversity*; Springer: Dordrecht, The Netherlands, 2004; pp. 49–68. [CrossRef]

8. Frajman, B.; Oxelman, B. Reticulate phylogenetics and phytogeographical structure of Heliosperma (Sileneae, Caryophyllaceae) inferred from chloroplast and nuclear DNA sequences. *Mol. Phylogenetics Evol.* **2007**, *43*, 140–155. [CrossRef]

9. Surina, B.; Schönswetter, P.; Schneeweiss, G.M. Quaternary range dynamics of ecologically divergent species (*Edraianthus serpyllifolius* and *E. tenuifolius*, Campanulaceae) within the *Balkan refugium*. *J. Biogeogr.* **2011**, *38*, 1381–1393. [CrossRef]

10. Kutnjak, D.; Kuttner, M.; Niketić, M.; Dullinger, S.; Schönswetter, P.; Frajman, B. Escaping to the summits: Phylogeography and predicted range dynamics of *Cerastium dinaricum*, an endangered high mountain plant endemic to the western Balkan Peninsula. *Mol. Phylogenetics Evol.* **2014**, *78*, 365–374. [CrossRef]

11. Pauli, H.; Gottfried, M.; Dullinger, S.; Abdaladze, O.; Akhalkatsi, M.; Alonso, J.L.B.; Coldea, G.; Dick, J.; Erschbamer, B.; Calzado, R.F.; et al. Recent Plant Diversity Changes on Europe's Mountain Summits. *Science* **2012**, *336*, 353–355. [CrossRef]

12. Frankham, R.; Ballou, J.D.; Briscoe, D.A.; McInnes, K.H. *Introduction to Conservation Genetics*; Cambridge University Press: Cambridge, UK; New York, NY, USA, 2002.

13. Meiners, J.; Winkelmann, T. Evaluation of reproductive barriers and realisation of interspecific hybridisations depending on genetic distances between species in the genus Helleborus. *Plant Boil.* **2012**, *14*, 576–585. [CrossRef]

14. Euro+Med (2006-): Euro+Med PlantBase-the information resource for Euro-Mediterranean Plant Diversity. Available online: http://ww2.bgbm.org/EuroPlusMed/ (accessed on 2 May 2020).

15. Balázs, V.L.; Filep, R.; Ambrus, T.; Kocsis, M.; Farkas, Á.; Stranczinger, S.; Papp, N. Ethnobotanical, historical and histological evaluation of Helleborus L. genetic resources used in veterinary and human ethnomedicine. *Genet. Resour. Crop. Evol.* **2020**, *67*, 781–797. [CrossRef]

16. Čakar, J.; Parić, A.; Vidic, D.; Haverić, A.; Haverić, S.; Maksimović, M.; Bajrović, K. Antioxidant and antiproliferative activities of Helleborus odorus Waldst. & Kit, H. multifidus Vis. and H. hercegovinus Martinis. *Nat. Prod. Res.* **2011**, *25*, 1969–1974. [CrossRef]
17. Banuls, L.M.Y.; Katz, A.; Miklos, W.; Cimmino, A.; Tal, D.M.; Ainbinder, E.; Zehl, M.; Urban, E.; Evidente, A.; Kopp, B.; et al. Hellebrin and its aglycone form hellebrigenin display similar in vitro growth inhibitory effects in cancer cells and binding profiles to the alpha subunits of the Na+/K+-ATPase. *Mol. Cancer* **2013**, *12*, 33. [CrossRef]
18. Reed, D.H.; Frankham, R. Society for Conservation Biology Correlation between Fitness and Genetic Diversity. *Conserv. Biol.* **2003**, *17*, 230–237. [CrossRef]
19. Szczecińska, M.; Sramkó, G.; Wołosz, K.; Sawicki, J. Genetic Diversity and Population Structure of the Rare and Endangered Plant Species Pulsatilla patens (L.) Mill in East Central Europe. *PLoS ONE* **2016**, *11*, e0151730. [CrossRef]
20. Reddy, M.P.; Sarla, N.; Siddiq, E. Inter simple sequence repeat (ISSR) polymorphism and its application in plant breeding. *Euphytica* **2002**, *128*, 9–17. [CrossRef]
21. Mondini, L.; Noorani, A.; Pagnotta, M.A. Assessing Plant Genetic Diversity by Molecular Tools. *Diversity* **2009**, *1*, 19–35. [CrossRef]
22. Yeh, F.C.; Yang, R.C.; Boyle, T.B.J.; Ye, Z.H.; Mao, J.X. *POPGENE, the User-Friendly Shareware for Population Genetic Analysis*; University of Alberta: Edmonton, AB, Canada, 1997.
23. Vekemans, X.; Beauwens, T.; Lemaire, M.; Roldan-Ruiz, I. Data from amplified fragment length polymorphism (AFLP) markers show indication of size homoplasy and of a relationship between degree of homoplasy and fragment size. *Mol. Ecol.* **2002**, *11*, 139–151. [CrossRef]
24. Peakall, R.; Smouse, P.E. GenAlEx 6: Genetic analysis in Excel. Population genetic software for teaching and research. *Mol. Ecol. Notes* **2006**, *6*, 288–295. [CrossRef]
25. Peakall, R.; Smouse, P.E. GenAlEx 6.5: Genetic analysis in Excel. Population genetic software for teaching and research—An update. *Bioinform* **2012**, *28*, 2537–2539. [CrossRef]
26. Lynch, M.; Milligan, B.G. Analysis of population genetic structure with RAPD markers. *Mol. Ecol.* **1994**, *3*, 91–99. [CrossRef]
27. Hampl, V.; Pavlíček, A.; Flegr, J. Construction and bootstrap analysis of DNA fingerprinting-based phylogenetic trees with the freeware program FreeTree: Application to trichomonad parasites. *Int. J. Syst. Evol. Microbiol.* **2001**, *51*, 731–735. [CrossRef]
28. Rambaut, A. *FigTree, v1.3.1*; beast: UK, 2010.
29. Huff, D.R.; Peakall, R.; Smouse, P.E. RAPD variation within and among natural populations of outcrossing buffalograss [Buchloë dactyloides (Nutt.) Engelm.]. *Theor. Appl. Genet.* **1993**, *86*, 927–934. [CrossRef]
30. Vavrek, M.J. Fossil: Palaeoecological and palaeogeographical analysis tools. *Palaeontol. Electron.* **2011**, *14*, 16.
31. Oksanen, A.J.; Blanchet, F.G.; Friendly, M.; Kindt, R.; Legendre, P.; Mcglinn, D.; Minchin, P.R.; Hara, R.B.O.; Simpson, G.L.; Solymos, P.; et al. *vegan: Community Ecology Package, version 2.5.2*; CRAN, R Core Team: Cary, CA, USA, 2018.
32. Pritchard, J.K.; Stephens, M.; Donnelly, P. Inference of population structure using multilocus genotype data. *Genetics* **2000**, *155*, 945–959.
33. Evanno, G.; Regnaut, S.; Goudet, J. Detecting the number of clusters of individuals using the software structure: A simulation study. *Mol. Ecol.* **2005**, *14*, 2611–2620. [CrossRef]
34. Earl, D.; Vonholdt, B.M. Structure Harvester: A website and program for visualizing Structure output and implementing the Evanno method. *Conserv. Genet. Resour.* **2011**, *4*, 359–361. [CrossRef]
35. Remington, D.L.; Thornsberry, J.M.; Matsuoka, Y.; Wilson, L.M.; Whitt, S.R.; Doebley, J.; Kresovich, S.; Goodman, M.M.; Buckler, E.S. Structure of linkage disequilibrium and phenotypic associations in the maize genome. *Proc. Natl. Acad. Sci. USA* **2001**, *98*, 11479–11484. [CrossRef]
36. Chamberlain, S.; Boettiger, C. R Python, and Ruby clients for GBIF species occurrence data. *PeerJ Preprints* **2017**, *5*, e3304v1. [CrossRef]
37. Robertson, M.P.; Visser, V.; Hui, C. Biogeo: An R package for assessing and improving data quality of occurrence record datasets. *Ecography* **2016**, *39*, 394–401. [CrossRef]
38. Hijmans, R.J.; Cameron, S.E.; Parra, J.L.; Jones, P.G.; Jarvis, A. Very high resolution interpolated climate surfaces for global land areas. *Int. J. Clim.* **2005**, *25*, 1965–1978. [CrossRef]

39. Karger, D.N.; Conrad, O.; Böhner, J.; Kawohl, T.; Kreft, H.; Soria-Auza, R.W.; Zimmermann, N.E.; Linder, H.P.; Kessler, M. Climatologies at high resolution for the earth's land surface areas. *Sci. Data* **2017**, *4*, 1–20. [CrossRef] [PubMed]

40. CHELSA Database. Available online: http://chelsa-climate.org/ (accessed on 2 May 2020).

41. Morales-Barbero, J.; Vega-Álvarez, J. Input matters matter: Bioclimatic consistency to map more reliable species distribution models. *Methods Ecol. Evol.* **2018**, *10*, 212–224. [CrossRef]

42. Araujo, M.B.; Anderson, R.P.; Barbosa, A.M.; Beale, C.M.; Dormann, C.F.; Early, R.; Garcia, R.A.; Guisan, A.; Maiorano, L.; Naimi, B.; et al. Standards for distribution models in biodiversity assessments. *Sci. Adv.* **2019**, *5*, eaat4858. [CrossRef]

43. Jarvis, A.; Reuter, H.I.; Nelson, A.; Guevara, E. Hole-filled SRTM for the Globe Version 3, from the CGIAR-CSI SRTM 90m Database 2006. Available online: http//srtm.csi.cgiar.org (accessed on 31 April 2020).

44. Hijmans, R.; van Etten, J. Raster: Geographic Data Analysis and Modeling. *R Package Version* **2014**, *517*, 2–12.

45. Evans, J.S. *spatialEco: Spatial Analysis and Modelling Utilities, version1.3-1*; CRAN, R Core Team: Cary, CA, USA, 2020.

46. Bornovas, I.; Rondogianni-Tsiambaou, T. *Geological Map of Greece Scale 1:500,000*, 2nd ed.; Inst. of Geology and Mineral Exploration, Division of General Geology and Economic Geology: Athens, Greece, 1983.

47. Dormann, C.F.; Elith, J.; Bacher, S.; Buchmann, C.; Carl, G.; Carré, G.; Márquez, J.R.G.; Gruber, B.; Lafourcade, B.; Leitão, P.J.; et al. Collinearity: A review of methods to deal with it and a simulation study evaluating their performance. *Ecography* **2012**, *36*, 27–46. [CrossRef]

48. Araujo, M.B.; New, M. Ensemble forecasting of species distributions. *Trends Ecol. Evol.* **2007**, *22*, 42–47. [CrossRef]

49. Thuiller, W.; Lafourcade, B.; Engler, R.; Araujo, M.B. BIOMOD—A platform for ensemble forecasting of species distributions. *Ecography* **2009**, *32*, 369–373. [CrossRef]

50. Ferrier, S.; Manion, G.; Elith, J.; Richardson, K. Using generalized dissimilarity modelling to analyse and predict patterns of beta diversity in regional biodiversity assessment. *Divers. Distrib.* **2007**, *13*, 252–264. [CrossRef]

51. Fitzpatrick, M.C.; Keller, S.R. Ecological genomics meets community-level modelling of biodiversity: Mapping the genomic landscape of current and future environmental adaptation. *Ecol. Lett.* **2014**, *18*, 1–16. [CrossRef]

52. Manion, G.; Lisk, M.; Ferrier, S.; Nieto-Lugible, D.; Mokany, K.; Fitzpatrick, M.C. *gdm: Generalized Dissimilarity Modeling, version 1.3.7*; CRAN, R Core Team: Cary, CA, USA, 2018.

53. R Core Development Team. *R: A language and Environment for Statistical Computing*; R Foundation for Statistical Computing: Vienna, Austria, 2019.

54. Rahbek, C. The elevational gradient of species richness: A uniform pattern? *Ecography* **1995**, *18*, 200–205. [CrossRef]

55. Hoffmann, A.A.; Sgrò, C.M. Climate change and evolutionary adaptation. *Nature* **2011**, *470*, 479–485. [CrossRef]

56. Pauls, S.U.; Nowak, C.; Bálint, M.; Pfenninger, M. The impact of global climate change on genetic diversity within populations and species. *Mol. Ecol.* **2012**, *22*, 925–946. [CrossRef] [PubMed]

57. Nelson, M.F.; Anderson, N.O. How many marker loci are necessary? Analysis of dominant marker data sets using two popular population genetic algorithms. *Ecol. Evol.* **2013**, *3*, 3455–3470. [CrossRef] [PubMed]

58. Nybom, H. Comparison of different nuclear DNA markers for estimating intraspecific genetic diversity in plants. *Mol. Ecol.* **2004**, *13*, 1143–1155. [CrossRef]

59. Hogbin, P.M.; Peakall, R. Evaluation of the Contribution of Genetic Research to the Management of the Endangered Plant *Zieria prostrata*. *Conserv. Boil.* **1999**, *13*, 514–522. [CrossRef]

60. Ellstrand, N.C.; Elam, D.R. Population Genetic Consequences of Small Population Size: Implications for Plant Conservation. *Annu. Rev. Ecol. Syst.* **1993**, *24*, 217–242. [CrossRef]

61. Slatkin, M.; Excoffier, L. Serial Founder Effects during Range Expansion: A Spatial Analog of Genetic Drift. *Genetics* **2012**, *191*, 171–181. [CrossRef]

62. Grdiša, M.; Radosavljević, I.; Liber, Z.; Stefkov, G.; Ralli, P.; Chatzopoulou, P.S.; Carović-Stanko, K.; Satovic, Z. Divergent selection and genetic structure of *Sideritis scardica* populations from southern Balkan Peninsula as revealed by AFLP fingerprinting. *Sci. Rep.* **2019**, *9*, 1–14. [CrossRef]

63. Larena, B.G.; Aguilar, J.F.; Feliner, G.N. Glacial-induced altitudinal migrations in Armeria (Plumbaginaceae) inferred from patterns of chloroplast DNA haplotype sharing. *Mol. Ecol.* **2002**, *11*, 1965–1974. [CrossRef]

64. Kramp, K.; Huck, S.; Niketić, M.; Tomović, G.; Schmitt, T. Multiple glacial refugia and complex postglacial range shifts of the obligatory woodland plant *Polygonatum verticillatum* (Convallariaceae). *Plant Boil.* **2009**, *11*, 392–404. [CrossRef] [PubMed]

65. Strid, A. The mountain flora of Greece with special reference to the Anatolian element. *Proc. R. Soc. Edinburgh. Sect. B. Boil. Sci.* **1986**, *89*, 59–68. [CrossRef]

66. Feidas, H.; Karagiannidis, A.; Keppas, S.; Vaitis, M.; Kontos, T.; Zanis, P.; Melas, D.; Anadranistakis, E. Modeling and mapping temperature and precipitation climate data in Greece using topographical and geographical parameters. *Theor. Appl. Clim.* **2013**, *118*, 133–146. [CrossRef]

67. Parmesan, C. Ecological and Evolutionary Responses to Recent Climate Change. *Annu. Rev. Ecol. Evol. Syst.* **2006**, *37*, 637–669. [CrossRef]

68. McInerny, G.; Turner, J.; Wong, H.; Travis, J.M.; Benton, T.G. How range shifts induced by climate change affect neutral evolution. *Proc. R. Soc. B Boil. Sci.* **2009**, *276*, 1527–1534. [CrossRef]

69. Riddle, B.R.; Dawson, M.N.; Hadly, E.A.; Hafner, D.J.; Hickerson, M.J.; Mantooth, S.J.; Yoder, A.D. The role of molecular genetics in sculpting the future of integrative biogeography. *Prog. Phys. Geogr. Earth Environ.* **2008**, *32*, 173–202. [CrossRef]

70. Scoble, J.; Lowe, A. A case for incorporating phylogeography and landscape genetics into species distribution modelling approaches to improve climate adaptation and conservation planning. *Divers. Distrib.* **2010**, *16*, 343–353. [CrossRef]

71. Eckert, C.G.; Samis, K.E.; Lougheed, S.C. Genetic variation across species' geographical ranges: The central–marginal hypothesis and beyond. *Mol. Ecol.* **2008**, *17*, 1170–1188. [CrossRef]

72. Sagarin, R.; Gaines, S.D. The 'abundant centre' distribution: To what extent is it a biogeographical rule? *Ecol. Lett.* **2002**, *5*, 137–147. [CrossRef]

Article

National Set of MAES Indicators in Greece: Ecosystem Services and Management Implications

Ioannis P. Kokkoris [1], Georgios Mallinis [2], Eleni S. Bekri [3], Vassiliki Vlami [4], Stamatis Zogaris [5], Irene Chrysafis [2], Ioannis Mitsopoulos [6] and Panayotis Dimopoulos [1,*]

[1] Department of Biology, Laboratory of Botany, University of Patras, 26504 Patras, Greece; ipkokkoris@upatras.gr

[2] Forest Remote Sensing and Geospatial Analysis Laboratory, Democritus University of Thrace, 68200 Orestiada, Greece; gmallin@fmenr.duth.gr (G.M.); echrysaf@fmenr.duth.gr (I.C.)

[3] Department of Civil Engineering, Environmental Engineering Laboratory, University of Patras, 26504 Patras, Greece; ebekri@upatras.gr

[4] Department of Environmental Engineering, University of Patras, 30100 Agrinion, Greece; vasvlami@upatras.gr

[5] Hellenic Centre for Marine Research, Institute of Marine Biological Resources and Inland Waters, 19013 Anavissos, Greece; zogaris@hcmr.gr

[6] Ministry of Environment & Energy, Directorate of Biodiversity and Natural Environment Management, 11526 Athens, Greece; i.mitsopoulos@prv.ypeka.gr

* Correspondence: pdimopoulos@upatras.gr; Tel.: +30-2610-996777

Received: 21 April 2020; Accepted: 21 May 2020; Published: 24 May 2020

Abstract: Research Highlights: The developed National Set of Indicators for the Mapping and Assessment of Ecosystems and their Services (MAES) implementation in Greece at the national level sets the official, national basis on which future studies will be conducted for MAES reporting for the achievement of targets within the National and the European Union (EU) biodiversity Strategy. Background and Objectives: Greece is currently developing and implementing a MAES nation-wide program based on the region's unique characteristics following the proposed methodologies by the European Commission, in the frame of the LIFE-IP 4 NATURA project (Integrated actions for the conservation and management of Natura 2000 sites, species, habitats and ecosystems in Greece). In this paper, we present the steps followed to compile standardized MAES indicators for Greece that include: (a) collection and review of the available MAES-related datasets, (b) shortcomings and limitations encountered and overcome, (c) identification of data gaps and (d) assumptions and framework setting. Correspondence to EU and National Strategies and Policies are also examined to provide an initial guidance for detailed thematic studies. Materials and Methods: We followed the requirements of the EU MAES framework for ecosystem services and ecosystem condition indicator selection. Ecosystem services reported under the selected indicators were assigned following the Common International Classification of Ecosystem Services. Spatial analysis techniques were applied to create relevant thematic maps. Results: A set of 40 MAES indicators was drafted, distributed in six general indicator groups, i.e., Biodiversity, Environmental quality, Food, material and energy, Forestry, Recreation and Water resources. The protocols for the development and implementation of an indicator were also drafted and adopted for future MAES studies in Greece, providing guidance for adaptive development and adding extra indicators when and where needed. Thematic maps representing ecosystem services (ES) bundles and ES hotspots were also created to identify areas of ES importance and simultaneously communicate the results at the national and regional levels.

Keywords: biodiversity; forestry; agriculture; recreation; water resources; natural resources management; adaptive monitoring; EU Green deal

---

125

1. Introduction

Reaching the end of the European Union (EU) Biodiversity Strategy to 2020 [1], and seven years after the publication of the multi-cited master document for Mapping and Assessment of Ecosystem and their Services (MAES) in the EU [2], many EU Member States (MS) have developed methods (e.g., References [3–7]) and conducted case studies (e.g., References [8–10]) towards the implementation of Action 5 of Target 2 of the EU Biodiversity Strategy. From 2015 to 2018, the Horizon 2020 Coordination and Support Action, ESMERALDA (Enhancing ecoSysteM sERvices mApping for poLicy and Decision mAking), aimed at developing guidance and a flexible methodology to support the EU member states in the MAES implementation; more specifically, the main objective of ESMERALDA was to provide guidance for integrated mapping and assessment of ecosystem services (ES) that can be used for sustainable decision-making in policy, business, society, practice and science at EU, national and regional levels [11,12]. In parallel, a collection of papers by Burkhard and Maes [13] provides a comprehensive guidance for MAES implementation and in combination with the ever-updating MAES Explorer online [14,15], that supports ecosystem ES research and proposes systematic ways of assessment, mapping and reporting for biophysical, economic and social aspects of ES applications (as well as for their possible wider integration). Furthermore, an operational framework for integrated MAES, developed by Burkhard et al. [16], builds on the MAES common assessment framework [2] and re-organizes it on the basis of specific, practical steps needed to be followed to ensure an integrated result, at EU and national levels.

For graphical representation and mapping of ES, a tiered approach is proposed by Grêt-Regamey et al. [17], and updated in Burkhard and Maes [13] (Chapter 5.6.1), to support MAES studies at standardized scales of detail and data availability. Thus, it is obvious that the methods and tools available for operationalizing Action 5 of Target 2 of the EU Biodiversity across and within MS should be urgently elaborated. However, Albert et al. [18] highlighted that even with this guidance, national implementation of MAES requires the development of adapted sets of indicators that are most applicable to each respective context [19]. Due to this need, a debate on national indicators [20,21] and how they can be incorporated into policy, planning and management [22–25] is still ongoing.

Indicators are also fundamental elements for ES monetary valuation and Natural Capital (NC) accounting. Some have continued to argue that monetary valuation of ES and NC is inappropriate, and we should preserve and protect nature strictly 'for its own sake', for its 'intrinsic values' [26,27]. Costanza et al. [28] point out that this perspective is itself an implicit valuation: it is simply arguing that nature is more valuable than any possible alternative. While in many cases this may be true, society has made decisions implying that this is not always the case [29]. Through our historical and current interactions with the environment, in order to develop infrastructure and produce the goods necessary for contemporary life and well-being, we exploit ecosystems and impact NC. Thus, being more explicit about the value of ES and NC can help society make better decisions in the many cases in which trade-offs and complex conflicts exist [30,31].

Simultaneously, everything in applied science, management and decision-making is data-dependent, as well as in need of interpretation. Throughout the relevant literature (e.g., References [7,32–34]), it is highlighted that the crucial step for a successful MAES implementation is the identification, selection, elaboration and/or development of the appropriate indicators which capture in space and time ecosystems' performance regarding their condition and the multitude of services and benefits they provide. Selecting the appropriate indicators is identified as one of the fundamental steps of the operational framework as proposed by Burkhard et al. [16] and comprises a selection of (a) indicators for ecosystem condition (EC) and (b) indicators for ES.

The purpose of indicators is to measure and ascribe a value to the various dimensions [35,36] of the complex ES concept [37]. Simultaneously, the indicator and indicandum (i.e., the phenomenon of interest reflected by the indicator) should be correlated with one another and the variance should be low [33]. It is clear that a common set of indicators cannot be applied across all the different ecosystem types; however, a standardized way of reporting ecosystem condition and ecosystem services at local, regional and national levels is important to support decision-making and strategic planning at MS and EU levels [2,38]. MAES indicators should have specific characteristics in order to support robust ES studies and should be applicable for policy-relevant interpretation, with the capacity to inform a broad array of policies related to the use, conservation and preservation of natural resources [38].

During the past five years, various studies guided by the MAES conceptual framework [2] have been conducted in Greece, providing information for different types of ecosystems and applying different methodologies and tools for mapping and assessing ES at local (e.g., Reference [39]), regional [40] and national levels [41–44]. In 2017, a group of scientists who believed in the importance of the MAES implementation (forming the Hellenic Ecosystem Partnership—HESP [45]) drafted the National Agenda for the MAES implementation in Greece [34] and set an Action Plan to 2020, including short- and mid-term objectives. Development and testing of a national set of indicators, by the end of 2020, is one of the mid-term objectives of the Agenda's Action Plan needed to be accomplished for further MAES implementation in Greece. The Life Integrated Project with the acronym "LIFE-IP 4 Natura" [46], led by the Hellenic Ministry of Environment and Energy, incorporates MAES implementation Actions at national, as well as at local (case-study) levels. These important developments towards standardizing MAES applications will also support NC accounting based on the System of Environmental Economic Accounting (SEEA)—Experimental Ecosystem Accounting [47].

This work aims to (a) collect and review all available data from national, regional and local authorities, which can be used to identify, assess and map ES at the national scale, (b) evaluate the potential usability of these data directly, after sorting and processing, or consider part of them as inappropriate for further use, and (c) identify ES data hot-spots and data-scarce areas in Greece. The final goal of the study is to provide a pre-defined National Set of Indicators for the MAES implementation at the national level. This set of Indicators will form the official, national basis on which future studies will be conducted for MAES, reporting towards supporting the targets of the National and the EU biodiversity Strategy.

2. Materials and Methods

2.1. Preparatory Actions

The initial overview of potential national indicators involved the following steps:

i. Identification of ecosystem types (terrestrial): using the Ecosystem type map of Europe [48], the Corine Land Cover dataset [49], the Natura 2000 Standard Data Forms and monitoring results [50,51] and spatial data for habitat types [52], and following the European Environment Agency (EEA) guidelines [53] and recent survey results [42,43], we have identified and classified the ecosystem types present in the Greek territory; a detailed ecosystem type mapping for Greece, i.e., MAES level 3 (sub-categories of MAES level 2 [2] ecosystem types in Greece) is an ongoing procedure of the LIFE-IP 4 NATURA project, summarized in Section 2.2.

ii. Literature review for ES indicators in Greece: we reviewed the provided list of primary and secondary ES indicators used for the site-level assessment of ES supply at mountainous Natura 2000 sites in Greece [42]. The overview on ecosystem condition indicators was based on the EU ecosystem condition assessment framework [38] and recent work for Greece [43].

iii. Exploration of data availability and quality: to identify and select possible indicators for (a) mapping and assessing ecosystem condition and ES, and (b) establishing a national set of indicators for future MAES studies, we revised and assessed (regarding their thematic detail and spatial scale) all available datasets for terrestrial ecosystems (based on the review by Dimopoulos et al. [34] and on datasets freely available by state authorities (Supplementary Table S1)), updated by the input of recent datasets and information from the relevant national and regional offices/authorities, after an official request by the Hellenic Ministry of Environment and Energy (leader of the LIFE-IP 4 Natura project).

iv.   Response to policy needs: using the guidance provided by Maes et al. [2,38] and in combination with the targets of (a) the EU Green Deal [54], (b) the Greek Biodiversity Strategy [55], (c) the Forest Strategy, (d) the Regional Policy, (e) the Agricultural Policy, (f) the Climate Policy and (g) the Water Policy, each indicator was evaluated and assigned with a "1" (yes) or "0" (no) mark regarding its direct relation with the above-mentioned strategies and policies.

v.   Supporting valuation and NC assessments: using a simple scaling method, experts from the LIFE-IP 4 NATURA consortium rated each indicator's dataset for its ability to support valuation and NC studies (i.e., 0—not relevant, 1—very low, 2—low, 3—medium, 4—high, 5—ready for use). This allowed the co-authors to decide on final indicators and to explore the uncertainty, data availability and applicability.

2.2. Drafting the National Set of Indicators

The selection of indicators to be drafted for the National Set was based on the requirements of the MAES indicator framework for ecosystem condition [38] (Table 1). These are considered as appropriate also for ES indicators, since ES are directly dependent on the condition of ecosystems (e.g., References [56–58]). Each indicator has been assessed for its compliance with the predefined requirements.

Table 1. Requirements for the Mapping and Assessment of Ecosystem and their Services (MAES) indicator framework on ecosystem condition assessment [38]. Numbers in column "Code" are used to assign the requirements with each one of the assessed indicators for the present study.

Requirements	Description	Code
Scientifically sound	Indicators should be based on the best available knowledge while giving a good representation of the ecosystem characteristics addressed.	1
Supporting environmental legislation	Indicators should support the implementation of environmental legislation in the European Union (EU).	2
Policy-relevant	Indicators should be policy-relevant: they have multiple policy uses and can support a policy narrative which links pressures, ecosystem condition, ecosystem services and policy objectives.	3
Include habitat and species conservation status	The conservation status of habitats and species (and in particular the parameters "area" and "structure and function") reported under Art.17 of the EU Habitats Directive should constitute a major indicator for assessing ecosystem condition.	4
Include soil-related information	Terrestrial ecosystems are not in good condition if their soils are not in good condition. Specific indicators which assess the condition of soils should therefore be included.	5
Applicable for natural capital accounts	The indicator framework should support the development and testing of ecosystem extent and condition accounts.	6
Spatially explicit	Ecosystem condition is not equal across space. Different spatial gradients of pressures and differences in the response of ecosystems to pressures result in spatial variance of ecosystem condition, which needs to be acknowledged in the indicator selection.	7
Baseline	Indicators should be measurable relative to a baseline year (e.g., 2010).	8
Sensitive to change	Indicators should be able to detect change over time.	9

The Common International Classification of Ecosystem Services (CICES ver. 5.1) [59] was used to assign ES reported under the selected indicators, with the international and EU practice of reporting ES. The selected indicators are also assigned to the corresponding categories of ES indicators, proposed in the second MAES report [32], followed by the designation of the relevant terrestrial MAES level 2 ecosystem type (i.e., Urban, Cropland, Grassland, Woodland and forest, Heathland and shrub, Sparsely vegetated land, Wetlands, Rivers and Lakes).

Moreover, each indicator is selected and tagged on the basis of its potential for application at the different tier assessments, i.e., tier 1 approach relies on widely available data and can be used to provide a rough overview of ES, tier 2 approach includes more specific information for the case study area, while tier 3 is the most data- and resource-demanding approach and is appropriate for large-scale and highly detailed assessments [13,17]. The steps followed to draft the final set of MAES indicators are presented in Figure 1.

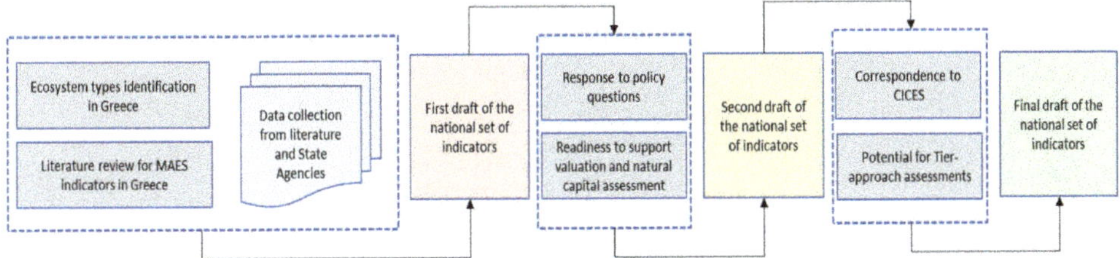

Figure 1. Steps followed for drafting the National Set of MAES Indicators. CICES: Common International Classification of Ecosystem Services.

Analysis of indicator-related data availability in space, time and scale has also been conducted and represented in thematic maps, identifying areas with data gaps, as well as areas where MAES implementation can be operationalized.

The above-mentioned procedure is the core part of the MAES implementation in Greece under the LIFE-IP 4 NATURA project and the relevant procedure flowchart is presented in Figure 2.

Development and Implementation of Indicators' Protocols

Following Kettunen et al. [60], Ferrari and Geneletti [61] and Nedkov et al. [62], we propose structured protocols for (a) the development and (b) the implementation of MAES indicators in Greece. This is to provide guidance on how to develop and implement indicators at various scales in future MAES studies. More precisely:

a. Development of the indicator protocol fields: (i) Indicator name (name of the index for national use), (ii) Definition (summary definition of the index), (iii) Description (summary description of the index), (iv) Application (for ES and/or EC assessments), (v) Use and interpretation (local, regional, national), (vi) Measurement units (e.g., m^3/ha), (vii) Data source (e.g., Ministry of Environment and Energy), (viii) Calculation method (detailed description of the method used to calculate the index).

b. Implementation of the indicator protocol fields: (i) Indicator name (name of the index for national use), (ii) Responsible, coordinating authority for the implementation (authority/organization, etc., and person responsible for indicator calculation and communication), (iii) Use and interpretation (includes: key questions to which the indicator responds, (iv) Users of the indicator, (v) Appropriate scale for the implementation, (vi) Potential for aggregation (interpretation of increasing or decreasing trends, possible causes for these trends), (vii) Impact of indicator alterations/change to management, (viii) Measurement unit, (ix) Data source, (x) Calculation method, (xi) Presentation of the index (maps, graphs, tables, etc., appropriate for communication purposes), (xii) Limitations of index utility and accuracy, (xiii) Update/revision of the index (frequency and procedure), (xiv) Relative indicators (report relevant indices if available), (xv) Additional information and comments (free text).

The proposed development protocol was used for the indicators of the present study.

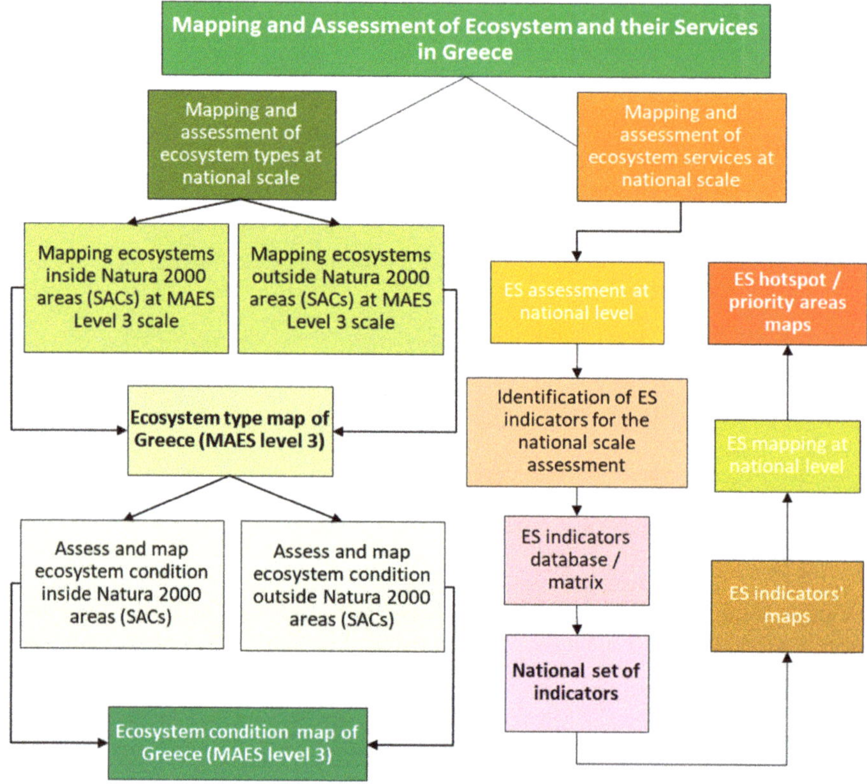

Figure 2. Flowchart for the MAES implementation in Greece, during the LIFE-IP 4 NATURA project (Integrated actions for the conservation and management of Natura 2000 sites, species, habitats and ecosystems in Greece). SAC: Special Areas of Conservation, ES: ecosystem services.

2.3. Thematic Maps

2.3.1. ES Indicators Bundles

To pinpoint ES hotspot areas, as well as areas where multiple ES are likely to be supplied simultaneously (ES bundles) in space and/or time, thematic maps have been prepared, by depicting indicator-related data overlaps; by this, areas of ES importance (high to low) are identified and the relevant priority MAES maps (priority areas maps) have been drafted. To identify areas where multiple ES are simultaneously supplied (or potentially supplied) in spatial and/or temporal terms, we used spatial analysis, and layer-overlay techniques, at the 10 × 10 km EEA reference grid for Greece [63], using Geographic Information Systems; thus, we highlighted areas where ES bundles are present and further study is needed at a more detailed level (i.e., regional and local levels). This methodology is based on the following assumptions: (a) each indicator layer was considered as of equal importance and its presence in each grid cell is assigned to a binary value, i.e., 0: not present, 1: present, in cell, (b) summing all values from the overlaying layers in each cell, represents each cell's importance for ES (higher values at cell represent higher importance and vice versa), (c) the available, spatial-referenced datasets were used for: (i) forest-management studies, (ii) wind-energy stations, (iii) hydroelectric-energy stations, (iv) solar-energy stations, (v) national forests, (vi) Natura 2000 areas and/or wildlife refuges, (vii) "blue flag" beaches, (viii) thermal springs, (ix) inland boating and rafting, (x) mountain shelters, (xi) cropland area and (xii) water resources, (d) these datasets were used as single or combined categorical variables, by recording their presence in each 10 × 10 EEA reference grid cell of Greece. By this, a matrix calculating the presence of the various datasets per grid cell has been created. The categorization of the total sum of the various datasets' availability in each grid cell, using GIS software, resulted in a thematic ES bundles national map, based on MAES-related available datasets for Greece. To support MAES knowledge at the administrative, regional-management level, the representativity of ES bundles at the regional (NUTS 2) level has also been prepared and presented using the analytics and visualization Tableau online platform (ver. 2020.1) [64] for the six general groups (i.e., Biodiversity, Environmental quality, Food, material and energy, Forestry, Recreation, Water resources) of the proposed set of indicators. The area of each region is also integrated in the analysis in order to encapsulate the relative importance of each region regarding the potential ES provision.

2.3.2. ES Hotspots

Using the 10 × 10 km EEA reference grid, ES hotspots are also depicted in thematic maps, representing areas at which a particular ecosystem service is provided (potential supply) in large proportions [65]. Large proportions refer to a high spatial density and extent of ES indicator spatial data.

2.3.3. Compliance with the MAES Indicator Framework

Maps of compliancy with the indicators' requirements [38] and as presented in Table 1 are also drafted using the total number of requirements simultaneously provided at each 10 × 10 km EEA reference grid cell.

2.3.4. Data Gaps

Since this exercise depicts current knowledge and is based on the best available datasets, data gaps must sometimes be overlooked to expedite applications. However, defining data gaps is crucial for concrete management and decision-making; thus, data-gap maps have also been drafted on the basis of the 10 × 10 km EEA reference grid. The overlay method, as described above, was also used to pinpoint the data-scarce areas by creating the relevant readiness maps for the MAES implementation in Greece, regarding (a) the indicator groups and (b) indicator groups' cumulative importance, in each region of Greece (NUTS 2).

2.3.5. Relevance to Policies

Indicator relevance with national strategies and polices (i.e., biodiversity strategy, forest strategy, regional policy, agricultural policy, climate policy and water policy) and with the EU "Green deal" set of transformative policies (i.e., clean energy, sustainable industry, building and renovating, sustainable mobility, biodiversity, "from farm to fork", eliminating pollution) is also considered. A thematic representation of the indicators relevance (as approached in Section 2.1, iv) was drafted to initially assess the indicators' potential use for implementing and evaluating various policies, across the Greek terrestrial territory.

3. Results

3.1. Classification Scheme of Terrestrial Ecosystem Types in Greece

The results of the drafted classification scheme for terrestrial ecosystem types in Greece proposes 28 MAES level 3 ecosystem types, included in eight MAES level 2 categories. More precisely, 110 habitat types are distributed along 19 MAES level 3 ecosystem types:

- Grassland: 13 habitat types in Natural grasslands.
- Woodland and forest: (i) seven habitat types in Temperate deciduous forests, (ii) seven habitat types in Mediterranean deciduous forests, (iii) five habitat types in Floodplain forests (Riparian forest/Fluvial forest), (iv) six habitat types in Temperate mountainous coniferous forests, (v) four

habitat types in Mediterranean coniferous forests, (vi) four habitat types in Mediterranean sclerophyllous forests and (vii) one habitat type in Mixed Forest.

- Heathland and shrub: (i) five habitat types in Moors and heathland and (ii) 10 habitat types in Sclerophyllous vegetation.
- Sparsely vegetated land: (i) 11 habitat types in Sparsely vegetated areas and (ii) eight habitat types in Beaches, dunes, sands.
- Wetlands: (i) three habitat types in Inland freshwater marshes, (ii) six habitat types in Inland saline marshes, (iii) four habitat types in Peat bogs and (iv) eight habitat types in Marine wetlands.
- Rivers and lakes: (i) five habitat types in Rivers and (ii) four habitat types in Lakes.

A detailed crosswalk among habitat types and ecosystem type levels is presented in Table 2. This typology will be used for the national scale MAES level 3 ecosystem type mapping and will provide a base-line map for the MAES implementation in Greece, under the LIFE-IP 4 Natura project.

3.2. Indicators for ES and EC Assessments

3.2.1. Selection of Indicators

A set of 40 indicators (Table 3) which comprises six groups of indicators, i.e., (i) Biodiversity, (ii) Environmental quality, (iii) Food, material and energy, (iv) Forestry, (v) Recreation and (vi) Water resources, has been drafted for Greece. The selected indicators cover all the three CICES sections as follows:

- Provisioning services: 15 indicators, corresponding to 10 CICES codes.
- Regulating and maintenance services: 18 indicators, corresponding to 22 CICES codes.
- Cultural services: 6 indicators, corresponding to 3 CICES codes.

Most of the indicators can be used for ES assessments, as well as for EC assessments; more precisely, 33 indicators comply for ES assessments (nine of them are exclusively applicable for ES assessments), while 30 indicators are suitable for EC assessments (four of them are exclusively applicable for EC assessments) (Table 3).

Figure 3 and Supplementary Table S1, present the correlations among ecosystem types and the proposed national set of MAES indicators. Regarding their utility per MAES level 2 ecosystem type, the proposed indicators are distributed as follows:

- Urban: 9 indicators (two from Biodiversity, three from Environmental quality, two from Food, material and energy and two from Recreation).
- Cropland: 11 indicators (four from Biodiversity, three from Environmental quality, two from Food, material and energy and two from Recreation).
- Grassland: 11 indicators (three from Biodiversity, three from Environmental quality, two from Food, material and energy, one from Forestry and two from Recreation).
- Woodland and forest: 20 indicators (three from Biodiversity, eight from Environmental quality, one from Food, material and energy, five from forestry and three from Recreation).
- Heathland and shrub: 17 indicators (four from Biodiversity, six from Environmental quality, three from Food, material and energy, two from forestry and two from Recreation).
- Sparsely vegetated land: 11 indicators (three from Biodiversity, three from Environmental quality, two from Food, material and energy, five from forestry and three from Recreation).
- Wetlands: 8 indicators (three from Biodiversity, three from Environmental quality and two from Recreation).
- Rivers and lakes: 15 indicators (three from Biodiversity, one from Environmental quality, one from Food, material and energy, three from Recreation and seven from Water resources).
- Marine inlets and transitional waters: 7 indicators (two from Biodiversity, three from Environmental quality and two from recreation).

Table 2. Classification of the ecosystem types (MAES level 3), for mapping and assessment in terrestrial areas of Greece.

MAES Ecosystem Category (Level 1)	Ecosystem Type for Mapping and Assessment (Level 2)	Ecosystem Type for Mapping and Assessment in Greece (Level 3)	Habitat Type Codes
Terrestrial	Urban	Dense to medium dense Urban Fabric (IM.D. 30%–100% + industrial, commercial, public, military and private units)	-
		Low-density Urban Fabric (IM.D. 0%–30%)	-
		Other/Transport	-
	Cropland	Arable land	-
		Permanent crops	-
		Heterogeneous agricultural areas	-
	Grassland	Managed grassland	-
		Natural grasslands prevailingly without trees and scrubs (T.C.D. < 30%)	6110 *, 6170, 6220 *, 6230 *, 6290, 62A0, 62D0, 6420, 6430, 6510, 651A, G628, G645
		Natural grasslands with trees and scrubs (T.C.D. > 30%)	6110 *, 6170, 6220 *, 6230 *, 6290, 62A0, 62D0, 6420, 6430, 6510, 651A, G628, G645
	Woodland and forest	Temperate deciduous forests	9110, 9130, 9140, 9150, 9180, G91K, G91L
		Mediterranean deciduous forests	91M0, 9280, 9250, 9310, 9350, 9260, 925A
		Floodplain forests (Riparian forest/Fluvial forest)	92A0, 92C0, 92D0, 91E0 *, 91F0
		Temperate mountainous coniferous forests	9530 *, 951B, 91BA, 91CA, 9410, 95A0
		Mediterranean coniferous forests	2270, 9540, 9560, 9290
		Mediterranean sclerophyllous forests	9340, 934A, 9320, 9370
		Mixed Forest	9270
	Heathland and shrub	Moors and heathland	4060, 4090, 5360, 5420, 5430
		Sclerophyllous vegetation	2250 *, 5110, 5150, 5160, 5210, 5230, 5310, 5330, 5340, 5350
	Sparsely vegetated land	Sparsely vegetated areas	8130, 8140, 8210, 8220, 8230, 8310, 8320, 8330, 2240, 2260, 9620
		Beaches, dunes, sands	1210, 1240, 1410, 2110, 2120, 2220, 2230, 2210
		Bare rock	-
		Burnt areas	-
		Glaciers and perpetual snow	-
		Mines, dumps, land without current use	-
	Wetlands	Inland freshwater marshes	72A0, 72B0, 2190
		Inland saline marshes	1310, 1410, 1420, 1430, 1510, 1440
		Peat bogs	7140, 7210, 7220, 7230
		Marine wetlands	1110, 1120, 1130, 1150, 1160, 1170, 1180, 1310
Freshwater	Rivers and lakes	Rivers	3240, 3250, 3260, 3280, 3290
		Lakes	3130, 3140, 3150, 3170 *

IM.D.: Impervious Degree, T.C.D.: Total Canopy Density, *: Habitat types of conservation priority in Europe.

Table 3. National Set of indicators for the MAES implementation in Greece, their correspondence with the Common International Classification of Ecosystem Services (CICES) sections and codes and their utility for ES and/or EC assessments.

Indicator Group	Indicator Name	CICES Section	CICES Code	ES Indicator	EC Indicator	MAES Framework Requirements (See Table 1)
Biodiversity	Diversity of agro-ecosystems with natural ecosystems (IB1)	Regulating and Maintenance	2.2.2.1, 2.2.2.3, 5.1.2.1, 5.2.2.1	Yes	Yes	1, 3, 8, 9
	Floristic diversity (IB2)		2.2.2.3	Yes	Yes	1, 2, 3, 4, 6, 7, 8, 9
	Micro-refugia of floristic and endemic diversity (IB3)		2.2.2.3	Yes	Yes	1, 2, 3, 7, 8, 9
	Network of crop limits with natural vegetation (IB4)		2.1.2.3, 2.2.2.1, 2.2.2.3, 5.1.2.1	Yes	Yes	1, 3, 8, 9
	Total biodiversity (IB5)		2.2.2.3	Yes	Yes	1, 2, 3, 4, 5, 6, 7, 8, 9
Environmental quality	Change of upper forest limits (IE1)	Regulating and Maintenance	-	No	Yes	1, 2, 3, 5, 7, 8, 9
	Conservation status at various scales (IE2)		-	No	Yes	1, 2, 3, 4, 5, 7, 8, 9
	Forest fires (density) (IE3)		-	No	Yes	1, 3, 5, 7, 8, 9
	Forest fires (frequency) (IE4)		-	No	Yes	1, 3, 5, 7, 8, 9
	Fractional vegetation cover (IE5)				Yes	1, 2, 3, 5, 7, 8, 9
	Natural regeneration (for woodland and forest) (IE6)		2.2.2.3, 2.2.6.1, 5.1.2.1, 5.2.2.1	Yes	Yes	1, 2, 3, 4, 6, 7, 8, 9
	Riparian area alteration (IE7)		2.2.1.3, 2.2.1.5, 2.2.2.3, 5.1.2.1, 5.2.1.2, 5.2.2.1	Yes	Yes	1, 2, 3, 4, 5, 6, 7, 8, 9
	Soil corrosivity (IE8)		2.2.1.1, 2.2.1.2, 2.2.4.1, 2.2.4.2, 5.2.1.1	Yes	Yes	1, 2, 3, 5, 7, 8, 9
	Urban green space (IE9)		2.1.2.2, 2.1.2.3, 2.2.1.4, 2.2.2.2, 2.2.2.3, 2.2.6.2, 5.2.1.3, 5.2.2.1	Yes	Yes	1, 3, 6, 7, 8, 9
	Urban temperature (IE10)		5.2.2.1	Yes	Yes	1, 3, 7, 8, 9
Food, material and energy	Cropland area (IM1)	Provisioning	1.1.1.1	Yes	No	1, 3, 6, 7, 8, 9
	Cropland efficiency (IM2)		1.1.1.1	Yes	Yes	1, 3, 6, 7, 8, 9
	Firewood energy value (IM3)		1.1.6.3	Yes	No	1, 3, 6, 7, 8, 9
	Solar energy (IM4)		4.3.2.4	Yes	No	1, 2, 3, 6, 7, 8, 9
	Water for energy production (IM5)		4.2.1.3	Yes	No	1, 2, 3, 6, 7, 8, 9
	Wind energy (IM6)		4.3.2.3	Yes	No	1, 2, 3, 6, 7, 8, 9

Table 3. *Cont.*

Indicator Group	Indicator Name	CICES Section	CICES Code	ES Indicator	EC Indicator	MAES Framework Requirements (See Table 1)
Forestry	Available firewood (IF1)	Provisioning	1.1.5.3	Yes	Yes	1, 3, 6, 7, 8, 9
	Pasture productivity (IF2)		1.1.5.1	Yes	Yes	1, 3, 6, 7, 8, 9
	Standing volume (IF3)		1.1.5.2	Yes	Yes	1, 3, 6, 7, 8, 9
	Technical wood (IF4)		1.1.5.2	Yes	Yes	1, 3, 6, 7, 8, 9
	Total annual increment (IF5)		1.1.5.2	Yes	Yes	1, 3, 6, 7, 8, 9
	Wood harvest (IF6)		1.1.5.2	Yes	Yes	1, 3, 6, 7, 8, 9
Recreation	National forests (IR1)	Cultural	6.2.2.1	Yes	No	1, 2, 3, 6, 7, 8, 9
	Inland waters rafting and boating (IR2)		6.1.1.1	Yes	No	1, 3, 6, 7, 8
	Organized beaches (IR3)		6.1.2.1	Yes	Yes	1, 3, 6, 7, 8
	Thermal springs (IR4)		6.1.2.1	Yes	No	1, 3, 6, 7, 8
	Trial walking systems (IR5)		6.1.1.1	Yes	No	1, 3, 6, 7, 8
	Visitors' preferences (IR6)		6.1.2.1	Yes	Yes	1, 3, 6, 7, 8
Water resources	Ability to satisfy demand by water use (IW1)	Provisioning	4.2.1.1, 4.2.1.2	Yes	Yes	1, 3, 6, 7, 8
	Chemical condition of surface water (for rivers and lakes) (IW2)	Regulating and Maintenance	2.2.5.1, 5.2.2.1	Yes	Yes	1, 2, 3, 4, 5, 6, 7, 8, 9
	Chemical condition of underground water (IW3)	Regulating and Maintenance	2.2.5.1, 5.2.2.1	Yes	Yes	1, 2, 3, 4, 5, 6, 7, 8, 9
	Demand (total use) (IW4)	Provisioning	4.2.1.1, 4.2.1.2	Yes	Yes	1, 3, 6, 7, 8
	Ecological condition of surface water (for rivers and lakes) (IW5)	Regulating and Maintenance	4.2.2.4_W1	Yes	Yes	1, 2, 3, 4, 5, 6, 7, 8, 9
	Quantity of underground water (IW6)	Provisioning	4.2.2.4_W2	Yes	Yes	1, 3, 6, 7, 8, 9
	Water exploitation index—WEI (IW7)	Provisioning	4.2.2.4_W3	Yes	Yes	1, 3, 6, 7, 8

W1, W2, W3: working code for Greece (following the guidelines of the CICES regarding the creation of additional working codes).

3.2.2. Data Quality and Suitability for Tiers

Due to the methodology applied for the indicators' selection, most indicators are suitable for tier 1 and tier 2 mapping (except for the indicators "Conservation status at various scales", "Demand (total use)" of water, "Water exploitation index—WEI" and "Ability to satisfy demand by water use", which are applicable for national-scale mapping). This is because spatial information for each indicator is provided at the level of the management unit (e.g., available stock for each managed forest class, cultivation and product type by each land parcel, kilometers of trails with georeferenced trace, etc.). However, most of the indicators cannot be used for tier 3 ES assessments and further elaboration (e.g., modelling) or field work is needed (e.g., for biodiversity assessment at the local level). Supplementary Table S1 (a) provides in detail all the available information for the National Set of Indicators and highlights their readiness for each tier-oriented study by a characteristic color coding, (b) simultaneously summarizes all the available knowledge from official designated sources of information and (c) is drafted as a baseline for the future steps of the MAES implementation in Greece.

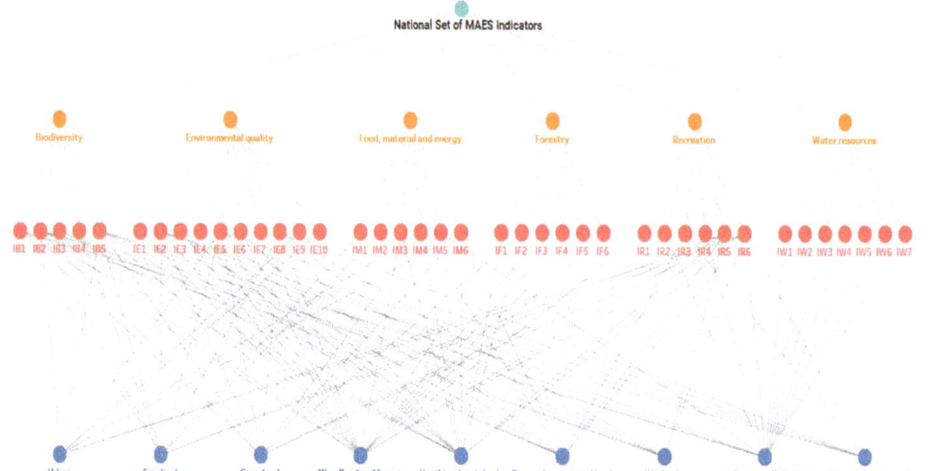

Figure 3. National set of indicators for Greece and their relationship with ecosystem types. IB: Biodiversity indicator, IE: Environmental quality indicator, IM: Food, material and energy indicator, IF: Forestry indicator, IR: Recreation indicator, IW: Water resources indicator (for a detailed abbreviations' reference, e.g., IB1, IB2, etc., see Table 3).

3.2.3. Indicator Development and Implementation

The drafted pre-defined protocols for the development and implementation of an indicator are presented in Supplementary Tables S2 and S3, respectively.

3.3. Thematic Mapping

3.3.1. MAES-Related Dataset Availability and Potential ES Bundles and Hotspots

ES Bundles

The thematic representation of co-existing MAES-related datasets, with spatial reference, at the 10×10 km EEA reference grid-cell level (Figure 4), highlights that the available, ready-for-use information follows a terrain-, relief-oriented pattern, as follows: (i) cells of mountain ranges, especially in the northern Peloponnese and further north, contain information from five or more of the available datasets (yellow to red colors), an area accounting for 751 out of 2215 cells (34% of the total extent). This is due to a lack of typical (timber-oriented) forest management and subsequently, to a lack of the relevant datasets, at the mountainous, forest and woodland areas of north-central Peloponnese and further south. (ii) Plain areas are dominated by cultivated land (designated by the "cropland area" dataset), e.g., western Peloponnese and Thessaly, and represent main agricultural areas of the country (dominated by greenish cells), (iii) island areas (Ionian Islands, Crete and Aegean Islands) are also dominated by cells of four or less MAES-related datasets. The distribution of the cumulative presence of MAES-related data in the grid cells throughout Greece is given in Table 4. These results pinpoint serious data gaps, since only twelve datasets are available to be directly used for the national scale assessment. The most complete and detailed datasets are those for cropland and forestry. The detailed information at each cell is provided in Supplementary Table S4.

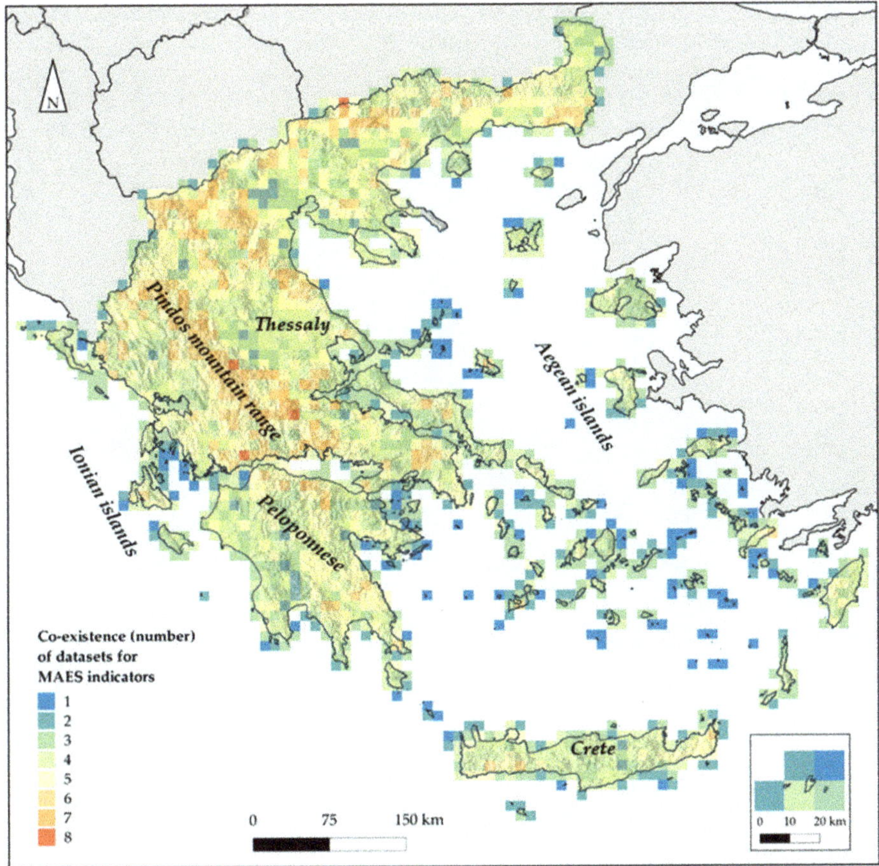

Figure 4. Ecosystem services bundles' map of Greece: a thematic representation of the co-existence of ES indicators at 10 × 10 km EEA reference grid cell level. Indicators taken into account and rated: (i) forest management studies, (ii) wind energy stations, (iii) hydroelectric energy stations, (iv) solar energy stations, (v) national forests, (vi) Natura 2000 areas and/or wildlife refuges, (vii) "blue flag" beaches, (viii) thermal springs, (ix) inland waters boating and rafting, (x) mountain shelters, (xi) cropland area and (xii) water resources.

Table 4. Distribution of the cumulative presence of MAES-related data in the terrestrial 10 × 10 European Environment Agency (EEA) reference grid cells of Greece.

Number of MAES-Related Datasets	Number of Corresponding Cells	% of Total Cell Number
0	14	0.63
1	98	4.42
2	166	7.49
3	474	21.41
4	712	32.14
5	482	21.76
6	195	8.80
7	69	3.12
8	5	0.23
Total	**2215**	**100**

ES bundling regions of Greece were also identified by analyzing the representativity of the various cell categories (i.e., cells rated from 0 to 6, representing the utility of zero to six ecosystem indicator groups) in each of the 13 administrative regions of Greece (NUTS 2 level) (i.e., Attica, North Aegean, South Aegean, Crete, Eastern Macedonia and Thrace, Central Macedonia, Western Macedonia, Epirus, Thessaly, Ionian Islands, Western Greece, Central Greece and Peloponnese). By this, for each indicator group, readiness-data-gap maps have been drafted for the MAES implementation in Greece (Figure 5), as well as for their cumulative importance (Figure 6). These maps integrate each region's area as a parameter. In Figure 6, pie slices represent the percentage (%) of EEA reference grid cells that include data for the application of zero, one, or more (numbers on the map) MAES indicator groups in each region of Greece. Pie size represents the relative percentage weighted by the NUTS 2 area size (i.e., number of 10 × 10 EEA reference grid cells included in the region). Grey tones also represent regions' area size (i.e., darker = bigger, lighter = smaller).

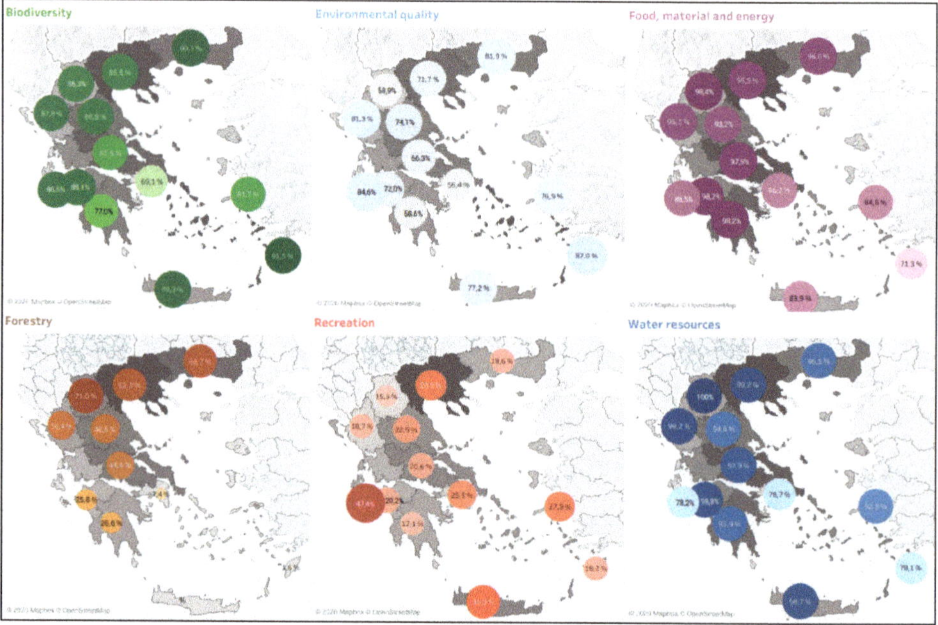

Figure 5. Readiness map for the MAES implementation in Greece, regarding the indicator groups of (i) Biodiversity, (ii) Environmental quality, (iii) Food material and Energy, (iv) Forestry, (v) Recreation and (vi) Water resources. Pie size represents the percentage (%) of EEA reference grid cells that include data for the application of each indicator group in each region of Greece (NUTS 2). Greyscale shading per region refers to the number of cells in each region with data of each indicator group (darker shades represent regions with a higher total number of cells with relevant data). Data gaps for the particular indicators can be inferred from both percentage and relative shadedness. This combination provides a proxy of relative "readiness" for MAES studies.

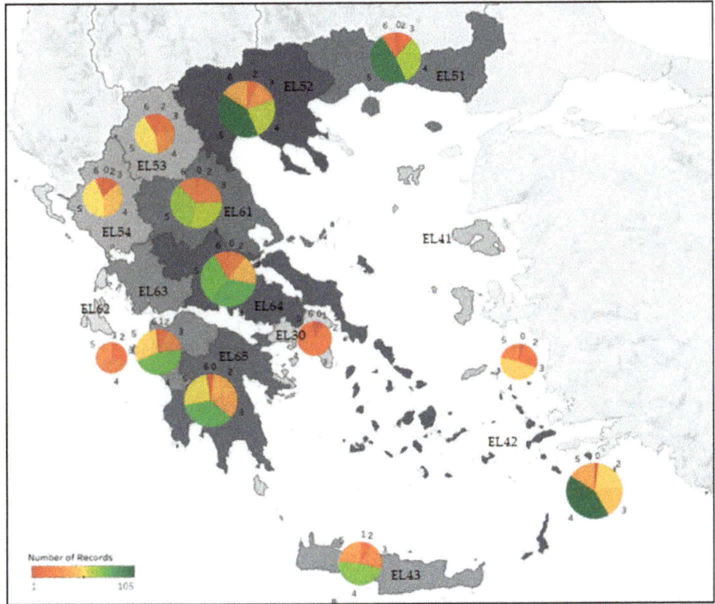

Figure 6. Readiness map for the MAES implementation in Greece. Pie slices represent the percentage of EEA reference grid cells that include data for the application of zero, one, or more MAES indicators (pie slice numbers) in each region of Greece (NUTS 2). Pie size represents the relative percentage weighted by NUTS 2 area size (i.e., number of 10 × 10 EEA reference grid cells included in the region). Grayscale shading represents the summation of the number of one or more indicator groups' data within the regions' grid cells. NUTS 2 Regions: Attica (EL30), North Aegean (EL41), South Aegean (EL42), Crete (EL43), Eastern Macedonia and Thrace (EL51), Central Macedonia (EL52), Western Macedonia (EL53), Epirus (EL54), Thessaly (EL61), Ionian Islands (EL62), Western Greece (EL63), Central Greece (EL64), Peloponnese (EL65).

The results of these maps suggest the following for the different indicator groups:

a. Biodiversity: most administrative regions are covered by an adequate number of biodiversity-related datasets (their majority is rated above 80%). The regions of South Aegean, East Macedonia and Thrace and Crete are covered at highest rates (91.5%, 90.7% and 89.3%, respectively). The region of Attica, hosting the metropolitan center and capital of Greece—Athens, is in the last position of the classification, with 69.1% of its area covered by biodiversity-related datasets. Overall, area-size differentiations are not of significant importance.

b. Environmental quality: in most regions (nine out of thirteen), more than 70% of their surface is covered with data relevant to environmental quality. The region of Attica, including Athens, ranks in the last position of the list. Area-size differentiations are not of significant importance.

c. Food, material and energy: all regions are considered to have adequate spatial coverage regarding the available datasets for food, material and energy. Area-size differentiations are not of significant importance.

d. Forestry: regions of northern Greece, Western Macedonia, East Macedonia and Thrace and Central Macedonia are the first in the list with 71.0%, 63.7% and 62.2% cover of their surface, respectively. The Metropolitan region of Attica and island regions are placed last in the list: Attica 7.4%, South Aegean 1.6%, North Aegean and the Ionian Islands at 0%. Area size is considered important and follows the pattern of area covered by the datasets.

e.   Recreation: most areas covered by recreation-related data are present in the Ionian Islands (47.4%), Crete (34.9%) and Central Macedonia (29.5%). Regions' area size is important; for example, for the Ionian Islands region, it is highlighted that 47.4% of the area cover is of high importance regarding the relatively small area of the region.

f.   Water resources: all regions are relatively well covered by water resources-related datasets and this is due mainly to the recent monitoring and assessment networks established for the EU Water Framework Directive. Although not apparent at this regional scale analysis, some local fine-scale data is still poorly compiled or surveyed, due to the country's high geographical fragmentation (many small river basins and small isolated wetlands). Area-size differentiations among regions are not of significant importance.

Central Macedonia, South Aegean and East Macedonia and Thrace are depicted (Figure 6) as the regions with larger areas, where, simultaneously, more than 4 indicators can be assessed. Regions of Central Greece, Peloponnese, Western Greece and Crete, follow. Epirus, North Aegean and Ionian Islands are depicted as of low importance in comparison with other regions, when considering areal extent (ha) for interpretation purposes. Area size is important for the interpretation of the results and highlights that larger area (e.g., Central Macedonia, Central Greece) is not a parameter linked to the available data diversity (e.g., Epirus, Ionian Islands, Attica).

ES Hotspots

Screening of the spatial distribution and density of the available datasets revealed areas which can be considered as ES hotspots. The following results were pinpointed for the three major ES categories (CICES sections):

a.   Provisioning services: (i) 39% of the terrestrial area is covered by agricultural areas, Thessaly, Western Greece, Central Macedonia, East Macedonia and Thrace and Crete can be considered as hotspots of agriculture-related provisioning services, (ii) approximately 15% of the country is under forest management control for timber and other forest products: Pindos, Rhodopi and Evros mountain ranges are considered as hotspots, (iii) most wind parks are developed in Central Greece, followed by Crete and East Macedonia and Thrace, and a local hotspot can be identified in Southern Evvoia (Central Greece), (iv) proposed and developed hydroelectric power plants from small dams hotspots are located throughout the mountain range of Pindos and continue southwards to the mountains of Northern Peloponnese: secondary hotspots are located lengthwise of north Greek borders, (v) a hotspot of solar energy production is identified at the surrounding area of Ptolemaida (Western Macedonia), while most solar energy units are developed in Thessaly and Central Greece.

b.   Regulating and maintenance services: Most high-altitude mountains of Greece, wetland areas, lakes and numerous coastal areas can be perceived as hotspots for regulating and maintenance services. Although some regions are characterized by relative higher proportional area coverage by Natura 2000 sites and wildlife refuges than others (e.g., East Macedonia and Thrace versus Attica, respectively), the spatial distribution pattern of these hotspot areas designates them as of equal importance for every region.

c.   Cultural services: Using the currently limited number of indicators the spatial pattern of the related information pinpoints Central Macedonia, Crete and the Ionian Islands as hotspot regions for cultural ecosystem services, hosting a dense concentration of high-quality mountain and coastal areas for recreational and spiritual activities/interactions (i.e., identified by a necessarily limited number of proxy indicators, being mountain shelters, "blue flag" beaches, thermal springs, fresh-water sports). This initial review shows that the restricted number of country-wide Cultural Ecosystem Services indicators should be increased and further developed.

Supplementary Figure S1 presents the thematic representation of the above-mentioned spatial distribution of ES-related datasets, for the hotspot identification.

Relevance to Policies

The screening assessment to identify relevance among the adopted indicators and national and EU strategies and policies came up with the following results (Figures 7 and 8):

a.   Biodiversity strategy: 21 relevant indicators, including 5 from Biodiversity, 8 from Environmental quality, 1 from Forestry, 4 from Recreation and 3 from Water resources groups.
b.   Forest strategy: 25 relevant indicators, including 3 from Biodiversity, 7 from Environmental quality, 1 from Food, material and energy, 6 from Forestry, 4 from Recreation and 4 from Water resources groups.
c.   Regional policy: 19 relevant indicators, including 7 from Environmental quality, 2 from Food, material and energy, 5 from Recreation and 5 from Water resources groups.
d.   Agricultural policy: 21 relevant indicators, including 4 from Biodiversity, 4 from Environmental quality, 4 from Food, material and energy, 1 from Forestry, 1 from Recreation and 7 from Water resources groups.
e.   Climate policy: 26 relevant indicators, including 2 from Biodiversity, 9 from Environmental quality, 4 from Food, material and energy, 4 from Forestry, 2 from Recreation and 5 from Water resources groups.
f.   EU Green Deal: All (40) proposed indicators are relevant to the EU Green Deal policy. However, specific relevance is present among indicators and the thematic targets of the EU Green Deals.

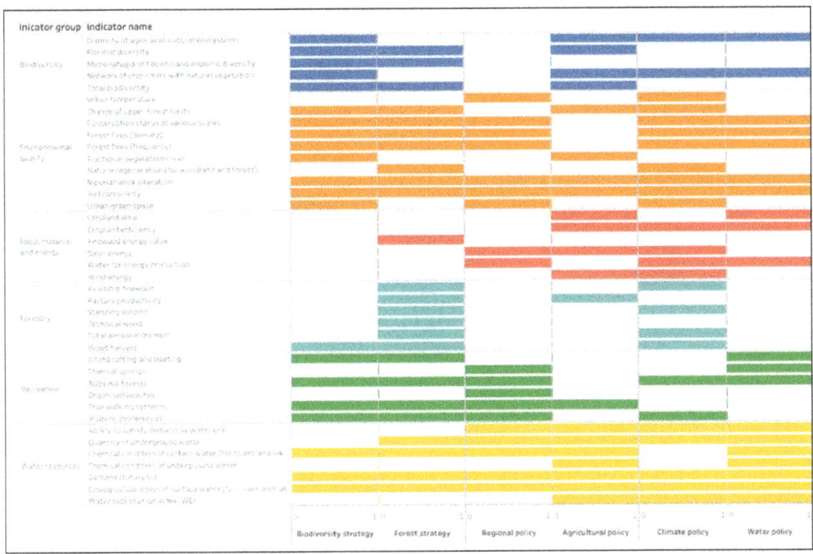

Figure 7. Correspondence matrix among indicators and national policies.

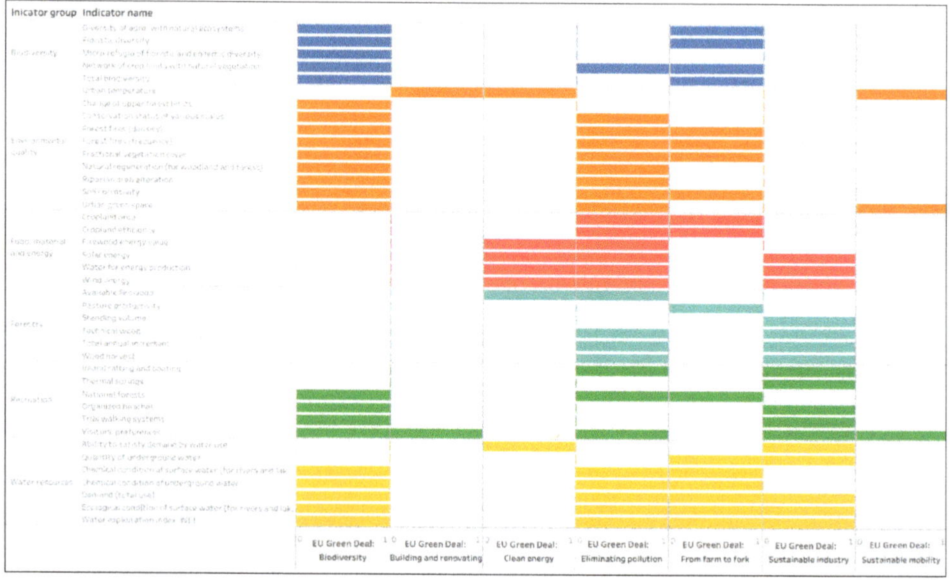

Figure 8. Correspondence matrix among indicators and the EU Green deal.

4. Discussion

Inventory, classification and standardized description of ES is the basis for any effort to measure, map or value them, and it is also the basis of being transparent regarding our methods and findings, so that we may effectively communicate and critique results [66]. However, the sheer number of ES and the correspondingly high number of indicators and large amounts of data needed for their assessment and valuation requires a pragmatic approach, such as the one proposed in the MAES indicators report, which focuses on using indicators supported by the available data [33]. At the same time, other criteria, such as validity and policy relevance, also need to be fulfilled regardless of data availability: ignore these and the usefulness of the indicators is seriously compromised [37,38,67,68]. Heink et al. [33] claim that (a) it is essential to show how indicators are related to goals and to conceptual frameworks and (b) the validity and relevance of indicators is just as important as data availability, as well as to the extent that the policy makers understand them.

The present study complies with the above-mentioned requirements and follows the CSLF (Credibility, Salience, Legitimacy, Feasibility) approach, which provides criteria for developing ecosystem service indicators [69]. Feasibility is of high importance in national ES assessments, since MS should (a) rapidly proceed with the baseline MAES status in their territory, (b) identify issues of importance regarding data gaps, management practices, trade-offs and/or sector policy conflicts and (c) move forward to detailed management-oriented and applied studies. Using available and feasible indicators and identifying data gaps avoids "paralysis by analysis" situations; taking such initiatives to apply indicators in practice also requires studying data gaps, risks, cultural idiosyncrasies and possible dysfunctions involved [70].

Our proposed MAES indicator framework does have some limitations since we are at the beginning of its state-wide development in Greece. Some of the available data are based on an uneven sampling distribution, and there are data-scarcity and data-consistency issues in many areas [34,41,71]; for example, higher quality information is available for some high-profile protected areas relative to the smaller and newer ones (e.g., Reference [72]), and this is broadly evident in our ecosystem bundles analysis (Figure 4). Moreover, work in the marine sector cannot at this time be fully integrated in this assessment framework since heterogeneity in data requirements, wide knowledge gaps and methodological limitations exist [34,73]. Although there are numerous challenges in applying policy-relevant ecosystems services with standardized methods in marine ecosystems, these problems are actively being explored [74–76]. Especially, in the Mediterranean MS, assessment approaches require careful considerations due to the diversity and intricacy of natural variability and cultural complexities [77]. A particular challenge among the ES types are Cultural Ecosystem Services, especially their streamlined assessment at the national and regional scales [78,79]. In our proposed MAES application, these are treated primarily within recreational values since it is difficult to survey other cultural services with any consistency on a state-wide scale (e.g., Reference [21]). CES are important in providing various methods to engage stakeholder and local applications (where conflict areas may also be identified) (e.g., References [80,81]). Obviously, we do not mean to degrade or denigrate other types of cultural ecosystem services [82,83], and important aspects such as landscape quality, including aesthetics and other landscape-scale attributes, are being actively investigated (e.g., References [84–86]). Many new methods and tools for landscape-scale analyses (e.g., References [87,88]) and approaches incorporating assessments of non-material benefits to societal well-being are being developed and new indicators should be incorporated in our adaptive national framework in the near future.

Despite some limitations, based on the results of the study, we consider that the available data in Greece are appropriate for national scale MAES studies (tier 3 and tier 2) and are sufficiently informative to support decision-making for many forms of natural resources management. Especially within the Natura 2000 sites (27.5% of the terrestrial territory), data availability, administrative support and human resources are considered as of high capacity and the potential is high to prioritize the implementation of MAES studies within protected areas in the near future. In the frame of the LIFE-IP 4 Natura Project, detailed MAES studies will be implemented in selected sites, and the proposed indicators can be modified or altered to comply with local-level needs [89]. Also, due to a variety of recently applied policy-relevant monitoring programs promoted in Greece, such as water body ecological quality monitoring applications [90,91], the quality and quantity of perpetually updated data should rapidly increase. The drafted set of indicators and their spatial representation within an adaptive framework provide an adequate guide to support case-study sites' selection and future ameliorations of assessment accuracy and precision.

Moreover, it was considered important to develop a standardized reporting methodology to EU authorities based on the 10 × 10 km EEA reference grid. Indicators' thematic maps on ES, data gaps and compliance with policy-related requirements produced in this study serve as an appropriate way of reporting the MS status on ES. Using the common ES coding provided by the CICES system, relevant maps at the EU level can be produced and in this way, we can identify ES bundles and hotspot areas across EU territory.

The next steps should be to conduct a nation-wide field assessment of ES and EC, using a web-based platform that has been developed by the University of Patras, for the LIFE-IP 4 NATURA project [92]. Plots within each EAA reference grid cell and for each ecosystem type are collected and this plots' dataset will: (a) support data validation and update, (b) thematic representation of ES actual and potential supply, based on field data (c) spatial trim literature ES supply and demand data and (d) assist the identification of ecosystems in degraded condition. This action, combined with the National Set of Indicators assessments, is expected to provide the best available knowledge for ecosystem types and their services in Greece and thus support decision- and policy-making with robust scientific information. Elaborated trade-off analysis is also needed and highlighted as a next step for the LIFE-IP 4 NATURA project, at least at the national scale. Fully developed trade-off analyses at this preliminary stage of MAES applications should proceed with extra care, since some types of ES, such as cultural services, may not be adequately represented by the current indicator set in all ecosystem types. However, displaying ES indicators makes trade-offs explicit and this may help facilitate management plans and project planning decisions; also, through stakeholder participation [28], this approach should promote sound value judgments at national and regional scales.

To summarize, drafting and implementing a National Set of Indicators in each MS is crucial at the country, as well as at the EU level, and provides (complementary to MAES): (a) assistance for implementing other Actions of the EU Biodiversity Strategy, (b) guidance on how to use information on ecosystem services in impact assessments or for policies needs and (c) the existing link of biodiversity and ecosystem condition to ecosystem services and human well-being. However, this set of indicators is not a "passe-partout" for the MAES implementation, and modifications and alterations should be welcome, assessed and considered for use at different scales. In fact, and as stated by Costanza et al. [28], "there is not one right way to assess and value ecosystem services. There is, however, a wrong way, that is, not to do it at all". We consider this must be an adaptive process that should involve incremental steps towards evolution, amelioration and frequent review of the efficiency, consistency and usefulness of the first set of chosen indicators.

5. Conclusions

This study introduced the National Set of MAES indicators for Greece, a set of 40 indicators, aiming to support and promote ES and EC assessments throughout the Greek territory. The methodology complies with EU MAES frameworks and the ES classification system and through this, contributes to standardized ES reporting purposes at the MS level. This work is a synthesis of a multidisciplinary review of the available datasets in Greece, evaluated by experts representing national authorities, various university departments and research centers; thus, ensuring administrative validity and a broad scientific background. The results presented at the national and the regional level (NUTS 2 administrative regions) of Greece, highlight areas where: (a) MAES studies can be directly applied by implementing the proposed indicators and (b) MAES preparatory actions are needed (e.g., field surveys, data collection from non-digital/non-digitized sources, modelling) due to the explored data gaps. Finally, this set of indicators forms the official national basis on which future studies will be conducted for MAES reporting and implementation towards supporting the targets of the National and the EU Strategies and Policies.

Supplementary Materials: The following are available online at http://www.mdpi.com/1999-4907/11/5/595/s1: Figure S1: Thematic representation of the available, ES-related datasets, for the hotspot identification. Table S1: Correlations among the proposed national set of MAES indicators, ecosystem types, scale adequacy, tiers, EU Green Deal targets, National Strategies and Policies—data sources are also provided. Table S2: Pre-defined indicator development protocol. Table S3: Pre-defined indicator implementation protocol. Table S4: Correlation matrix among EEA 10 × 10 reference grid cells, NUTS 2 regions, general group of indicators, EU Green Deal Targets, National Strategies and Policies.

Author Contributions: Conceptualization, I.P.K. and P.D.; methodology, I.P.K., G.M. and P.D.; software, I.P.K.; validation, I.P.K, E.S.B., G.M., V.V., S.Z. and P.D.; formal analysis, I.P.K., G.M., E.S.B., I.C. and P.D.; investigation, I.P.K., G.M., E.S.B., V.V., S.Z. and P.D.; resources, I.P.K., G.M., I.M. and P.D.; data curation, I.P.K., G.M., E.S.B., V.V. and I.C.; writing—original draft preparation, I.P.K., G.M., E.S.B., V.V. and P.D.; writing—review and editing, I.P.K., G.M., V.V., S.Z. and P.D.; visualization, I.P.K.; supervision, I.P.K. and P.D.; project administration, I.P.K. and P.D.; funding acquisition, I.M., G.M. and P.D. All authors have read and agreed to the published version of the manuscript.

Funding: This research was funded by the European Commission LIFE Integrated Project, LIFE-IP 4 NATURA "Integrated Actions for the Conservation and Management of Natura 2000 sites, species, habitats and ecosystems in Greece", Grant Number: LIFE 16 IPE/GR/000002.

Acknowledgments: The study was conducted within the framework of the National Mapping and Assessment of Ecosystem and their Services project, under the LIFE-IP 4 NATURA project. We would like to thank the Hellenic Ministry of the Environment and Energy and the Hellenic Ministry of Rural Development and Food, for the provision of nation-wide datasets for forestry, agriculture and mapping material. We also thank Constantia Patelodimou for all the assistance in this work, especially on database handling and visualizations under the Tableau online platform.

Conflicts of Interest: The authors declare no conflict of interest. The funder had no role in the design of the study; in the collection, analyses, or interpretation of data; in the writing of the manuscript, or in the decision to publish the results.

References

1. European Commission. *Our Life Insurance, Our Natural Capital: An EU Biodiversity Strategy to Communication from the Commission to the European Parliament, the Council the Economic and Social Committee and the Committee of the Regions*; Directorate-General for Environment: Brussels, Belgium, 2011.
2. Maes, J.; Teller, A.; Erhard, M.; Liquete, C.; Braat, L.; Berry, P.; Egoh, B.; Puydarrieus, P.; Fiorina, C.; Santos, F.; et al. *Mapping and Assessment of Ecosystem and Their Services. An Analytical Framework for Ecosystem Assessments under Action 5 of the EU Biodiversity Strategy to 2020*; Publications Office of the European Union: Luxemburg, 2013; ISBN 9789279293696.
3. Haines-Young, R.; Paterson, J.; Potschin, M.; Wilson, A.; Kass, G. *The UK NEA Scenarios: Development of Storylines and Analysis of Outcomes*; The UK National Ecosystem Assessment Technical Report; United Nations Environment Programme World Conservation Monitoring Centre (UNEP-WCMC): Cambridge, UK, 2011.
4. Spanish National Ecosystem Assessment. *Ecosystems and Biodiversity for Human Wellbeing: Synthesis of Key Findings*; Ministerio de Agricultura, Alimentación y Medio Ambiente: Madrid, Spain, 2014.
5. Bratanova-Doncheva, S.; Chipev, N.; Gocheva, K.; Stoyan, V.; Fikova, R. Methodological framework for assessment and mapping of ecosystem condition and ecosystem services in Bulgaria. In *Conceptual Bases and Principles of Application*; Digital Illusions: Clorindm, Argentina, 2017; ISBN 978-619-7379-21-1.
6. Grunewald, K.; Herold, H.; Marzelli, S.; Meinel, G.; Richter, B.; Syrbe, R.-U.; Walz, U. Assessment of ecosystem services at the national level in Germany—Illustration of the concept and the development of indicators by way of the example wood provision. *Ecol. Indic.* **2016**, *70*, 181–195. [CrossRef]
7. Grunewald, K.; Syrbe, R.U.; Walz, U.; Richter, B.; Meinel, G.; Herold, H.; Marzelli, S. Germany's Ecosystem Services—State of the Indicator Development for a Nationwide Assessment and Monitoring. *One Ecosyst.* **2017**, *2*, e14021. [CrossRef]
8. Henke, J.M.; Petropoulos, G.P. A GIS-based exploration of the relationships between human health, social deprivation and ecosystem services: The case of Wales, UK. *Appl. Geogr.* **2013**, *45*, 77–88. [CrossRef]
9. Palomo, I.; Martín-López, B.; Zorrilla-Miras, P.; Del Amo, D.G.; Montes, C. Deliberative mapping of ecosystem services within and around Doñana National Park (SW Spain) in relation to land use change. *Reg. Environ. Chang.* **2013**, *14*, 237–251. [CrossRef]
10. Cortinovis, C.; Geneletti, D. Ecosystem services in urban plans: What is there, and what is still needed for better decisions. *Land Use Policy* **2018**, *70*, 298–312. [CrossRef]
11. Burkhard, B.; Maes, J.; Potschin-Young, M.; Santos-Martín, F.; Geneletti, D.; Stoev, P.; Kopperoinen, L.; Adamescu, C.; Esmail, B.A.; Arany, I.; et al. Mapping and assessing ecosystem services in the EU—Lessons learned from the ESMERALDA approach of integration. *One Ecosyst.* **2018**, *3*, e29153. [CrossRef]
12. Burkhard, B.; Sapundzhieva, A.; Kuzmova, I.; Maes, J.; Geneletti, D.; Adem Esmail, B.; Potschin-Young, M.; Santos-Martin, F.; Mulders, S.; Kopperoinen, L.; et al. *Action 5 Implementation Plan. Deliverable D1.7 EU Horizon 2020 ESMERALDA Project*; Grant agreement No. 642007; Leipzig University: Hannover, Germany, 2018.
13. Burkhard, B.; Maes, J. (Eds.) *Mapping Ecosystem Services*; Pensoft Publishers: Sofia, Bulgaria, 2017.
14. ESMERALDA MAES Explorer. Available online: http://www.maes-explorer.eu (accessed on 12 April 2020).
15. Santos-Martin, F.; Viinikka, A.; Mononen, L.; Brander, L.; Vihervaara, P.; Liekens, I.; Potschin-Young, M. Creating an operational database for Ecosystems Services Mapping and Assessment Methods. *One Ecosyst.* **2018**, *3*, e26719. [CrossRef]
16. Burkhard, B.; Santos-Martín, F.; Nedkov, S.; Maes, J. An operational framework for integrated Mapping and Assessment of Ecosystems and their Services (MAES). *One Ecosyst.* **2018**, *3*, e22831. [CrossRef]
17. Grêt-Regamey, A.; Weibel, B.; Kienast, F.; Rabe, S.-E.; Zulian, G. A tiered approach for mapping ecosystem services. *Ecosyst. Serv.* **2015**, *13*, 16–27. [CrossRef]
18. Albert, C.; Bonn, A.; Burkhard, B.; Daube, S.; Dietrich, K.; Engels, B.; Frommer, J.; Götzl, M.; Grêt-Regamey, A.; Job-Hoben, B.; et al. Towards a national set of ecosystem service indicators: Insights from Germany. *Ecol. Indic.* **2016**, *61*, 38–48. [CrossRef]
19. Maes, J.; Egoh, B.; Willemen, L.; Liquete, C.; Vihervaara, P.; Schägner, J.P.; Grizzetti, B.; Drakou, E.; La Notte, A.; Zulian, G.; et al. Mapping ecosystem services for policy support and decision making in the European Union. *Ecosyst. Serv.* **2012**, *1*, 31–39. [CrossRef]

20. Mononen, L.; Auvinen, A.-P.; Ahokumpu, A.-L.; Rönkä, M.; Aarras, N.; Tolvanen, H.; Kamppinen, M.; Viirret, E.; Kumpula, T.; Vihervaara, P. National ecosystem service indicators: Measures of social–ecological sustainability. *Ecol. Indic.* **2016**, *61*, 27–37. [CrossRef]

21. Tratalos, J.; Haines-Young, R.; Potschin, M.; Fish, R.; Church, A. Cultural ecosystem services in the UK: Lessons on designing indicators to inform management and policy. *Ecol. Indic.* **2016**, *61*, 63–73. [CrossRef]

22. Albert, C.; Hauck, J.; Buhr, N.; Von Haaren, C. What ecosystem services information do users want? Investigating interests and requirements among landscape and regional planners in Germany. *Landsc. Ecol.* **2014**, *29*, 1301–1313. [CrossRef]

23. Albert, C.; Galler, C.; Hermes, J.; Neuendorf, F.; Von Haaren, C.; Lovett, A. Applying ecosystem services indicators in landscape planning and management: The ES-in-Planning framework. *Ecol. Indic.* **2016**, *61*, 100–113. [CrossRef]

24. Hauck, J.; Schweppe-Kraft, B.; Albert, C.; Görg, C.; Jensen, R.; Fürst, C.; Maes, J.; Ring, I.; Hönigová, I.; Burkhard, B.; et al. The Promise of the Ecosystem Services Concept for Planning and Decision-Making. *GAIA Ecol. Perspect. Sci. Soc.* **2013**, *22*, 232–236. [CrossRef]

25. Hayek, U.W.; Teich, M.; Klein, T.; Grêt-Regamey, A. Bringing ecosystem services indicators into spatial planning practice: Lessons from collaborative development of a web-based visualization platform. *Ecol. Indic.* **2016**, *61*, 90–99. [CrossRef]

26. Chee, Y.E. An ecological perspective on the valuation of ecosystem services. *Biol. Conserv.* **2004**, *120*, 549–565. [CrossRef]

27. McCauley, D.J. Selling out on nature. *Nature* **2006**, *443*, 27–28. [CrossRef]

28. Costanza, R.; De Groot, R.; Braat, L.; Kubiszewski, I.; Kubiszewski, I.; Sutton, P.; Farber, S.; Grasso, M. Twenty years of ecosystem services: How far have we come and how far do we still need to go? *Ecosyst. Serv.* **2017**, *28*, 1–16. [CrossRef]

29. Russell-Smith, J.; Lindenmayer, D.; Kubiszewski, I.; Green, P.; Costanza, R.; Campbell, A. Moving beyond evidence-free environmental policy. *Front. Ecol. Environ.* **2015**, *13*, 441–448. [CrossRef]

30. De Groot, R.; Fisher, B.; Christie, M.; Aronson, J.; Braat, L.; Gowdy, J.; Haines-Young, R.; Maltby, E.; Neuville, A.; Polasky, S.; et al. Integrating the ecological and economic dimensions in biodiversity and ecosystem service valuation. In *The Economics of Ecosystems and Biodiversity: Ecological and Economic Foundations*; Earthscan: London, UK; Washington, DC, USA, 2012; ISBN 9781849775489.

31. Braat, L.C.; De Groot, R. The ecosystem services agenda: bridging the worlds of natural science and economics, conservation and development, and public and private policy. *Ecosyst. Serv.* **2012**, *1*, 4–15. [CrossRef]

32. Maes, J.; Teller, A.; Erhard, M.; Liquete, C.; Braat, L.; Berry, P.; Egoh, B.; Puydarrieux, P.; Fiorina, C.; Santos, F. *Mapping and Assessment of Ecosystems and Their Services. Indicators for Ecosystem Assessments under Action 5 of the EU Biodiversity Strategy to 2020*; Publications office of the European Union: Luxemburg, 2014; ISBN 9789279361616.

33. Heink, U.; Hauck, J.; Jax, K.; Sukopp, U. Requirements for the selection of ecosystem service indicators – The case of MAES indicators. *Ecol. Indic.* **2016**, *61*, 18–26. [CrossRef]

34. Dimopoulos, P.; Drakou, E.; Kokkoris, I.; Katsanevakis, S.; Kallimanis, A.; Tsiafouli, M.; Bormpoudakis, D.; Kormas, K.; Arends, J. The need for the implementation of an Ecosystem Services assessment in Greece: Drafting the national agenda. *One Ecosyst.* **2017**, *2*, e13714. [CrossRef]

35. Smart, J.J.C.; Hempel, C.G. Foundations of the Unity of Science. Volume II, no. 7: Fundamentals of Concept Formation in Empirical Science. *Philos. Rev.* **1953**, *62*, 473. [CrossRef]

36. Williams, D.G.; Babbie, E.R. The Practice of Social Research. *Contemp. Sociol. A J. Rev.* **1976**, *5*, 163. [CrossRef]

37. Heink, U.; Kowarik, I. What criteria should be used to select biodiversity indicators? *Biodivers. Conserv.* **2010**, *19*, 3769–3797. [CrossRef]

38. Maes, J.; Teller, A.; Erhard, M.; Grizzetti, B.; Barredo, J.I.; Paracchini, M.L.; Condé, S.; Somma, F.; Orgiazzi, A.; Jones, A.; et al. *Mapping and Assessment of Ecosystems and their Services: An Analytical Framework for Ecosystem Condition*; Publications office of the European Union: Luxemburg, 2018; ISBN 978-92-79-74288-0.

39. Kokkoris, I.P.; Bekri, E.S.; Skuras, D.; Vlami, V.; Zogaris, S.; Maroulis, G.; Dimopoulos, D.; Dimopoulos, P. Integrating MAES implementation into protected area management under climate change: A fine-scale application in Greece. *Sci. Total. Environ.* **2019**, *695*, 133530. [CrossRef]

40. Hatziiordanou, L.; Fitoka, E.; Hadjicharalampous, E.; Votsi, N.; Palaskas, D.; Malak, D. Indicators for mapping and assessment of ecosystem condition and of the ecosystem service habitat maintenance in support of the EU Biodiversity Strategy to 2020. *One Ecosyst.* **2019**, *4*, e32704. [CrossRef]

41. Vlami, V.; Kokkoris, I.P.; Zogaris, S.; Cartalis, C.; Kehayias, G.; Dimopoulos, P. Cultural landscapes and attributes of "culturalness" in protected areas: An exploratory assessment in Greece. *Sci. Total. Environ.* **2017**, *595*, 229–243. [CrossRef]

42. Kokkoris, I.P.; Drakou, E.; Maes, J.; Dimopoulos, P. Ecosystem services supply in protected mountains of Greece: Setting the baseline for conservation management. *Int. J. Biodivers. Sci. Ecosyst. Serv. Manag.* **2017**, *14*, 45–59. [CrossRef]

43. Kokkoris, I.; Dimopoulos, P.; Xystrakis, F.; Tsiripidis, I. National scale ecosystem condition assessment with emphasis on forest types in Greece. *One Ecosyst.* **2018**, *3*, e25434. [CrossRef]

44. Nikolaidou, C.; Votsi, N.-E.; Sgardelis, S.; Halley, J.; Pantis, J.; Tsiafouli, M. Ecosystem Service capacity is higher in areas of multiple designation types. *One Ecosyst.* **2017**, *2*, e13718. [CrossRef]

45. Hellenic Ecosystem Services Partnership (HESP). Available online: https://www.es-partnership.org/community/regional-chapters/south-east-europe/greece-hesp/ (accessed on 12 April 2020).

46. LIFE-IP 4 NATURA. Available online: https://edozoume.gr/en/ (accessed on 12 April 2020).

47. United Nations. European Commission; Food and Agricultural Organization of the United Nations; Organization for Economic Co-operation and Development. In *World Bank System of Environmental-Economic Accounting 2012: Experimental Ecosystem Accounting*; United Nations: New York, NY, USA, 2014; ISBN 9789210559263.

48. Ecosystem Types of Europe. Available online: https://www.eea.europa.eu/data-and-maps/data/ecosystem-types-of-europe-1 (accessed on 12 April 2020).

49. CORINE Land Cover. Available online: https://land.copernicus.eu/pan-european/corine-land-cover (accessed on 12 April 2020).

50. *Interpretation Manual of European Union Habitats–EUR28*; European Commission, DG Environment: Brussels, Belgium, 2013; ISBN HAB 96/2 FINAL—EN Version EUR 15.

51. Dimopoulos, P.; Bergmeier, E.; Theodoropoulos, K.; Fischer, P.; Tsiafouli, M. *Monitoring Guide for Habitat types and Plant Taxa in Natura 2000 Sites with Management Bodies*; University of Ioannina, Ministry of the Environment: Agrinio, Greece, 2005; ISBN 960-233-168-2.

52. Development of Large Scale (1:5000) Spatial Data Infrastructure for Terrestrial Areas Protected under the «Natura 2000» Network at a National Scale 2016. Available online: http://www.ypeka.gr/LinkClick.aspx?fileticket=txXACYhOLPI%3d&tabid=37&language=el-GR (accessed on 10 April 2018).

53. Linkages of Species and Habitat Types to MAES Ecosystems. Available online: https://www.eea.europa.eu/data-and-maps/data/linkages-of-species-and-habitat#tab-metadata (accessed on 12 April 2020).

54. The European Green Deal. Communication from the Commission to the European Parliament, the European Council, the Council, the European Economic and Social Committee and the Committee of the Regions. Brussels. 2019. Available online: https://ec.europa.eu/info/sites/info/files/european-green-deal-communication_en.pdf (accessed on 12 April 2020).

55. *National Biodiversity Strategy and Action Plan*; Hellenic Ministry of Environment and Climate Change: Athens, Greece, 2014; ISBN 978-960-7284-33-4.

56. Smith, A.; Harrison, P.; Soba, M.P.; Archaux, F.; Blicharska, M.; Egoh, B.; Erős, T.; Domenech, N.F.; György, Á.I.; Haines-Young, R.; et al. How natural capital delivers ecosystem services: A typology derived from a systematic review. *Ecosyst. Serv.* **2017**, *26*, 111–126. [CrossRef]

57. Pouso, S.; Uyarra, M.C.; Borja, A. Recreational fishers' perceptions and behaviour towards cultural ecosystem services in response to the Nerbioi estuary ecosystem restoration. *Estuarine, Coast. Shelf Sci.* **2018**, *208*, 96–106. [CrossRef]

58. Grizzetti, B.; Liquete, C.; Pistocchi, A.; Vigiak, O.; Zulian, G.; Bouraoui, F.; De Roo, A.; Cardoso, A. Relationship between ecological condition and ecosystem services in European rivers, lakes and coastal waters. *Sci. Total. Environ.* **2019**, *671*, 452–465. [CrossRef]

59. Haines-Young, R.; Potschin, M. CICES Towards a Common Classification of Ecosystem Services. Available online: https://cices.eu/ (accessed on 12 April 2020).

60. Kettunen, M.; Ministers, N.C.O.; Vihervaara, P.; Kinnunen, S.; D'Amato, D.; Badura, T.; Argimon, M.; Brink, P.T. Socio-economic importance of ecosystem services in the Nordic Countries. In *Socio-Economic Importance of Ecosystem Services in the Nordic Countries*; Nordic Council of Ministers: Copenhagen, Denmark, 2012.

61. Ferrari, M.; Geneletti, D. Mapping and assessing multiple ecosystem services in an alpine region: A study in Trentino, Italy. *Ann. Bot.* **2014**, *4*, 65–71.

62. Nedkov, S.; Borisova, B.; Koulov, B.; Zhiyanski, M.; Bratanova-Doncheva, S.; Nikolova, M.; Kroumova, J. Towards integrated mapping and assessment of ecosystems and their services in Bulgaria: The Central Balkan case study. *One Ecosyst.* **2018**, *3*, e25428. [CrossRef]

63. EEA Reference Grid. Available online: https://www.eea.europa.eu/data-and-maps/data/eea-reference-grids-2 (accessed on 12 April 2020).

64. Tableau Online 2020.1. Available online: https://www.tableau.com/products/cloud-bi (accessed on 12 April 2020).

65. Egoh, B.; Reyers, B.; Rouget, M.; Richardson, D.M.; Le Maitre, D.C.; Van Jaarsveld, A.S. Mapping ecosystem services for planning and management. *Agric. Ecosyst. Environ.* **2008**, *127*, 135–140. [CrossRef]

66. Czúcz, B.; Arany, I.; Potschin-Young, M.; Bereczki, K.; Kertész, M.; Kiss, M.; Aszalós, R.; Haines-Young, R. Where concepts meet the real world: A systematic review of ecosystem service indicators and their classification using CICES. *Ecosyst. Serv.* **2018**, *29*, 145–157. [CrossRef]

67. Feld, C.; Sousa, J.P.; Da Silva, P.M.; Dawson, T. Indicators for biodiversity and ecosystem services: Towards an improved framework for ecosystems assessment. *Biodivers. Conserv.* **2010**, *19*, 2895–2919. [CrossRef]

68. Müller, F.; Burkhard, B. The indicator side of ecosystem services. *Ecosyst. Serv.* **2012**, *1*, 26–30. [CrossRef]

69. Van Oudenhoven, A.P.E.; Schröter, M.; Drakou, E.; Geijzendorffer, I.R.; Jacobs, S.; Van Bodegom, P.; Chazee, L.; Czúcz, B.; Grunewald, K.; Lillebø, A.; et al. Key criteria for developing ecosystem service indicators to inform decision making. *Ecol. Indic.* **2018**, *95*, 417–426. [CrossRef]

70. Van Reeth, W. *Ecosystem Service Indicators in Flanders: Are We Measuring What We Want to Manage?* Instituut voor Natuuren Bosonderzoek: Brussel, Belgium, 2014.

71. Zogaris, S.; Economou, A.N. The biogeographic characteristics of the river basins of Greece. In *The Rivers of Greece: Evolution, Current Status and Perspectives*; Skoulikidis, N., Karaouzas, I., Dimitriou, E., Eds.; Springer-Verlag: Berlin/Heidelberg Germany, 2017; pp. 53–95.

72. Tsianou, M.A.; Mazaris, A.D.; Kallimanis, A.S.; Deligioridi, P.-S.K.; Apostolopoulou, E.; Pantis, J.D. Identifying the criteria underlying the political decision for the prioritization of the Greek Natura 2000 conservation network. *Biol. Conserv.* **2013**, *166*, 103–110. [CrossRef]

73. Katsanevakis, S.; Sini, M.; Dailianis, T.; Gerovasileiou, V.; Koukourouvli, N.; Topouzelis, K.; Ragkousis, M. Identifying where vulnerable species occur in a data-poor context: Combining satellite imaging and underwater occupancy surveys. *Mar. Ecol. Prog. Ser.* **2017**, *577*, 17–32. [CrossRef]

74. Börger, T.; Beaumont, N.J.; Pendleton, L.; Boyle, K.J.; Cooper, P.; Fletcher, S.; Haab, T.; Hanemann, M.; Hooper, T.L.; Hussain, S.S.; et al. Incorporating ecosystem services in marine planning: The role of valuation. *Mar. Policy* **2014**, *46*, 161–170. [CrossRef]

75. Rodrigues, J.G.; Conides, A.; Rodriguez, S.R.; Raicevich, S.; Pita, P.; Kleisner, K.; Pita, C.; Lopes, P.; Roldán, V.A.; Ramos, S.; et al. Marine and Coastal Cultural Ecosystem Services: Knowledge gaps and research priorities. *One Ecosyst.* **2017**, *2*. [CrossRef]

76. Salomidi, M.; Katsanevakis, S.; Borja, A.; Braeckman, U.; Damalas, D.; Galparsoro, I.; Mifsud, R.; Mirto, S.; Pascual, M.; Pipitone, C.; et al. Assessment of goods and services, vulnerability, and conservation status of European seabed biotopes: A stepping stone towards ecosystem-based marine spatial management. *Mediterr. Mar. Sci.* **2012**, *13*, 49–88. [CrossRef]

77. Balzan, M.V.; Pinheiro, A.M.; Mascarenhas, A.; Morán-Ordóñez, A.; Ruiz-Frau, A.; Carvalho-Santos, C.; Vogiatzakis, I.; Arends, J.; Santana-Garcon, J.; Roces-Díaz, J.V.; et al. Improving ecosystem assessments in Mediterranean social-ecological systems: A DPSIR analysis. *Ecosyst. People* **2019**, *15*, 136–155. [CrossRef]

78. Satz, D.; Gould, R.K.; Chan, K.M.A.; Guerry, A.; Norton, B.; Satterfield, T.; Halpern, B.S.; Levine, J.; Woodside, U.; Hannahs, N.; et al. The Challenges of Incorporating Cultural Ecosystem Services into Environmental Assessment. *Ambio* **2013**, *42*, 675–684. [CrossRef]

79. Jaligot, R.; Hasler, S.; Chenal, J. National assessment of cultural ecosystem services: Participatory mapping in Switzerland. *Ambio* **2019**, *48*, 1219–1233. [CrossRef] [PubMed]

80. Raum, S. A framework for integrating systematic stakeholder analysis in ecosystem services research: Stakeholder mapping for forest ecosystem services in the UK. *Ecosyst. Serv.* **2018**, *29*, 170–184. [CrossRef]

81. Vlami, V.; Danek, J.; Zogaris, S.; Gallou, E.; Kokkoris, I.P.; Kehayias, G.; Dimopoulos, P. Residents' Views on Landscape and Ecosystem Services during a Wind Farm Proposal in an Island Protected Area. *Sustainability* **2020**, *12*, 2442. [CrossRef]

82. Tengberg, A.; Fredholm, S.; Eliasson, I.; Knez, I.; Saltzman, K.; Wetterberg, O. Cultural ecosystem services provided by landscapes: Assessment of heritage values and identity. *Ecosyst. Serv.* **2012**, *2*, 14–26. [CrossRef]

83. Schaubroeck, T. The Concept of Cultural Ecosystem Services Should Not Be Abandoned. *BioScienece* **2019**, *69*, 585. [CrossRef]

84. Schaich, H.; Bieling, C.; Plieninger, T. Linking Ecosystem Services with Cultural Landscape Research. *GAIA Ecol. Perspect. Sci. Soc.* **2010**, *19*, 269–277. [CrossRef]

85. Ungaro, F.; Hafner, K.; Zasada, I.; Piorr, A. Mapping cultural ecosystem services: Connecting visual landscape quality to cost estimations for enhanced services provision. *Land Use Policy* **2016**, *54*, 399–412. [CrossRef]

86. Vlami, V.; Zogaris, S.; Djuma, H.; Kokkoris, I.P.; Kehayias, G.; Dimopoulos, P. A Field Method for Landscape Conservation Surveying: The Landscape Assessment Protocol (LAP). *Sustainability* **2019**, *11*, 2019. [CrossRef]

87. Lorilla, R.S.; Kalogirou, S.; Poirazidis, K.; Kefalas, G. Identifying spatial mismatches between the supply and demand of ecosystem services to achieve a sustainable management regime in the Ionian Islands (Western Greece). *Land Use Policy* **2019**, *88*, 104171. [CrossRef]

88. Lorilla, R.S.; Poirazidis, K.; Detsis, V.; Kalogirou, S.; Chalkias, C. Socio-ecological determinants of multiple ecosystem services on the Mediterranean landscapes of the Ionian Islands (Greece). *Ecol. Model.* **2020**, *422*, 108994. [CrossRef]

89. LIFE IP 4Natura—Integrated Actions for the Conservation and Management of Natura 2000 Sites, Species, Habitats and Ecosystems in Greece (LIFE16 IPE/GR/000002). Available online: https://ec.europa.eu/environment/life/project/Projects/index.cfm?fuseaction=search.dspPage&n_proj_id=6520 (accessed on 12 April 2020).

90. Economou, A.N.; Zogaris, S.; Vardakas, L.; Koutsikos, N.; Chatzinikolaou, Y.; Kommatas, D.; Kapakos, Y.; Giakoumi, S.; Oikonomou, E.; Tachos, V. Developing policy-relevant river fish monitoring in Greece: Insights from a nation-wide survey. *Mediterr. Mar. Sci.* **2016**, *17*, 302. [CrossRef]

91. Mentzafou, A.; Panagopoulos, Y.; Dimitriou, E. Designing the National Network for Automatic Monitoring of Water Quality Parameters in Greece. *Water* **2019**, *11*, 1310. [CrossRef]

92. MAES GR Tool: Mapping and Assessment of Ecosystems and their Services: A Dynamic Tool for Ecosystem Services. Available online: https://83.212.170.27:8085/login (accessed on 12 April 2020).

Article

Medicinal and Aromatic Lamiaceae Plants in Greece: Linking Diversity and Distribution Patterns with Ecosystem Services

Alexian Cheminal [1,2], Ioannis P. Kokkoris [1], Arne Strid [3] and Panayotis Dimopoulos [1,*]

[1] Department of Biology, Laboratory of Botany, University of Patras, 26504 Patras, Greece; alexian.cheminal@agrosupdijon.fr (A.C.); ipkokkoris@upatras.gr (I.P.K.)
[2] Higher National School of Agriculture and Food Sciences (AgroSup Dijon), 26, bd Docteur Petitjean-BP 87999, 21079 Dijon CEDEX, France
[3] 4 Bakkevej 6, DK-5853 Ørbæk, Denmark; arne.strid@youmail.dk
* Correspondence: pdimopoulos@upatras.gr; Tel.: +0030-261-099-6777

Received: 5 June 2020; Accepted: 9 June 2020; Published: 10 June 2020

Abstract: *Research Highlights:* This is the first review of existing knowledge on the Lamiaceae taxa of Greece, considering their distribution patterns and their linkage to the ecosystem services they may provide. *Background and Objectives:* While nature-based solutions are sought in many fields, the Lamiaceae family is well-known as an important ecosystem services provider. In Greece, this family counts 111 endemic taxa and the aim of the present study is to summarize their known occurrences, properties and chemical composition and analyze the correlations between these characteristics. *Materials and Methods:* After reviewing all available literature on the studied taxa, statistical and GIS spatial analyses were conducted. *Results:* The known properties of the endemic Lamiaceae taxa refer mostly to medicinal and antimicrobial ones, but also concern nutritional and environmental aspects. Essential oils compositions with high concentrations in molecules of interest (e.g., carvacrol, caryphyllene oxide, etc.) have been found in some taxa, suggesting unexploited applications for these taxa. Distribution patterns show a higher concentration of endemic Lamiaceae on the island of Kriti and southern Peloponnisos; patterns of the endemics' properties are also highlighted in the biodiversity hotspot of Kriti. However, the lack of data for two thirds of the taxa, regarding their properties or specific distribution, shows a gap of knowledge. Results on endemic Lamiaceae properties and composition are correlated with the supply or potential supply of ecosystem services and the relevant hotspots have been identified. *Conclusions:* The Greek endemic Lamiaceae taxa are proved to be of great importance, regarding their chemical composition and the properties they confer. The distribution analysis suggests the existence of clustering patterns of plant species with common properties. Finally, this study highlights knowledge gaps that should be filled in order to ensure the conservation of the endemic Lamiaceae taxa and the preservation of the ecosystem services they provide or could potentially provide.

Keywords: biodiversity management; endemic taxa; Greek flora; knowledge gaps; MAES implementation

---

1. Introduction

The biological resources of medicinal and aromatic plants have been used extensively for health care and healing practices across history and cultures [1–6]. Hundreds of millions of people, especially in developing countries and regions, collect plant and animal material to fulfil their needs for substances for personal uses or for trade as a complementary or primary income [7,8]. Characteristically, the World Health Organization reports that medicinal and aromatic plants still form the basis of traditional or indigenous health systems of the populations in most of the developing countries [9]. The collection and use of nature products and especially medicinal and aromatic plants is a common practice also in developed countries for cultural reasons as well as for trade commodities that meet the demand of often distant markets [10]. Moreover, medicinal and aromatic plants represent the largest natural resource (in terms of taxa number) used for its properties and compounds, especially in the growing, international market of plant-based cosmetics, spices, medicine and health products [11].

Subsequently, intensive pressure on natural resources is placed on the populations of the medicinal and aromatic plants, due to increasing demand, most of which are still collected in the wild. Uncontrolled overexploitation of wild plants, their habitat loss and alteration are the main reasons why medicinal plants, their study, evaluation, utilization and conservation have become essential parts of research programs [12] and increasing recommendations by many agencies that wild species should be brought into cultivation [13–15]. Although cultivation of medicinal plants can reduce the harvesting pressures on wild populations, it could also result to habitat and ecosystem degradation, genetic diversity reduction and the loss of incentives for the conservation of wild populations [16]. On the other hand, wild plant populations are affected by disturbance processes and positive links are identified among medicinal plant diversity and disturbance factors; the example of *Arnica montana* is typical, where traditional grazing practices in European meadows supports the conservation of rare plant populations [17,18].

The trade-off documentation (e.g., among conservation and production benefits) for medicinal and aromatic plant exploitation is crucial for supporting management decisions and policy-making processes on species conservation options and actions needed to be implemented [19–21]. In situ and ex situ conservation (field, seed and in vitro collections), which are complementary conservation strategies, are being implemented in Europe and other continents in the world for plant genetic resources in general, and medicinal and aromatic plants species in particular [22]. During recent decades, the demand and need for natural products and their substances, instead of safety-questionable synthetic compounds, has guided numerous studies regarding wild plant chemical composition, properties and uses [23–25].

The ecosystem services (ES) approach [26–28] recognizes that humans, characterized by their cultural and economic diversity, form an integral component of the natural environment. This strong relationship among human activities and ecosystems is the core of the sustainable development framework and highlights the dependence of human society on ecosystems [10,29], as well as that human and ecosystem "prosperity" need to be jointly assessed [30]. When the human condition and the condition of the ecosystem are favorable or improving jointly, then sustainable society functioning should be achieved [31,32]. Plant and animal population and distribution trends are indicators of sustainable resource use. Subsequently, a sustainable harvesting system, including medicinal and aromatic plants, proposes the collection of plant material from a certain area without impact on the structure and functions of the harvested plant population [33,34].

The capacity of nature to provide medicinal and aromatic plant resources depends on the species richness of medicinal/aromatic plants. Several areas in Europe and Central Asia are characterized by high medicinal plant species richness, including the Mediterranean region, the Alps and the Pyrenees, the Massif Central in France, the Balkan Peninsula, the Crimean Peninsula and the Carpathian Mountains [35]. Bogers et al. [36] concluded that the use of naturally available resources is of high priority in areas where climatic conditions and environmental attributes permit it, and Greece is considered as one of the globally important places where medicinal and aromatic plants constitute an important natural resource [37,38]. An outstanding example of a plant family with important medicinal and aromatic plants in Greece is the Lamiaceae, colloquially known as the "mint family"; it includes many ethnobotanically renowned species such as sages, thymes, lavenders, "mountain teas" and oreganos [39].

Under the key targets of the Mapping and Assessment of Ecosystems and their Services (MAES) implementation in Greece [40] and in the frame of the currently in progress Flora of Greece project, this paper provides an overview of the endemic Lamiaceae medicinal and aromatic plants of Greece, by using all the available literature data for each taxon's uses and attributes. More precisely, this work aims to: (a) provide a catalogue of the endemic Lamiaceae medicinal and aromatic plant species and subspecies (taxa) with country-wide distributional data, assigned to ecosystem types and followed by their properties and components characteristics, (b) summarize the distribution patterns for Lamiaceae medicinal and aromatic plants in Greece, (c) highlight diversity and endemism hotspots for the Lamiaceae plants, (d) support MAES implementation, by assigning each taxon to the actual or potential supply of one or more ecosystem services and (e) pinpoint data gaps and further steps needed for the sustainable exploitation and management of medicinal and aromatic plants' wild populations.

2. Materials and Methods

The Lamiaceae taxa present in Greece and their status as Greek endemics and/or range-restricted are identified and designated following Dimopoulos et al. [41,42] and the related Flora of Greece Web database portal [43]. In the present study, taxa are defined as comprising species and subspecies. Separation of the terrestrial Greek territory into floristic regions follows Strid and Tan [44] and the specific biogeographical region names are used. The implementation of the study consists of the following eight steps (Figure 1):

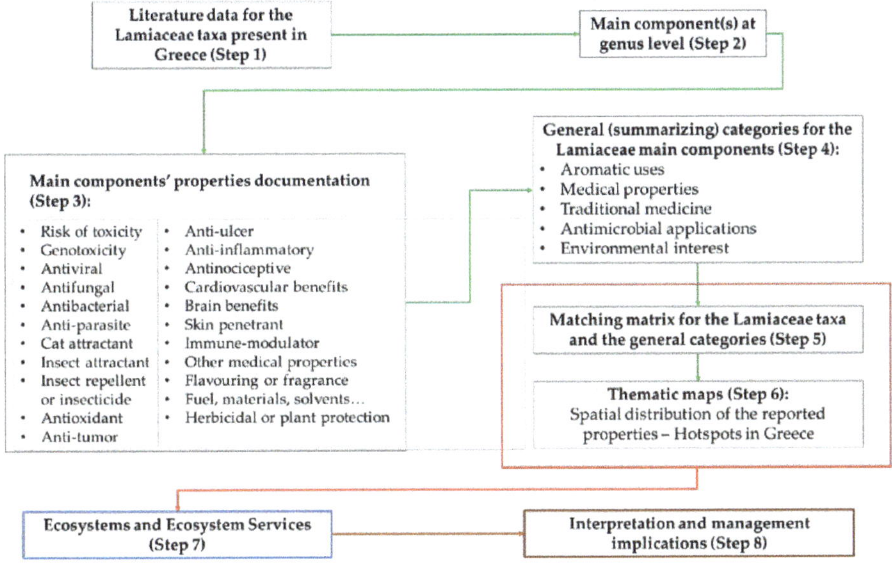

Figure 1. Flowchart of the study.

Step 1: Data on the existing knowledge for each endemic Lamiaceae taxon are collected by using "taxon name" and/or "Lamiaceae + Greece" as key words in the Google Scholar, Science direct, Scopus and Web of Science databases. The goal was to gather all existing literature for each Lamiaceae taxon referring to: (a) the chemical constituents of its essential oils, (b) its medicinal uses, (c) its environmental value, (d) its culinary uses and (e) the last discoveries about its biochemical, potential applications. In addition, the knowledge about the endemic taxa has been completed by data concerning closely related, non-endemic species (e.g., *Lamium garganicum* subsp. *striatum* (Sm.) Hayek, *Satureja montana* subsp. *pisidia* (Wettst.) Šilic, *Teucrium chamaedrys* subsp. *lydium* O. Schwarz). As a result, 741 papers have been assessed, including 356 studies dealing with molecules' properties;

Step 2: The gathered information was filtered on the basis of the main component (e.g., 1.8-cineole, alpha-cadinol, alpha-copaene, alpha-pinene, bêta-caryophyllene, carvacrol, etc.) at the genus level;

Step 3: Main components were assigned to one or more properties (i.e., risk of toxicity, genotoxicity, antiviral, antifungal, antibacterial, anti-parasite, cat attractant, insect attractant, insect repellent or insecticide, antioxidant, anti-tumor, anti-ulcer, anti-inflammatory, antinociceptive, cardiovascular benefits, brain benefits, skin penetrant, immune-modulator, other medical properties, flavoring or fragrance, fuel, materials, solvents, herbicidal or plant protection;

Step 4: Properties identified in Step 3 were assigned under five general categories, i.e., (i) aromatic uses: this category points out the species with known uses as aromatic oils, food, spices or cosmetics; (ii) medical properties: this category refers to plant extracts that have been scientifically proved to possess properties of medical interest (antioxidant, antitumor, antidiabetic, anti-leukemic, immuno-stimulating, etc.), some of them may already be exploited and commercialized for medical purposes (treatments, essential oil, etc.); (iii) traditional medicine: this category includes plants that have been used for a long time by the locals, and have been empirically known for their medical properties; (iv) antimicrobial applications: this category refers to plant extracts presenting antibacterial, antifungal, anti-yeast or antiviral potential applications; and (v) environmental interest: this category includes plants presenting ornamental or landscaping interests, or plants which can have beneficial applications when used or utilized as a living organism (educational value in botanical gardens and protected areas, ecotourism value, habitat indicator, habitat or soil restoration, honey plant, etc.) and/or as a natural extract (insecticidal, herbicidal, etc.);

Step 5: For identifying the properties per taxon, a matching table (matrix) for the main components, the relevant properties and the general categories was drafted for each taxon;

Step 6: Using distribution data for each taxon, thematic maps were prepared focusing on the spatial distribution of the endemic and non-endemic Lamiaceae taxa and of their properties. Gradients patterns and heatmaps are used to distinguish hotspots and data gaps for each map. This analysis includes endemic and non-endemic Lamiaceae taxa of Greece, aiming to compare with the taxa of interest (i.e., endemic);

Step 7: Correspondence to the Common International Classification of Ecosystem Services (CICES ver. 5.1) [45] sections (i.e., Provisioning, Regulating and Maintenance, Cultural) and the relevant codes is also presented and assigned to the relevant categories of the IPEBS [46], Millenium Ecosystem Assessment (MA) [30] and TEEB [47]. Ecosystem types supporting these ES are highlighted and ES bundles regarding the identified ES are depicted in thematic maps;

Step 8: Results interpretation and management implications, suggestions for future actions and support to policy making.

3. Results

3.1. Flora Statistics and Distribution

A list of 414 Lamiaceae taxa, including 111 Greek endemics, has been exported from Dimopoulos et al. [41,42] and the Flora of Greece Web online database [43]. These 111 taxa belong to 88 species, from 19 genera. *Stachys* is the richest in endemics genus with 24 endemic taxa (18 species), followed by *Scutellaria* (13 taxa from 7 species) and *Teucrium* (12 taxa from 11 species). The endemism rates (i.e., count of endemic taxa/total count of taxa) at the family and the genus level in relation to their species and taxa are as follows:

i.   Family level: 26.8% (111 out of 414 taxa) when considering the taxa and 34% (88 of the 262 species) when considering the species;

ii.   Genus level: *Origanum* is characterized by the highest endemism rate (70%) with 7 endemic species out of 10, followed by *Satureja* (60%), *Nepeta* and *Teucrium* (58%) (Figure 2a). *Nepeta* presents the highest rate with 10 endemic taxa out of 18 (56%), mostly due to the numerous endemic *Nepeta argolica* subspecies, followed by *Origanum* (54%) and *Scutellaria* (46%) (Figure 2b).

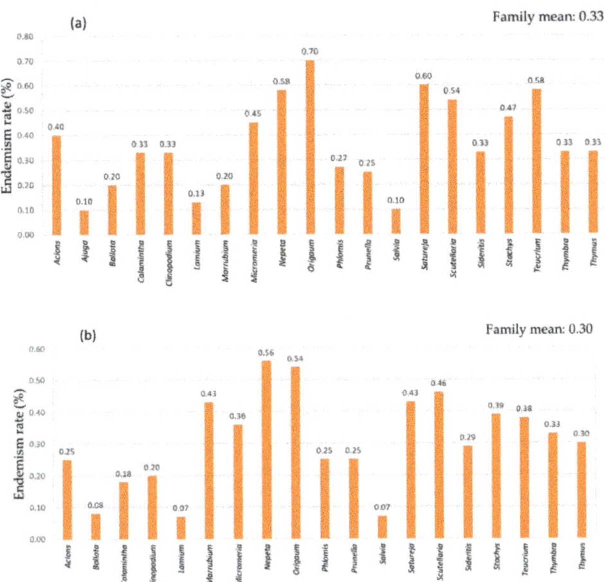

Figure 2. Endemism rates of Lamiaceae genera, at their species (**a**) and taxon (**b**) level.

The distribution patterns of the Lamiaceae in Greece for total taxon and endemic taxon richness, as thematically represented on gradient maps, highlight the following:

a. Areas rich in Lamiaceae (orange to red cells) are located in northern and southern Pindos (NPi, SPi), Stereas Ellas (StE) and Peloponnisos (Pe), North Central Greece (NC) and Kriti (Kr) (Figure 3);

b. Hotspot regions of Lamiaceae endemics are located on the Kriti (Kr) and Karpathos (Kp) islands and in Peloponnisos (Pe), especially in the southeast part of the region; Mt Athos (NE) can also be considered as a local hotspot (Figure 4).

3.2. Main Components and Properties

3.2.1. Main Components

For 57 out of the 111 endemic Lamiaceae taxa, the components have been detected from previous studies, testing a total of 134 samples of these taxa (Table 1). When focusing on the major component of each of these 57 taxa, we find that the more frequent one is carvacrol, detected in 12 taxa and especially in *Satureja* with 4 taxa, followed by thymol (6 taxa, including 3 *Satureja* and 2 *Thymus*), alpha-pinene (5 taxa, including 4 *Sideritis* and 1 *Phlomis*) and p-cymene (5 taxa, including 2 *Origanum* and 3 *Thymus*). Other molecules have been also detected and some have only been found as main components for two taxa (six molecules including bêta-elemene, delta-cadinene, linalool, nepetalactone derivatives) or for one taxon (11 molecules including (E)-nerolidol, alpha-copaene, gamma-terpinene and spathulenol, among others). For some taxa, the content of metabolites like iroids, flavonoids or neoclerodanes has been investigated, especially for eight taxa for which these lists are the only available information about their essential oils' composition (including two *Marrubium* taxa). For four other taxa, only one to two compounds have been detected, in order to compare these taxa with others in their own genus. However, the composition of the essential oils for 54 endemic taxa are still unknown and could only be deduced from the composition of similar species belonging to the same genus (e.g., *Acinos, Ballota, Clinopodium*).

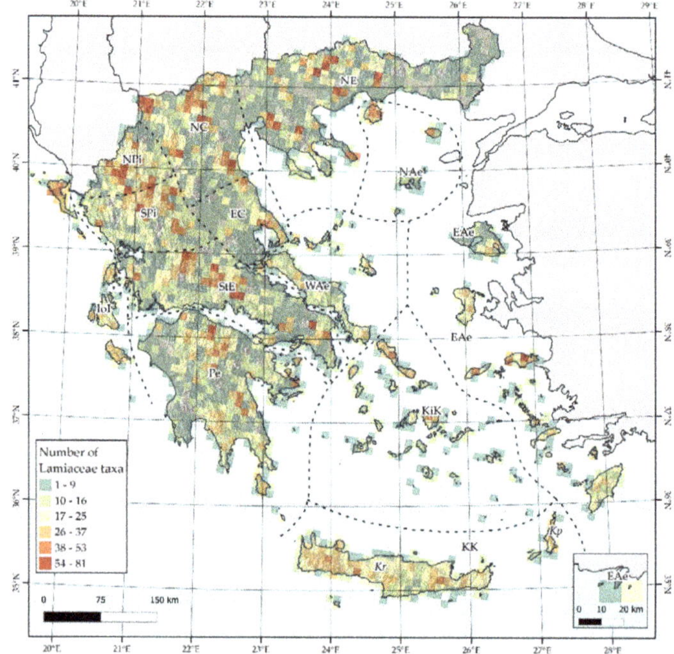

Figure 3. Distribution of the Lamiaceae taxa in Greece (10 × 10 km, EEA reference grid). Dotted lines delineate the floristic regions of Greece [44].

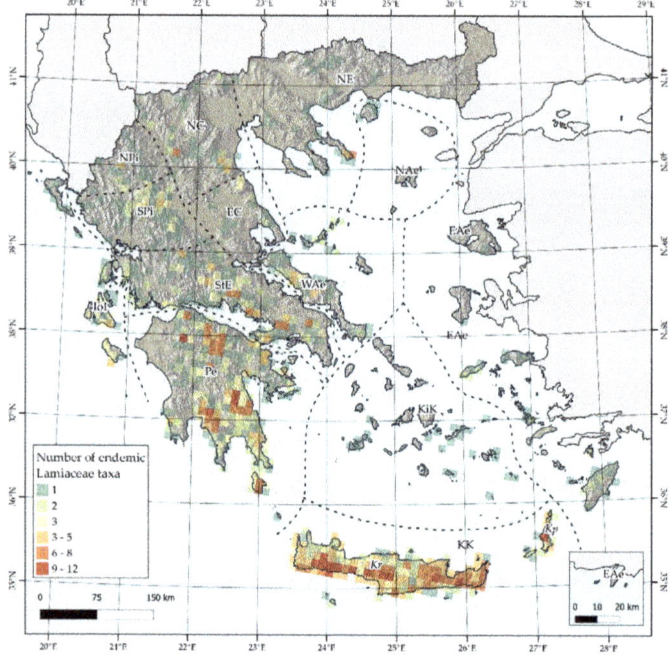

Figure 4. Distribution of the endemic Lamiaceae taxa in Greece (10x10 km, EEA reference grid). Dotted lines delineate the floristic regions of Greece [44].

Table 1. Main components of Greek endemic Lamiaceae, assigned to number of taxa and genus/genera represented. Genera in bold characters represent the genus with most taxa per main component.

Main Components	Number of Taxa	Genus (Genera) Represented
(E)-caryophyllene	4	*Satureja,* ***Stachys***
(E)-nerolidol	1	*Stachys*
1,8-cineole	5	*Nepeta*
Alpha-cadinol	1	*Stachys*
Alpha-copaene	1	*Stachys*
Apha-pinene	5	*Phlomis,* ***Sideritis***
Bêta-caryophyllene	2	*Sideritis, Teucrium*
Bêta-copaene	1	*Sideritis*
Bêta-elemene	2	*Stachys*
Carvacrol	12	*Origanum,* ***Satureja,*** *Sideritis, Stachys, Teucrium, Thymbra, Thymus*
Caryophyllene oxide	4	*Nepeta,* ***Stachys***
Delta-cadinene	2	*Stachys*
Gamma-terpinene	1	*Satureja*
Geraniol	1	*Thymus*
Germacrene D	4	*Phlomis,* ***Teucrium,*** *Thymus*
Isoabienol	3	***Sideritis,*** *Stachys*
Limonene	1	*Stachys*
Linalool	2	*Scutellaria, Thymus*
Nepetalactone derivatives	2	*Nepeta*
p-cymene	5	*Origanum,* ***Thymus***
Piperitenone oxide	1	*Calamintha*
Piperitone oxide	2	*Calamintha*
Spathulenol	1	*Satureja*
Terpinen-4-ol	1	*Origanum*
Thymol	6	*Origanum,* ***Satureja,*** *Thymus*
Viridiflorol	1	*Stachys*
Only flavonoid, iroid, phenolic or neoclerodane contents available	8	***Marrubium,*** *Scutellaria,* ***Stachys,*** *Teucrium*
Only one or two compounds known	4	*Origanum, Scutellaria, Teucrium, Thymus*
Unknown	54	*Acinos, Ballota, Clinopodium, Lamium, Micromeria, Prunella, Salvia, Scutellaria, …*

3.2.2. Properties

The review of the existing scientific literature enabled this study to identify noticeable properties for the components (Table 2) of 37 Lamiaceae taxa among the 111 Greek endemics and assign them to general property categories (Tables S1 and S2 of the supplement). In addition, two more types of information enriched the review: (i) some studies suggested properties, without scientific evidence or proof, but considered their traditional use, and (ii) certain properties proved to exist within a species, which have been attributed as suggested properties for closely related subspecies (i.e., the properties suggested by scientists without proven scientific demonstration). In the latter case, for example, *Nepeta argolica* subsp. *dirphya* (Boiss.) Strid & Kit Tan has been documented as antibacterial, which suggests that the endemics *Nepeta argolica* Bory & Chaub. subsp. *argolica* and *Nepeta argolica* subsp. *malacotrichos* (Baden) Strid & Kit Tan could also possess this property. With the inclusion of these different sources, 74 taxa out of the 111 have been assigned to one or several general property categories (66%). At the same time, more than one third of the total number of taxa still has not been studied with regard to their properties; an additional one third has only been proved to possess a single property. Some more documented taxa are suspected to have two properties (24%) or more (12%).

Table 2. Correspondence matrix among components and properties. x = proved property; ? = suspected or weak property; No = property proved to be absent; green cells = most investigated property for that substance.

Components	Risk of Toxicity	Genotoxicity	Antiviral	Antifungal	Antibacterial	Anti-Parasite	Cat Attractant	Insect Attractant	Insect Repellent or Insecticide	Antioxidant	Anti-Tumor	Anti-Ulcer	Anti-Inflammatory	Antinociceptive	Cardiovascular Benefits	Brain Benefits	Skin Penetrant	Immune-Modulator	Other Medical Properties	Flavouring or Fragrance	Fuel, Materials, Solvents	Herbicidal or Plant Protection
(E)-caryophyllene								x	x		No		?	?								
(E)-nerolidol		?		x	x	x		x	x	x	x	x		x		x				x		?
1,8-cineole		No		x	x				x	x	x		x	x	x	x	x		x		x	x
Alpha-cadinol					?	?			x				?						?			
Alpha-copaene		No		?				x		x	x											
Alpha-pinene		No		x	x			x	x	x	x	x	x	x		x			x	x	x	
Beta-caryophyllene		No			x			x	x	x	x	x	x	x	x	x	x	x	x	x		x
Beta-copaene																						
Beta-elemene				x	?					x	x		x						x			
Carvacrol	?	No	X	x	x	x			x	x	x		x	x	x	x		x	x		x	x
Caryophyllene oxide	x	No		x	x	x			x	x	x		x	x	x	x						x
Delta cadinene				?	x				?											x		?
Gamma-terpinene					x					x			x	x							x	
Geraniol	?			x	x			x	x	x	x	x	x	x	x				x		x	
Germacrene D					?			x		x												
Isoabienol					?															x		
Limonene	?			?	x				x	?	x		x	x	x	x		?	x	x	x	
Linalool				x	x				x	x				x		x						
Nepetalactone derivatives				x	x		x		x								x					
p-cymene					x				x	x	?		x	x								
Piperitenone oxide		x	X	x	x	x			x	x	x		x	x	x				x	x		
Piperitone oxide						?			?													
Spathulenol				x	x					x	x		x									
Terpinen-4-ol				x	x				x		x		x						x			
Thymol		No		x	x	x			x	x	?		x						x		x	
Viridiflorol				x	x				x	x			x									
Total of x	1	1	2	13	18	6	1	7	18	16	13	4	16	12	7	8	3	2	10	10	7	4
Total of ?	3	1		3	4	2			2	1	2		2	1				1	1			2
Total of No		7									1											
Total (x, ?, No)	4	9	2	16	22	8	1	7	20	17	16	4	18	13	7	8	3	3	11	10	7	6

Among the 74 taxa documented as of interest, the most recurrent property is the scientifically documented medicinal aspect; the literature review enabled to list 22 taxa as being of interest for human health, and 10 as suspected to be so. Nevertheless, commercial use as a medicinal drug has only been quoted for *Origanum dictamnus* L. Another main property is the antimicrobial aspect, proved to exist in 21 taxa, and suspected in 16 taxa. The more represented antimicrobial taxa belong to the *Sideritis* and *Stachys* genera. The environmental benefit by the Lamiaceae species is also frequently quoted (18 proved, 5 suspected), especially for *Scutellaria* or *Thymus* species, whose extracts can have insecticidal effects, or can be used for ornamental purposes. The aromatic aspect has mostly been quoted but not yet established, explaining the list of 20 suspected taxa. However, based on their culinary or cosmetical uses, 11 Lamiaceae taxa endemic to Greece have been proved to have this property. In addition to these four categories, *Origanum*, *Satureja* and *Sideritis* have been quoted among the seven proved and the six suspected taxa used in traditional medicine.

Gradient Maps

The thematic representation of the spatial distribution of each general property category per endemic taxon is depicted in Figure 5a–e; five gradient maps, one for each general property category, have been drafted, where the different hotspots are visible, according to the presence of endemic plants of interest for each property. The results of the analysis per category are as follows:

i.   Aromatic properties: mountainous areas of SW Kriti (KK) are considered as the main hotspot for endemic taxa with aromatic properties, followed by the mountains of southern Peloponnisos (Pe). Secondary hotspots can be considered the areas of the Pindos mountain range (NPi and SPi) and Mt Olimbos (NC) (Figure 5a);

ii.   Medical properties: mountainous areas of Greece are pinpointed as hotspots for endemic plants with medical properties. Kriti (KK) is the main hotspot (especially the eastern part of the island), followed by the mountains of Peloponnisos (Pe), Sterea Ellas (StE), southern Pindos (SPi) and Mt Olimbos (NC). Mt Athos (NE) is considered as a local hotspot (Figure 5b);

iii.   Antimicrobial properties: all the mountains of Peloponnisos (Pe) are hotspots for the Greek endemic taxa with antimicrobial properties. Kriti (KK) follows, with one or more Greek endemics with antimicrobial properties occurring almost throughout the island. Mountains of Sterea Ellas (StE) are also highlighted as important, alongside the local hotspots of Mt Olimbos (NC) and Mt Athos (NE) (Figure 5c);

iv.   Traditional medicine: Mt Olimbos (NC), Mt Parnon (Pe) and Mt Lefka Ori (KK) are the prevailing hotspots; other mountain ranges throughout Greece follow. It should be mentioned that the lack of well documented traditional uses of Greek endemic Lamiaceae in the literature is probably giving a weak signal with regard to the traditional uses of the Greek endemic Lamiaceae and thus biases the result of their spatial representation (Figure 5d);

v.   Environmental interest: Mt Taigetos (Pe), Mt Lefka Ori (KK), Mt Vourinos (NC) and the island of Kefallinia (IoI) are the main hotspots. A weak signal is also present for this category, due to the lack of extensive literature sources (Figure 5e).

The cumulative result of the above-mentioned categories is presented in the map of Figure 5f. Most significant hotspots of the co-existing Greek endemic Lamiaceae have been depicted and include Kriti (KK), Peloponnisos (Pe), southern and northern Pindos (SPi and NPi). Local hotspots are present throughout Greece, in the mainland and secondary to the islands, mainly at the Ionian Islands (IoI), and secondary in the Aegean (EAe, NAe) (Symi, Ikaria and Thasos). Mt Olimbos (NC) and Mt Athos (NE) are also pinpointed as cumulative local hotspots.

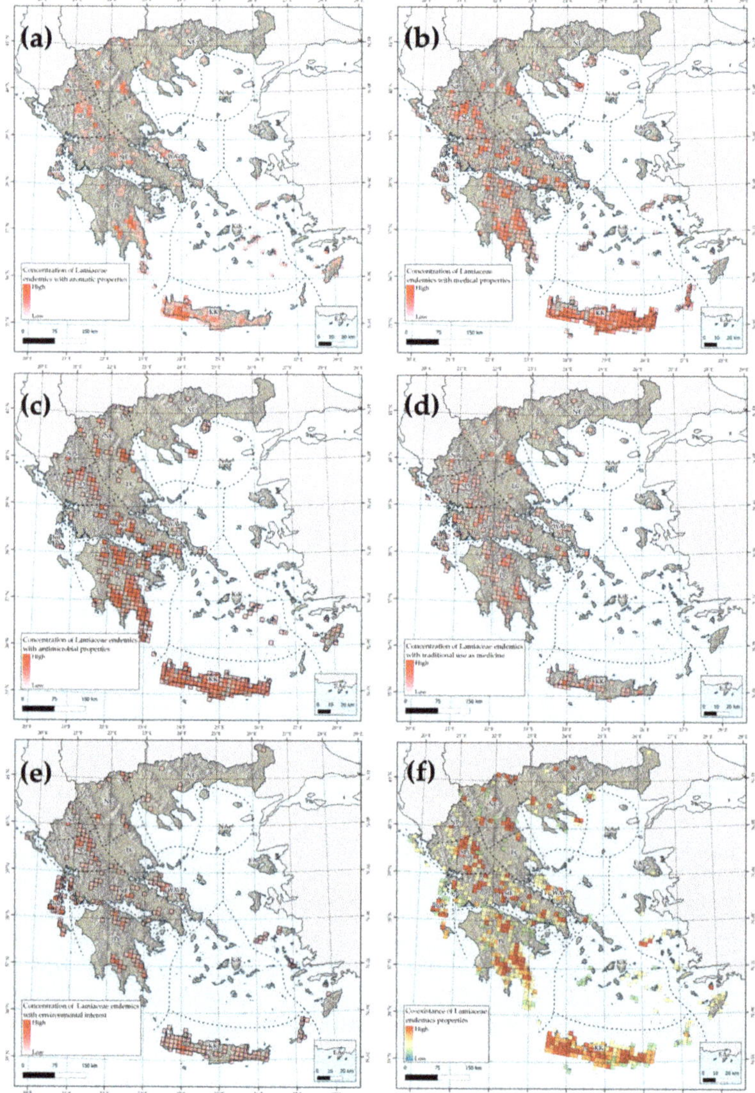

Figure 5. Distribution of the reported properties per endemic taxon using the 10x10 km EEA reference grid; (**a**) aromatic properties, (**b**) medical properties, (**c**) antimicrobial properties, (**d**) traditional medicine, (**e**) environmental interest, (**f**) cumulative representation of the different properties present per cell.

3.2.3. Ecosystems and Ecosystem Services

Habitat Categories and Ecosystem Types

The habitat preferences analysis of the Greek endemic Lamiaceae taxa assigned them to the different habitat categories [41,42] and their correspondence to the relevant MAES level ecosystem types [28] (Table 3) and highlighted the following:

i. The majority of the endemic Lamiaceae (41%, i.e., 45 taxa out of 111) occurs on cliffs, rocks, walls, boulder surfaces and ravines, a habitat category corresponding to the sparsely vegetated land MAES ecosystem type (level 2); 31 of these taxa (28% of the total) are exclusively found in this habitat category;

ii. A total of 31 taxa (28%) occur in xeric Mediterranean phrygana and grasslands (Mediterranean dwarf shrub formations, annual-rich pastures and lowland screes), a habitat category corresponding to the heathland and shrub MAES ecosystem type (level 2); 15 of these taxa (14% of the total) are exclusively found in this habitat category;

iii. A total of 30 taxa (28%) occur in high mountain vegetation (i.e., mountain- and oro-Mediterranean grasslands, screes and rocks, scrub above the tree line), a habitat category corresponding to the grasslands and sparsely vegetated land MAES ecosystem types (level 2); 20 of these taxa (18% of the total) are exclusively found in this habitat category;

iv. A total of 21 taxa (19%) occur in temperate and sub-Mediterranean grasslands, i.e., lowland to montane dry and mesic meadows and pastures, rock outcrops and stony ground, grassy non-ruderal verges and forest edges; these habitats correspond to the grasslands and sparsely vegetated land MAES ecosystem types (level 2); 12 of these taxa (11% of the total) are exclusively found in this habitat category;

v. A total of 13 taxa (12%) are found exclusively in woodlands and scrub, i.e., broadleaved and coniferous forests, riparian and mountain forests and scrubs, hedges and shady woodland margins; these habitats correspond to the woodland and forest MAES ecosystem types (level 2); two of these taxa (2% of the total) are exclusively found in this habitat category;

vi. Finally, two taxa (2%) are found in agricultural and ruderal habitats (fields, gardens and plantations, roadsides and trampled sites, frequently disturbed and pioneer habitats), habitats corresponding to the cropland and urban MAES ecosystem types (level 2); one of these taxa (1% of the total) is exclusively found in this habitat category.

Table 3. Number of Greek endemic Lamiaceae taxa per habitat category and Mapping and Assessment of Ecosystems and their Services (MAES) ecosystem type (level 2). Numbers in the parenthesis indicate number of taxa exclusively present in the corresponding habitat category.

Habitat Category [41,43]	MAES Ecosystem Type (Level 2) [28]	Number of Greek Endemic Lamiaceae Taxa
C: Cliffs, rocks, walls, ravines, boulders	Sparsely vegetated land	45 (31)
G: Temperate and sub-Mediterranean grasslands (lowland to montane dry and mesic meadows and pastures, rock outcrops and stony ground, grassy non-ruderal verges and forest edges)	Grasslands/Sparsely vegetated land	21 (12)
H: High mountain vegetation (mountain- and oro-Mediterranean grasslands, screes and rocks, scrub above the tree line)	Grasslands/Sparsely vegetated land	30 (20)
P: Xeric Mediterranean phrygana and grasslands (Mediterranean dwarf shrub formations, annual-rich pastures and lowland screes)	Heathland and shrubs	31 (15)
R: Agricultural and ruderal habitats (fields, gardens and plantations, roadsides and trampled sites, frequently disturbed and pioneer habitats)	Cropland/Urban	2 (1)
W: Woodlands and scrub (broadleaved and coniferous forest, riparian and mountain forest and scrub, hedges and shady woodland margins)	Woodland and forest	13 (2)

It is evident that sparsely vegetated land hosts most of the Greek endemic Lamiaceae and the majority of these taxa are exclusively present in this ecosystem type (i.e., 31 taxa) or are present in sparsely vegetated land and in grasslands (32 taxa), a total of 63 taxa. This fact pinpoints grasslands as the second richest ecosystem type in Greek endemic Lamiaceae; heathland and shrubs, woodland and forest follow.

Ecosystem Services

Assigning the general (summarizing) categories of the Lamiaceae endemics to the CICES categories and to the relevant ecosystem services categories of the MA and TEEB (Table 4), we present the following:

i. Aromatic uses correspond to (a) two CICES codes of provisioning services, (b) three IPEBS categories, (c) four MA categories and (d) three TEEB categories;

ii. Medical properties correspond to (a) one CICES code of provisioning services, (b) two IPEBS categories, (c) three MA categories and (d) two TEEB categories;

iii. Traditional medicine corresponds to (a) one CICES code of provisioning services and to three CICES codes of cultural services, (b) three IPEBS categories, (c) seven MA categories and (d) five TEEB categories;

iv. Antimicrobial applications correspond to (a) one CICES code of provisioning services, (b) one IPBES category, (c) three MA categories and (d) two TEEB categories;

v. Environmental interest corresponds to (a) three CICES codes for regulating and maintenance services and to four CICES codes of cultural services, (b) seven IPBES categories, (c) fourMA categories and (d) one TEEB category.

The localization of the areas of Greece where the highest numbers of endemic Lamiaceae have been recorded, i.e., different taxa records and different records from the same taxon, set the scientific basis for the ecosystem services hotspot areas and the ecosystem services cumulative importance (ecosystem services bundles) documentation. The results of this analysis are presented in Figure 6, combined with the delineated Natura 2000 sites to identify possible conservation needs and protection status. We pinpoint the following:

i. The island of Kriti is the main hotspot and especially its mountainous regions and mountain summits;

ii. The mountain summits of the southern Peloponnisos Peninsula (especially Mt Taigetos and Mt Parnon), Evvoia Island (Mt Dirfis), Mt Olimbos and Mt Athos (Chalkidiki peninsula) are assessed as local hotspots. Secondary local hotspots have been assessed on the mountains of northern Peloponnisos (Mt Chelmos, Mt Killini), Sterea Ellas (Mt Parnis, Mt Parnassos, Mt Giona) and of northern and southern Pindos (Mt Timfi, Mt Peristeri, Mt Tzoumerka);

iii. The extent of the Natura 2000 network in Greece covers all identified ecosystem services supply, or potential supply hotspots, as well as areas of lower importance based on the endemic Lamiaceae records.

Table 4. Correspondence of the general (summarizing) categories of the Lamiaceae Greek endemics to the CICES categories and the relevant ecosystem services categories of IPBES, MA and TEEB.

General Categories of Main Components for Lamiaceae Greek Endemics	CICES Section			IPBES Name (Code)	MA	TEEB
	Provisioning (Codes)	Regulating and Maintenance (Codes)	Cultural (Codes)			
Aromatic uses	1.1.5.1;1.1.5.2	-	-	12; 13; 14	Food; fiber, timber; ornamental; biochemical	Food; raw materials; medicinal resources
Medical properties	1.1.5.2	-	-	13; 14	Fiber, timber; ornamental; biochemical	Raw materials; medicinal resources
Traditional medicine	1.1.5.2	-	3.1.2.1;3.1.2.2; 3.1.2.3	6; 13; 15	Fiber, timber; ornamental; biochemical; Knowledge systems and educational values; cultural diversity; aesthetic values; spiritual and religious values	Raw materials; medicinal resources; information and cognitive development; inspiration for culture, art and design; aesthetic information
Antimicrobial applications	1.1.5.2	-	-	14	Fiber, timber; ornamental; biochemical	Raw materials; medicinal resources
Environmental interest	-	2.2.2.1; 2.2.2.3; 2.2.3.1;	3.1.2.1;3.1.2.2; 3.1.2.4; 3.2.1.3	1; 2; 6; 10; 13; 15; 17	Pest regulation; knowledge systems and educational values; cultural diversity; aesthetic values	Biological control

CICES class codes and names: 1.1.5.1—wild plants (terrestrial and aquatic, including fungi, algae) used for nutrition; 1.1.5.2—fibers and other materials from wild plants for direct use or processing (excluding genetic materials); 2.2.2.1—pollination (or "gamete" dispersal in a marine context); 2.2.3.1—pest control (including invasive species); 3.1.2.1—characteristics of living systems that enable scientific investigation or the creation of traditional ecological knowledge; 3.2.1.3—Elements of living systems used for entertainment or representation. IPEBS codes and names: 1—habitat creation and maintenance; 2—pollination and dispersal of seeds and other propagules; 6—regulation of freshwater quantity, location and timing; 10—regulation of organisms detrimental to humans; 12—food and feed; 13—materials and assistance; 14—medicinal, biochemical and genetic resources; 15—learning and inspiration; 17—supporting identities.

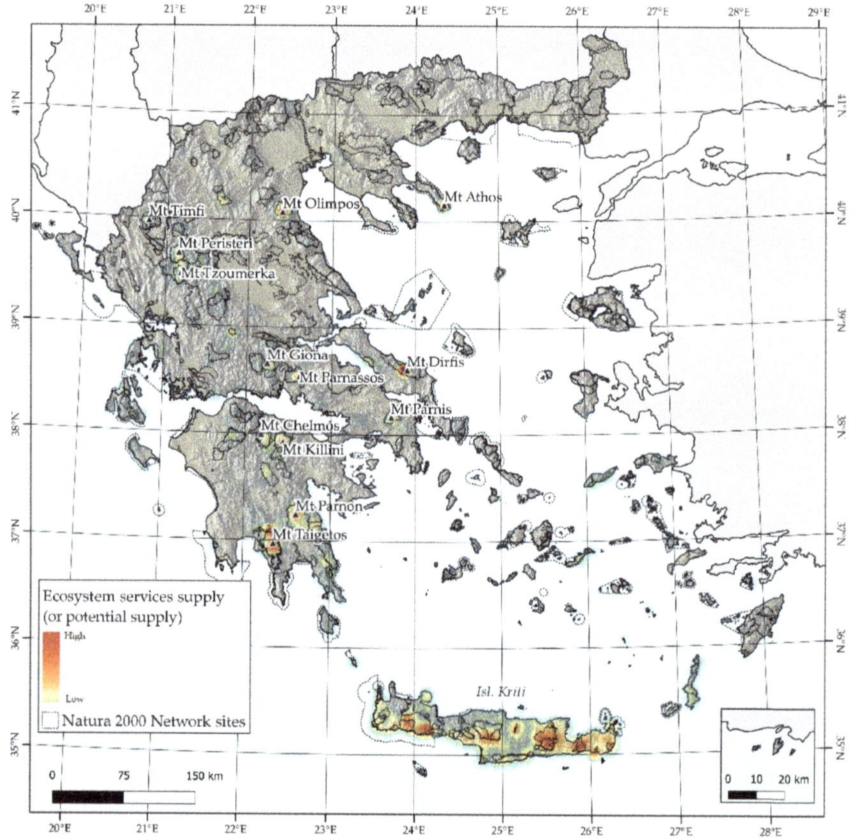

Figure 6. Ecosystem services hotspots in Greece, based on the recordings of the Greek endemic Lamiaceae. Natura 2000 sites in Greece are also depicted.

4. Discussion

Endemic taxa of Lamiaceae are distributed among 19 genera out of the 35 present in Greece. The most endemic taxon-rich genera are *Stachys*, *Scutellaria* and *Teucrium*, representing 45% of the endemic taxa of Lamiaceae, as well as 30% of the total Lamiaceae taxa in Greece. At the taxon (species and subspecies) and the species level, other genera such as *Nepeta*, *Origanum* and *Satureja* contain less taxa but a higher proportion of endemics. Thus, two groups of genera could be highlighted: (a) genera with a high number of endemics and (b) genera with a high proportion of endemics. These two groups could be used for plant conservation purposes, but also for management of commercial exploitation. More precisely, the main focus for the first group could target variations in the components proportions in each genus, whereas the second group would tend to offer more new active compounds or properties.

4.1. Properties and Components

Two thirds of the endemic taxa in Lamiaceae have no proven human-interest applications, according to the current state of knowledge. Nevertheless, other taxa included in the same genus or species, endemic or not, have already been investigated and documented as having properties of interest. As an example, from the review data extracted, *Teucrium montanum* subsp. *helianthemoides* (Adamović) Baden is a yet uninvestigated subspecies of *Teucrium montanum* L., whereas the latter has been quoted as having antifungal and antioxidant properties, and is already used in traditional medicine and for culinary purposes [48–50]. Moreover, for the currently documented species, studies have focused on certain properties, whereas others, not yet explored, might also be of interest. Medicinal and antimicrobial uses are the most recurrent properties studied. These properties can be proved by experiments led in a restricted time in the laboratory. The environmental aspects dealing with habitat restoration or indicators are rarely considered, and the social aspects as traditional remedies or garden plants are only approached through ethnobotany and ethnopharmacology.

The properties of the endemic Lamiaceae taxa are due to the presence of one or a combination of essential oil components. In order to obtain a better understanding of the properties and active compounds that the taxa could provide, a review of the molecules was necessary. Some already well-known essential oil components, such as carvacrol and thymol, as their precursors gamma-terpinene and p-cymene, were expected to be found and are part of many of the investigated species, as main or secondary compounds. However, other species proved to be rich in more uncommon components, e.g., *Thymus holosericeus* Čelak. for geraniol, *Stachys spruneri* Boiss. for limonene and *Calamintha cretica* (L.) Lam. for piperitenone oxide. This information could be of great importance to identify naturally provided sources of these compounds. Indeed, the three molecules previously quoted as examples have all been targeted as presenting positive attributes, i.e., geraniol could have applications in insect repellency and for anti-tumor treatments [51–54], limonene has significative antinociceptive effects [55] and piperitenone oxide is investigated for insecticidal uses and cardiovascular protection [56–58]. However, research gaps are still present, since each component has not yet been studied (e.g., beta-copaene, isoabienol), and some kinds of properties are still to be investigated for all the components (e.g., antiviral, immuno-stimulant, cardiovascular protection). It should be mentioned that the widespread molecules like carvacrol or caryophyllene oxide, which have already been deeply examined, have demonstrated efficiency in a wide range of applications [59,60].

4.2. Limitations of the Study

This study identified two main data gaps in the knowledge of the properties of the endemic Lamiaceae: (i) the first one is geographical and refers to the lack of surveys in some areas of Greece (shown as without-color patches on the thematic maps), either due to low biodiversity in these areas, or due to the better accessibility or reputation of other areas; (ii) the second one concerns the focus of the studies; knowledge is missing on the chemical composition and on the existence of properties in some taxa. In particular, the data gap concerns the choice of the studied aspects, i.e., some properties are more prone to be investigated than others, while other properties may still be discovered as they have been found in other taxa of the same genus. Certain knowledge gaps are also apparent at the links between the molecules and the properties, especially when more than one component needs to be combined to be more efficient. Subsequently, this demonstrates the importance of conducting further research to explore more about the endemic taxa, their specific distributions and their uses.

Other limitations include the following: (a) some species that are found in different locations may have different chemical compositions and properties depending on specific ecological conditions (e.g., altitude, soil, climate, season) [61–63], (b) some non-endemic taxa of particular genera have been proved to contain major compounds, which have not yet been detected in endemic ones, e.g., hexadecanoic acid in the *Scutellaria* genus and bêta-thujone in *Satureja* species [64,65], (c) the clusters produced by the spatial analysis are based on the restrictive hypothesis that the attributes of the taxa are the same at all times and all places. Future steps for improving the method used in this study could take into account more precise spatial and temporal data, especially since essential oils compositions have been proven to vary depending on the season [66–68]. Still, having a perfect description of the essential oils would necessitate handling too much data at once, e.g., plant organ considered, plant development stage, interactions with predators and with the ecosystem and modifications due to cultivation, which provide more limitations for a holistic approach and thus the segmentation of the analysis to specific-oriented parts.

4.3. Ecosystem Services and Management Implications

The Greek endemic taxa of Lamiaceae as ecosystem services providers are significant, since they can (a) support traditional harvesting methods and trade, and (b) act as genetic resources for reproductive materials of cultivated plants. Moreover, Heinrich [69] proposed that these results support guidance on modern and efficient research for bioactive components. Additionally, data collected during herbal market surveys, e.g., [5,70,71], regarding the origin and collection sites of each plant taxon can provide important ecological (e.g., population sites and size) and economic information (e.g., contribution to local economy). By this, management and policy decisions can be supported for the sustainable use of wild resources, as well as for the development of the cultivation of aromatic and medicinal plants, towards the protection of wild populations from overexploitation [5]. The need to protect natural populations has become much more urgent in recent years, since the overexploitation or even the fatal damage of aromatic plants (i.e., by inappropriate harvesting methods), especially the herbaceous ones, mainly for trade, is repeatedly reported by the state authorities (Forest Service). Overharvesting and lack of enforcement even in protected areas are serious problems for many targeted endemic species, e.g., *Origanum dictamnus* L. in Kriti [5] and the "mountain teas" (including *Sideritis* spp.) on the mainland.

As Solomou et al. [4] concluded, medicinal and aromatic plants have a crucial role in the utilization of the natural wealth and biodiversity conservation in the country; moreover, due to the selective and multifaceted biological activity of essential oils, considerable potential on the use of aromatic plants for novel applications in sustainable agriculture exists, valuable uses are possible and medicinal and aromatic plant diversity represents attainable new, environmentally and economically sustainable opportunities for agricultural areas.

The results revealed that all known endemic hotspots for Lamiaceae taxa in Greece are included in the Natura 2000 network. However, only four of the endemic plants of Lamiaceae (*Micromeria taygetea* P.H. Davis, *Nepeta argolica* subsp. *dirphya* (Boiss.) Strid & Kit Tan, *Nepeta sphaciotica* P.H. Davis, *Origanum dictamus* L.) are included in Annex II of the EU Habitats Directive (Dir. 92/43/EE) and monitored for their conservation status. Aromatic plants and especially the local endemics should be considered as an important and unique part of the country's natural capital; thus, quantitative data for the ecosystem services supply (or potential supply) by the plant species (e.g., of components, properties, material, spiritual, aesthetic importance) are needed to develop natural capital accounts and integrate them into local, regional and national decisions and policy-making processes.

5. Conclusions

The Lamiaceae family is represented in Greece by 111 endemic taxa. Most of them are located in southern Greece and Kriti. Many endemic taxa have not been studied in depth. However, according to existing documentation and the non-endemic plants assigned to the same genera, numerous and various properties are to be expected, ranging from medicinal applications to ethnobotanical and environmental interest. Moreover, their essential oils are rich in compounds that could be extracted for pharmaceutical uses, such as antimicrobials or insect repellents. From the ecosystem services point of view, the potential benefits of the endemic Lamiaceae are documented, including as molecule or other chemical properties providers. Additionally, this study also points out patterns of the endemics' properties in the biodiversity hotspot of Kriti. This information could be useful to concentrate measures for the sustainable exploitation of plants occurring in zones of special interest, but also to improve species and ecosystem-based conservation management.

Supplementary Materials: The following are available online at http://www.mdpi.com/1999-4907/11/6/661/s1, Table S1: Properties and components of the Greek endemic Lamiaceae; Table S2: Components correspondence to the general properties' categories.

Author Contributions: Conceptualization, A.C., I.P.K and P.D.; methodology, A.C., I.P.K. and P.D.; software, A.C. and I.P.K.; validation, A.C., I.P.K. and P.D.; formal analysis, A.C. and I.P.K..; investigation, A.C.; resources, A.C., P.D. and A.S.; data curation, A.C. and I.P.K.; writing—original draft preparation, A.C. and I.P.K.; writing—review and editing, A.C., I.P.K., P.D.; visualization, A.C. and I.P.K.; supervision, P.D.; All authors have read and agreed to the published version of the manuscript.

Funding: This research received no external funding.

Acknowledgments: This work was partially conducted during Alexian Cheminal's internship (March 23rd to August 30th, 2019, Department of Biology, University of Patras), as a student of the Higher National School of Agriculture and Food Sciences (AgroSup Dijon).

Conflicts of Interest: The authors declare no conflict of interest.

References

1. Lietava, J. Medicinal plants in a Middle Paleolithic grave Shanidar IV? *J. Ethnopharmacol.* **1992**, *35*, 263–266. [CrossRef]
2. Nunn, J.F. *Ancient Egyptian Medicine*; University of Oklahoma Press: Norman, OK, USA, 1996.
3. Kelly, K. *The History of Medicine*; Infobase Publishing: New York, NY, USA, 2009.
4. Solomou, A.D.; Martinos, K.; Skoufogianni, E.; Danalatos, N.G. Medicinal and Aromatic Plants Diversity in Greece and Their Future Prospects: A Review. *Agric. Sci.* **2016**, *4*, 9–20. [CrossRef]
5. Petrakou, K.; Iatrou, G.; Lamari, F.N. Ethnopharmacological survey of medicinal plants traded in herbal markets in the Peloponnisos, Greece. *J. Herb. Med.* **2020**, *19*, 100305. [CrossRef]
6. Bruneton, J. *Farmacognosia, Fitoquímica, Plantas Medicinales*, 2nd ed.; Editorial Acribia: Zaragoza, Spain, 2001.
7. Iqbal, M. *International Trade in Non-Wood Forest Products: An Overview*; FAO: Rome, Italy, 1993.
8. Walter, S. *Non Wood Forest Products in Africa: A Regional and National Overview*; FAO: Rome, Italy, 2001.
9. WHO. *Traditional Medicine Strategy 2014–2023*; WHO: Geneva, Switzerland, 2013.
10. Schippmann, U.; Leaman, D.; Cunningham, A.B. A comparison of cultivation and wild collection of medicinal and aromatic plants under sustainability aspects. In *Medicinal and Aromatic Plants*; Bogers, R., Craker, L., Lange, D., Eds.; Springer: Dordrecht, The Netherlands, 2006; pp. 75–95. ISBN 978-1-4020-5448-8.
11. Leaman, D.J. The International standard for sustainable wild collection of medicinal and aromatic plants (ISSC-MAP): Elements of ISSC-MAP resource assessment guidance relevant to cites NDF. In Proceedings of the International Expert Workshop on CITES Non-Detriment Findings Perennial Plant Working Group (Ornamentals, Medicinal and Aromatic Plants), Cancun, Mexico, 15 August 2008.
12. Barata, A.M.; Rocha, F.A.; Lopes, V.M.; Maia, J.; Dias, S.; Costa, M.; Salgueiro, L.; Morgado, J.; Bettencourt, E.; Delgado, F.; et al. Networking on conservation and use of Medicinal, Aromatic and Culinary Plants Genetic Resources in Portugal. *Acta Hortic.* **2011**, *925*, 21–35. [CrossRef]
13. World Health Organization (WHO). *Guidelines on the Conservation of Medicinal Plants*; WHO: Geneva, Switzerland, 1993.
14. Lambert, J.; Srivastava, J.; Vietmeyer, N. *Medicinal Plants: Rescuing a Global Heritage*; World Bank Tech. Pap. No 355; The World Bank: Washington, DC, USA, 1997; Available online: http://documents.worldbank.org/curated/en/538541468242090002/pdf/multi-page.pdf (accessed on 12 May 2020).
15. BAH. *Pflanzliche Arzneimittel Heute: Wissenschaftliche Erkenntnisse und Arzneirechtliche Rahmenbedingungen; Bestandsaufnahme und Perspektiven*, 4th ed.; Bundesfachverband der Arzneimittelhersteller (BAH): Bonn, Germany, 2004.
16. IUCN. *Assessing the Impacts of Commercial Captive Breeding and Artificial Propagation on Wild Species Conservation*; IUCN: Jacksonville, FL, USA, 2002.
17. Ellenberg, A. Assuming responsibility for a protected plant: WELEDA's endeavour to secure the firm's supply of Arnica montana. In Proceedings of the Medicinal Plant Trade in Europe: Conservation and Supply: Proceedings of the First International Symposium on the Conservation of Medicinal Plants in Trade in Europe, Royal Botanic Gardens, Kew, UK, 22–23 June 1998; Traffic Europe, Ed.; Traffic Europe: Brussels, Belgium, 1998; pp. 127–130.
18. Myklestad, Å.; Sætersdal, M. The importance of traditional meadow management techniques for conservation of vascular plant species richness in Norway. *Biol. Conserv.* **2004**, *118*, 133–139. [CrossRef]
19. Bodeker, G.; Bhat, K.K.; Jeffrey, B.; Vantomme, P. *Medicinal Plants for Forest Conservation and Health Care: Non-Wood Forest Products 11*; FAO: Rome, Italy, 1997.

20. Schippmann, U.; Leaman, D.J.; Cunningham, C.B. Impact of cultivation and gathering of medicinal plants in biodiversity: Global trends and issues. In *Biodiversity and the Ecosystem Approach in Agriculture, Forestry, and Fisheries*; FAO, Ed.; FAO: Rome, Italy, 2002; pp. 142–167.

21. Schippmann, U.; Leaman, D.J.; Cunningham, A.B.; Walter, S. Impact of cultivation and collection on the conservation of medicinal plants: Global trends and issues. In Proceedings of the Conservation, Cultivation and Sustainable Use of MAPs: A Proceedings of WOCMAP III: The IIIrd World Congress on Medicinal Aromatic Plants, Chiang Mai, Thailand, 3–7 February 2003; Jatisatienr, A., Paratasilpin, T., Elliott, S., Anusarnsunthorn, V., Wedge, D., Craker, L.E., Gardner, Z., Eds.; ISHS Acta Horticulturae: Leuven, Belgium, 2005; pp. 31–34.

22. Barata, A.M.; Rocha, F.; Lopes, V.; Carvalho, A.M. Conservation and sustainable uses of medicinal and aromatic plants genetic resources on the worldwide for human welfare. *Ind. Crops Prod.* **2016**, *88*, 11–18. [CrossRef]

23. Inatani, R.; Nakatani, N.; Fuwa, H. Antioxidative Effect of the Constituents of Rosemary (*Rosmarinus officinalis* L.) and Their Derivatives. *Agric. Biol. Chem.* **1983**, *47*, 521–528. [CrossRef]

24. Franz, C. Actual Problems on the Quality of Medicinal and Aromatic Plants. *Acta Hortic.* **1986**, 21–34. [CrossRef]

25. Piccaglia, R.; Marotti, M.; Giovanelli, E.; Deans, S.G.; Eaglesham, E. Antibacterial and antioxidant properties of Mediterranean aromatic plants. *Ind. Crops Prod.* **1993**, *2*, 47–50. [CrossRef]

26. Daily, G.C. *Nature's Services*; Island Press: Washington, DC, USA, 1997.

27. Costanza, R.; Arge, R.; De Groot, R.; Farberk, S.; Grasso, M.; Hannon, B.; Limburg, K.; Naeem, S.; O'Neill, R.V.; Paruelo, J.; et al. The value of the world's ecosystem services and natural capital. *Nature* **1997**, *387*, 253–260. [CrossRef]

28. Maes, J.; Teller, A.; Erhard, M.; Liquete, C.; Braat, L.; Berry, P.; Egoh, B.; Puydarrieus, P.; Fiorina, C.; Santos, F.; et al. *Mapping and Assessment of Ecosystem and Their Services. An Analytical Framework for Ecosystem Assessments under Action 5 of the EU Biodiversity Strategy to 2020*; Publications Office of the European Union: Luxemburg, 2013; ISBN 9789279293696.

29. Vlami, V.; Kokkoris, I.P.; Zogaris, S.; Cartalis, C.; Kehayias, G.; Dimopoulos, P. Cultural landscapes and attributes of "culturalness" in protected areas: An exploratory assessment in Greece. *Sci. Total Environ.* **2017**, *595*, 229–243. [CrossRef] [PubMed]

30. Millenium Ecosystem Assessment. *Ecosystems & Human Well-Being*; Island Press: Washington, DC, USA, 2005.

31. Wu, J. Landscape sustainability science: Ecosystem services and human well-being in changing landscapes. *Landsc. Ecol.* **2013**, *28*, 999–1023. [CrossRef]

32. Maes, J.; Teller, A.; Erhard, M.; Grizzetti, B.; Barredo, J.I.; Paracchini, M.L.; Condé, S.; Somma, F.; Orgiazzi, A.; Jones, A.; et al. *Mapping and Assessment of Ecosystems and Their Services: An Analytical Framework for Ecosystem Condition*; Publications Office of the European Union: Luxembourg, 2018.

33. Peters, C.M. *Sustainable Harvest of Non-Timber Plant Resources in Tropical Moist Forests: An Ecological Primer*; World Wildlife Fund: Washington, DC, USA, 1994.

34. Cunningham, A.B. *Applied Ethnobotany: People, Wild Plant Use and Conservation*; Earthscan: London, UK, 2001.

35. Allen, D.; Bilz, M.; Leaman, D.J.; Miller, R.M.; Timoshyna, A.; Window, J. *European Red List of Medicinal Plants*; Publications Office of the European Union: Luxemburg, 2014.

36. Bogers, R.J.; Craker, L.E.; Lange, D. *Medicinal and Aromatic Plants: Agricultural, Commercial, Ecological, Legal, Pharmacological and Social Aspects*; Bogers, R.J., Craker, L.E., Lange, D., Eds.; Springer: Wageningen, The Netherlands, 2006.

37. Lange, D. Trade in medicinal and aromatic plants: A financial instrument for nature conservation in Eastern and Southeast Europe. In *Financial Instruments for Nature Conservation in Central and Eastern Europe*; Heinze, B., Bäurle, G., Stolpe, G., Eds.; BfN—German Federal Agency for Nature Conservation: Bonn, Germany, 2001; pp. 157–171.

38. Mateescu, I.; Paun, L.; Popescu, S.; Roata, G.; Sidoroff, M. Medicinal and aromatic plants—A statistical study on the role of Phytotherapy in human health. *Bull. UASVM Anim. Sci. Biotechnol.* **2014**, *71*, 14–19.

39. Mulas, M. Traditional uses of Labiatae in the Mediterranean area. In Proceedings of the International Symposium on the Labiatae: Advances in Production, Biotechnology and Utilisation 723, Sanremo, Italy, 22–25 February 2006; pp. 25–32.

40. Hellenic Ministry of Environment and Energy. *LIFE IP 4Natura—Integrated Actions for the Conservation and Management of Natura 2000 Sites, Species, Habitats and Ecosystems in Greece. LIFE16 IPE/GR/000002.* 1 December 2017. Available online: https://ec.europa.eu/environment/life/project/Projects/index.cfm?fuseaction=search.dspPage&n_proj_id=6520 (accessed on 12 May 2020).

41. Dimopoulos, P.; Raus, T.; Bergmeier, E.; Constantinidis, T.; Iatrou, G.; Kokkini, S.; Strid, A.; Tzanoudakis, D. Vascular plants of Greece: An annotated checklist. *Englera* **2013**, *31*, 370. [CrossRef]

42. Dimopoulos, P.; Raus, T.; Bergmeier, E.; Constantinidis, T.; Iatrou, G.; Kokkini, S.; Strid, A.; Tzanoudakis, D. Vascular plants of Greece: An annotated checklist. Supplement. *Willdenowia* **2016**, *46*, 301–347. [CrossRef]

43. Dimopoulos, P.; Raus, T.; Strid, A. Flora of Greece Web: Vascular Plants of Greece an Annotated Checklist. Available online: http://portal.cybertaxonomy.org/flora-greece/intro (accessed on 12 May 2020).

44. Arne, S.; Tan, K. *Flora Hellenica*; Koeltz: Königstein, Germany, 1997.

45. Haines-Young, R.; Potschin, M. CICES Towards a Common Classification of Ecosystem Services. Available online: https://cices.eu/ (accessed on 12 April 2020).

46. IPEBS. *Global Assessment Report on Biodiversity and Ecosystem Services of the Intergovernmental Science-Policy Platform on Biodiversity and Ecosystem Services*; Brondizio, E.S., Settele, J., Díaz, S., Ngo, H., Eds.; IPBES Secretariat: Bonn, Germany, 2019.

47. TEEB. *The Economics of Ecosystems and Biodiversity: Mainstreaming the Economics of Nature: A Synthesis of the Approach, Conclusions and Recommendations of TEEB*; Progress Press: Birkirkara, Malta, 2010.

48. Panovska, T.K.; Kulevanova, S.; Stefova, M. In vitro antioxidant activity of some *Teucrium* species (Lamiaceae). *Acta Pharm.* **2005**, *55*, 207–214.

49. Čanadanović-Brunet, J.M.; Djilas, S.M.; Ćetković, G.S.; Tumbas, V.T.; Mandić, A.I.; Čanadanović, V.M. Antioxidant activities of different Teucrium montanum L. Extracts. *Int. J. Food Sci. Technol.* **2006**, *41*, 667–673. [CrossRef]

50. Vukovic, N.; Milosevic, T.; Sukdolak, S.; Solujic, S. Antimicrobial activities of essential oil and methanol extract of *Teucrium montanum*. *Evid. Based Complement. Alternat. Med.* **2007**, *4* (Suppl. 1), 17–20. [CrossRef] [PubMed]

51. Burke, Y.D.; Stark, M.J.; Roach, S.L.; Sen, S.E.; Crowell, P.L. Inhibition of pancreatic cancer growth by the dietary isoprenoids farnesol and geraniol. *Lipids* **1997**, *32*, 151. [CrossRef] [PubMed]

52. Carnesecchi, S.; Schneider, Y.; Ceraline, J.; Duranton, B.; Gosse, F.; Seiler, N.; Raul, F. Geraniol, a component of plant essential oils, inhibits growth and polyamine biosynthesis in human colon cancer cells. *J. Pharmacol. Exp. Ther.* **2001**, *298*, 197–200. [PubMed]

53. Duncan, R.E.; Lau, D.; El-Sohemy, A.; Archer, M.C. Geraniol and β-ionone inhibit proliferation, cell cycle progression, and cyclin-dependent kinase 2 activity in MCF-7 breast cancer cells independent of effects on HMG-CoA reductase activity. *Biochem. Pharmacol.* **2004**, *68*, 1739–1747. [CrossRef] [PubMed]

54. Müller, G.C.; Junnila, A.; Butler, J.; Kravchenko, V.D.; Revay, E.E.; Weiss, R.W.; Schlein, Y. Efficacy of the botanical repellents geraniol, linalool, and citronella against mosquitoes. *J. Vector Ecol.* **2009**, *34*, 2–8. [CrossRef] [PubMed]

55. Do Vale, T.G.; Furtado, E.C.; Santos, J.G., Jr.; Viana, G.S.B. Central effects of citral, myrcene and limonene, constituents of essential oil chemotypes from Lippia alba (mill.) N.E. Brown. *Phytomedicine* **2002**, *9*, 709–714. [CrossRef] [PubMed]

56. Božovic, M.; Pirolli, A.; Ragno, R. Mentha suaveolens Ehrh. (Lamiaceae) essential oil and its main constituent piperitenone oxide: Biological activities and chemistry. *Molecules* **2015**, *20*, 8605–8633. [CrossRef] [PubMed]

57. Lahlou, S.; Carneiro-Leão, R.F.L.; Leal-Cardoso, J.H.; Toscano, C.F. Cardiovascular effects of the essential oil of Mentha x villosa and its main constituent, piperitenone oxide, in normotensive anaesthetised rats: Role of the autonomic nervous system. *Planta Med.* **2001**, *67*, 638–643. [CrossRef] [PubMed]

58. Tripathi, A.K.; Prajapati, V.; Ahmad, A.; Aggarwal, K.K.; Khanuja, S.P.S. Piperitenone Oxide as Toxic, Repellent, and Reproduction Retardant toward Malarial Vector Anopheles stephensi (Diptera: Anophelinae). *J. Med. Entomol.* **2004**, *41*, 691–698. [CrossRef] [PubMed]

59. Fidyt, K.; Fiedorowicz, A.; Strządała, L.; Szumny, A. β-caryophyllene and β-caryophyllene oxide—Natural compounds of anticancer and analgesic properties. *Cancer Med.* **2016**, *5*, 3007–3017. [CrossRef] [PubMed]

60. Mohammedi, Z. Carvacrol: An update of biological activities and mechanism of action. *Open Access J. Chem.* **2017**, *1*, 53–62.

61. Jayanthy, A.; Kumar, P.U.; Remashree, A.B. Seasonal and Geographical Variations in Cellular Characters and Chemical Contents in *Desmodium gangeticum* (L.) DC.—An Ayurvedic Medicinal Plant. *Int. J. Herb. Med.* **2013**, *1*, 37.

62. Boulila, A.; Sanaa, A.; Ben Salem, I.; Rokbeni, N.; M'rabet, Y.; Hosni, K.; Fernandez, X. Antioxidant properties and phenolic variation in wild populations of *Marrubium vulgare* L. (Lamiaceae). *Ind. Crops Prod.* **2015**, *76*, 616–622. [CrossRef]

63. Lemos, M.F.; Lemos, M.F.; Pacheco, H.P.; Guimarães, A.C.; Fronza, M.; Endringer, D.C.; Scherer, R. Seasonal variation affects the composition and antibacterial and antioxidant activities of Thymus vulgaris. *Ind. Crops Prod.* **2017**, *95*, 543–548. [CrossRef]

64. Cicek, M.; Demirci, B.; Yilmaz, G.; Ketenoglu, O.; Baser, K.H.C. Composition of the essential oils of subspecies of scutellaria albida L. from Turkey. *J. Essent. Oil Res.* **2010**, *22*, 55–58. [CrossRef]

65. Glamoèlija, J.; Sokoviæ, M.; Vukojeviæ, J.; Milenkoviæ, I.; Van Griensven, L.J.L.D. Chemical composition and antifungal activities of essential oils of *Satureja thymbra* L. and *Salvia pomifera* ssp. *calycina* (Sm.) Hayek. *J. Essent. Oil Res.* **2006**, *18*, 115–117. [CrossRef]

66. Presti, M.L.; Crupi, M.L.; Costa, R.; Dugo, G.; Mondello, L.; Ragusa, S.; Santi, L. Seasonal variations of teucrium flavum l. essential oil. *J. Essent. Oil Res.* **2010**, *22*, 211–216. [CrossRef]

67. Piozzi, F.; Bruno, M. Diterpenoids from roots and aerial parts of the genus *Stachys*. *Rec. Nat. Prod.* **2011**, *5*, 1–11.

68. Pacifico, S.; Galasso, S.; Piccolella, S.; Kretschmer, N.; Pan, S.P.; Marciano, S.; Bauer, R.; Monaco, P. Seasonal variation in phenolic composition and antioxidant and anti-inflammatory activities of *Calamintha nepeta* (L.) Savi. *Food Res. Int.* **2015**, *69*, 121–132. [CrossRef]

69. Heinrich, M. Ethnopharmacology: Quo vadis? Challenges for the future. *Rev. Bras. Farmacogn.* **2014**, *24*, 99–102. [CrossRef]

70. Hanlidou, E.; Karousou, R.; Kleftoyanni, V.; Kokkini, S. The herbal market of Thessaloniki (N Greece) and its relation to the ethnobotanical tradition. *J. Ethnopharmacol.* **2004**, *91*, 281–299. [CrossRef] [PubMed]

71. Karousou, R.; Deirmentzoglou, S. The herbal market of Cyprus: Traditional links and cultural exchanges. *J. Ethnopharmacol.* **2011**, *133*, 191–203. [CrossRef] [PubMed]

Article

Integrating Plant Diversity Data into Mapping and Assessment of Ecosystem and Their Services (MAES) Implementation in Greece: Woodland and Forest Pilot

Konstantinos Kotsiras [1], Ioannis P. Kokkoris [1], Arne Strid [2] and Panayotis Dimopoulos [1,*]

[1] Department of Biology, Laboratory of Botany, University of Patras, 26504 Patras, Greece; cmng3151@upnet.gr (K.K.); ipkokkoris@upatras.gr (I.P.K.)
[2] Bakkevej 6, DK-5853 Ørbæk, Denmark; arne.strid@youmail.dk
* Correspondence: pdimopoulos@upatras.gr; Tel.: +30-261-099-6777

Received: 1 August 2020; Accepted: 29 August 2020; Published: 1 September 2020

Abstract: Research Highlights: This is the first approach that integrates biodiversity data into Mapping and Assessment of Ecosystem and their Services (MAES) implementation and natural capital accounting process, at the national scale, using an extensive vascular plant dataset for Greece. Background and Objectives: The study aims to support the MAES implementation in Greece, by assessing, as a pilot, the woodland and forest ecosystem type; the targets of the study are: (a) Identify and map ecosystem type extent; (b) identify ecosystem condition using biodiversity in terms of plant species richness (i.e., total, ecosystem exclusive, endemic, ecosystem exclusive endemic diversity); (c) develop ecosystem asset proxy indicators by combining ecosystem extent and ecosystem condition outcomes; (d) identify shortcomings; and (e) propose future steps and implications for the MAES implementation and natural capital accounting, based on biodiversity data. Materials and Methods: Following the national European Union's and United Nations System of Environmental Economic Accounts-Experimental Ecosystem Accounting (SEEA-EEA) guidelines and the adopted National Set of MAES Indicators, we developed a set of four proxy ecosystem asset indicators to assess ecosystem types with respect to ecosystem area extent and ecosystem condition. This was as interpreted by its plant diversity in terms of species richness (total, ecosystem exclusive, endemic, and ecosystem exclusive endemic diversity). Results: The results revealed that when indicators use well-developed biodiversity datasets, in combination with ecosystem extent data, they can provide the baseline for ecosystem condition assessment, ecosystem asset delineation, and support operational MAES studies. Conclusions: The relation among biodiversity, ecosystem condition, and ecosystem services is not a linear equation and detailed, fine-scale assessments are needed to identify and interpret all aspects of biodiversity. However, areas of importance are pinpointed throughout Greece, and guidance is provided for case-study selection, conservation strategy, and decision-making under the perspective of national and EU environmental policies.

Keywords: ecosystem condition; ecosystem extent; ecosystem asset; SEEA-EEA; Greek flora; national set of indicators; LIFE-IP 4 NATURA

---

1. Introduction

The benefits that derive from economic, social, cultural, or other human activities performed on ecosystems are defined as Ecosystem Services (ES)—which have been introduced in the scientific forefront, during the 1970s and 1980s [1–3], and have been established as an environmental advisory tool for policy- and decision-making during the 2000s [4–7]. Biodiversity is in the epicenter of the ES approach (see Millennium Ecosystem Assessment (MEA) [8], Mapping and Assessment of Ecosystem and their Services (MAES) [7,9], Intergovernmental Science-Policy Platform on Biodiversity and

Ecosystem Services (IPBES) [10]). It is considered as the cornerstone for future wellbeing, particularly with respect to equity and fairness in the socio-economic system, which guides the direct drivers of change [8]. In the European Union (EU), the EU Biodiversity Strategy for 2030 [11], and the EU Green Deal [12] highlight the important role of biodiversity, the need for its assessment and recording its spatial extent (i.e., mapping biodiversity and its attributes) to support conservation measures and management strategies needed for sustainable development. A growing concern on ES maintenance and its sustainable use is rising, since the documented biodiversity loss may affect ecosystem functioning and alter the provision of various ES threatening human wellbeing [1–3]. However, assessing and documenting the importance of ES delivery in the social sphere has proven difficult, with major challenges being the complexity of the topic and the availability of applicable approaches [13]. Moreover, the relationship between biodiversity and ES is considered as confused and is hindering the efforts for the development of coherent policy [14].

Braat and ten Brink [15] demonstrate how changes in biodiversity affect different types of ecosystem services, while Harrison et al. [16] in their literature review study found that the relationships between biodiversity attributes and ecosystem services are, in their majority, positively correlated. This is also highlighted by the work of Grunewald et al. [17], in which regulating ES is found to be usually positively correlated with higher biodiversity. However, detailed studies at various area-extent and dataset scales are needed to provide robust information; Steur et al. [18] studied plant diversity relationship with tropical forest ecosystem services, which were often found inconclusive, or showed both positive and negative correlations.

As the concept of ES gained popularity [6], the demand for appropriate indicators, quantification, and spatial localization methods has increased [19–21]. Various biodiversity indicators have been proposed to interpret ecosystem services provision at different ecosystems and at various spatial scales (e.g., [20,22]), as well as for ecosystem condition assessments (i.e., utilizing species diversity and abundance) as described by Maes et al. [23] in the analytical framework for mapping and assessment of ecosystem condition in the EU. In Europe, many EU Member States have already developed or are currently developing indicators to support MAES studies, some of which are focusing on different types of biodiversity and/or its spatial extent and distribution (e.g., [24–26]).

The need to measure ecosystems and their ES led to the broadly applicable system of the United Nations System of Environmental Economic Accounts-Experimental Ecosystem Accounting (SEEA-EEA) [27,28], which includes a set of accounts such as ecosystem extent, ecosystem services (supply and use), ecosystem assets, and biodiversity. The biodiversity accounts focus on species richness, abundance, and threats [29]. It is worth mentioning that biodiversity monitoring and biodiversity accounting systems have substantial differences; accounting methods are to be informative at an aggregated level with a limited set of indicators, that will capture biophysical information and at the same time the outcomes to be easily communicated to policy- and decision-makers [30].

In Greece, MAES implementation is already in progress, since 2018, via the LIFE-IP 4 NATURA project [31], which is coordinated by the Ministry of the Environment and Energy and incorporates a national set of indicators that includes six indicator groups, comprised by 40 indicators of which five are dealing with biodiversity, i.e., (i) diversity of agro-ecosystems with natural ecosystems, (ii) floristic diversity, (iii) micro-refugia of floristic and endemic diversity, (iv) network of crop limits with natural vegetation and (v) total biodiversity [26]. From the mentioned indicators, "floristic diversity" and "total biodiversity" are also considered as applicable for natural capital accounting, under the SEEA-EEA approach.

Woodland and forest is the dominant, natural terrestrial ecosystem type in Greece, covering ca. 29% [32] of the terrestrial area; woodland and forest ecosystem type is highlighted as of particular importance in Greece, including seven forest categories (Temperate deciduous forests, Mediterranean deciduous forests, Floodplain forests, Riparian forest/Fluvial forests, Temperate mountainous coniferous forests, Mediterranean coniferous forests, Mediterranean sclerophyllous forests, Mixed forests) [26,33], and distributed among 34 habitat types. Woodland and forests provide a variety of ecosystem services [34], which are proposed to be initially assessed by 20 indicators (three from Biodiversity, eight from Environmental Quality, one from Food, Material, and Energy, five from Forestry and three from Recreation indicator groups) [26].

This study aims to contribute to the efforts for overcoming the aforementioned challenges and support the MAES implementation in Greece, by assessing, as a pilot, woodland and forest ecosystem type; the EU and National guidelines for assessing and mapping ecosystem extent, ecosystem condition, and ecosystem asset, as described by SEEA-EEA and the relevant literature, e.g., [27,29,35,36] have been followed for the assessment. More precisely, the targets of the study are: (a) Identify and map ecosystem type extent; (b) identify ecosystem condition using biodiversity in terms of plant species richness (i.e., total, ecosystem exclusive, endemic, ecosystem exclusive endemic diversity); (c) develop ecosystem asset proxy indicators by combining ecosystem extent and ecosystem condition outcomes; (d) identify shortcomings; and (e) propose future steps and implications for the MAES implementation and natural capital accounting, based on biodiversity data.

2. Materials and Methods

Using the following guidance—(a) the analytical framework for mapping and assessment of ecosystems and their services in EU [7]; (b) the analytical framework for mapping and assessment of ecosystem condition in EU [23]; (c) SEEA and EU guidance on incorporating biodiversity into natural capital accounting (measuring the condition of ecosystem assets) [35,36]; and (d) the National Set of MAES Indicators for Greece [26]—we developed a species-richness based methodology to develop proxy indicators for ecosystem asset assessment and support the MAES implementation efforts in Greece. In general, the proposed methodology incorporates the ecosystem type area and the species richness in a given ecosystem type, at the 10 × 10 km European Environment Agency (EEA) reference grid [37] scale.

2.1. Datasets and Typology

For the analysis, we used the CORINE Land Cover (CLC) dataset for Greece [38] and the floristic records from the Flora of Greece Web project [39]. Species distribution, chorology, and habitat preferences follow Dimopoulos et al. [40,41] and the relevant information provided by the Flora of Greece Web portal [39]; the term 'species' includes both plant species and subspecies. Subsequently, a typology to assign species' habitats and MAES level 2 ecosystem types [7] has been developed following the concept by Kokkoris et al. [26].

2.2. Methodological Procedure

The methodological procedure of the present study consists of the following three steps:

Step 1: Ecosystem extent. It includes the identification of each ecosystem types' area spatial extent using the 10 × 10 km EEA reference grid for Greece [37] and based on the proposed typology for MAES [7], by which each CLC class is matched to one of the MAES level 2 ecosystem types and applied on the Greek terrestrial area using Geographic Information Systems (GIS). Subsequently, each grid cell includes one or more polygons of different ecosystem types. For the present study, we used two metrics for the ecosystem type extent per cell: (a) The actual area of the ecosystem type in the cell and (b) the relative area of the ecosystem type in the cell, with respect to the total area of the ecosystem type in the floristic region [39,42], where the cell belongs to. For example, and for interpretation purposes, let us assume that the cell of interest is as presented in Figure 1 and the ecosystem type of interest is D with the area at the given cell equal to d ha and its total area in the floristic region where the cell belongs to is x ha; thus, the relative ecosystem area extent for ecosystem type D in the given cell is d/x ha.

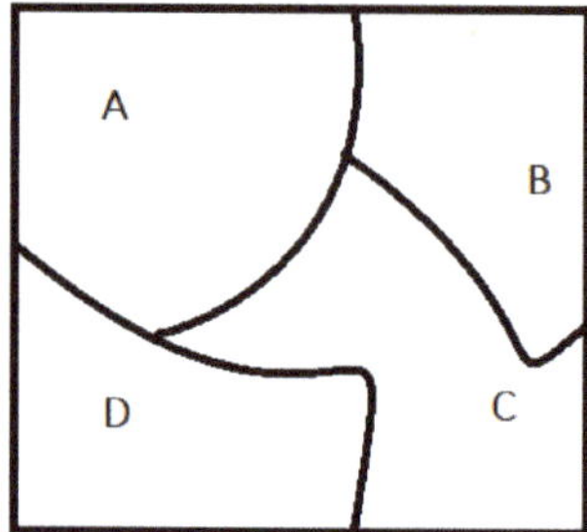

Figure 1. Graphical representation of a 10 × 10 EEA (Experimental Ecosystem Accounting) reference grid cell that includes four different ecosystem types, i.e., A, B, C, D.

By this, the normalized, relative area of each MAES level 2 ecosystem type [32] is calculated in each grid cell, with respect to the total area extent of the given ecosystem type in each floristic region of Greece [29,33] and all cells are scored in a common scale (0 to 1). In the present study, we followed this procedure for the woodland and forest ecosystems.

Step 2: Ecosystem condition. It includes the identification and assessment of plant diversity with respect to species richness at each grid cell, including only species assigned to the under-assessment ecosystem type. For example, for woodland and forest ecosystem type assessment, we analyzed only records from species present in woodland and forests. This analysis continues to a more detailed assessment using species exclusively present in the given ecosystem type. The same analysis is applied for endemic species, as well as for endemic species present only in the given ecosystem type. All calculations refer to normalized (0 to 1), relative species number with respect to each category's (i.e., total species richness, richness of species exclusively present in the ecosystem type, endemic species richness, richness of endemic species exclusively present in the ecosystem type) total species number in each floristic region of Greece. Table 1 provides a detailed description of the proposed calculations.

Table 1. Description of the proposed method for species-richness calculations for the four plant diversity categories.

Plant Diversity Categories	Calculation
Total species richness	$\dfrac{Number\ of\ species\ supported\ by\ the\ ecosystem\ at\ the\ cell\ level}{Number\ of\ species\ supported\ by\ the\ ecosystem\ in\ the\ floristic\ region}$
Richness of species exclusively present in the ecosystem type	$\dfrac{Number\ of\ species\ present\ exclusively\ in\ the\ ecosystem\ at\ the\ cell\ level}{Number\ of\ species\ present\ exclusively\ in\ the\ ecosystem\ in\ the\ floristic\ region}$
Endemic species richness	$\dfrac{Number\ of\ endemic\ species\ present\ in\ the\ ecosystem\ at\ the\ cell\ level}{Number\ of\ endemic\ species\ present\ in\ the\ ecosystem\ in\ the\ floristic\ region}$
Richness of endemic species exclusively present in the ecosystem type	$\dfrac{Number\ of\ endemic\ species\ exclusively\ present\ in\ the\ ecosystem\ at\ the\ cell\ level}{Number\ of\ endemic\ species\ exclusively\ present\ in\ the\ ecosystem\ at\ the\ floristic\ region}$

Step 3: Ecosystem asset proxy indicators. Normalized values of ecosystem type area extent and ecosystem condition (plant diversity) outcomes are combined per grid cell by summing their cell value. The sum has been subsequently normalized in a 0 to 1 scale, and the result is the proxy indicator for each one of the four possible combinations.

The abovementioned methodology for developing and assessing ecosystem asset proxy indicators is presented in Figure 2.

Thematic representation of the results has also been performed, by producing gradient maps in Geographic Information Systems (GIS), using a five-rating scale (i.e., very low, low, medium, high, very high). By this, areas of importance are highlighted, hotspots (i.e., areas where high concentration occurs of cells rated as "high" and/or "very high") are identified, and the results are better communicated to the non-expert community.

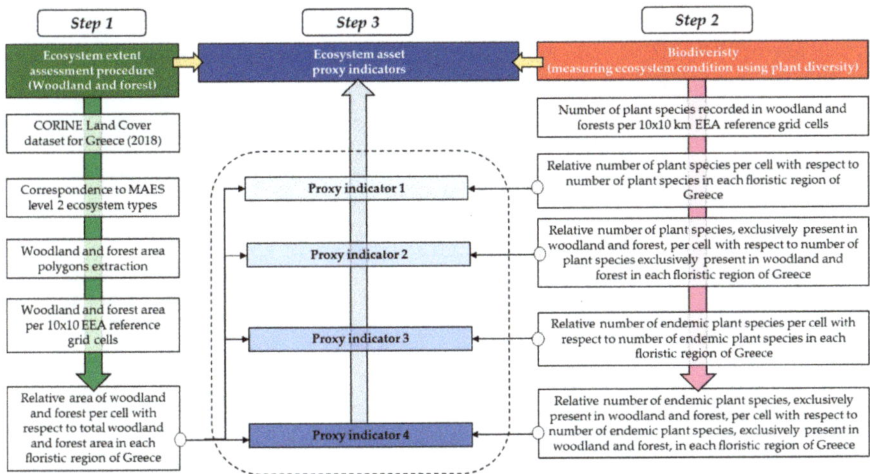

Figure 2. Methodological flowchart for developing and assessing ecosystem asset proxy indicators.

3. Results

3.1. Plant Habitats Categories to MAES Ecosystem Types Typology

The eight habitat categories for the vascular plants of Greece and as identified for the Greek flora by Dimopoulos et al. [40,41] have been assigned to the relevant MAES level 2 ecosystem types [7]. For most ecosystem types, including woodland and forests, the correspondence is straightforward; only 'heathland and shrub' and 'sparsely vegetated land' ecosystem types correspond to two different habitat categories, i.e., (a) Temperate and sub-Mediterranean grasslands (G), Xeric Mediterranean phrygana and grasslands (P) and (b) Cliffs, rocks, walls, ravines, boulders (C) and Coastal habitats (M), respectively. Subsequently, the total number of species has been assigned to each ecosystem type (Table 2). The woodland and forest ecosystem type hosts 1506 species out of 6760 species present in Greece [39], i.e., 22% of the Greek flora.

Table 2. Correspondence between ecosystem types (MAES level 2), Corine Land Cover classes ecosystem types, and habitats of the vascular plants of Greece. The total number of species present in each category is also presented.

Ecosystem Types (MAES Level 2) [7]	CORINE Land Cover Classes [7,43]	Habitats of Vascular Plants of Greece (Code) [39–41]	Plant Species (Number)
Cropland	2.1.1., 2.1.2., 2.1.3., 2.2.1., 2.2.2., 2.2.3., 2.4.1., 2.4.2., 2.4.3., 2.4.4.	Agricultural and ruderal habitats (R)	1868
Grassland	2.3.1., 3.2.1.	High mountain vegetation (H)	1385
Woodland and forest	3.1.1., 3.1.2., 3.2.4.	Woodlands and scrub (W)	1506
Heathland and shrub	3.2.2., 3.2.3	Temperate and sub-Mediterranean grasslands (G)	1927
		Xeric Mediterranean phrygana and grasslands (P)	1608
Sparsely vegetated land	3.3.2., 3.3.3., 3.3.4.	Cliffs, rocks, walls, ravines, boulders (C)	959
	3.3.1.	Coastal habitats (M)	483
Wetlands	4.1.1., 4.1.2., 4.2.1., 4.2.2.	Freshwater habitats (A)	931

3.2. Woodland and Forest Extent

Woodland and forest ecosystem type covers in total 40,735 km^2 throughout the Greek territory. Mainland Greece hosts 89.82% of the ecosystem type's area, while the remaining 10.18% is scattered throughout the island regions. More precisely and based on the floristic regions' division of Greece, North-East Greece (NE) hosts the 25.82% of its area, followed by North Central Greece (NC) (14.64%) and North Pindos (NPi) (12.10%). Floristic regions with the smallest area cover are Kiklades (KiK) (0.05%), North Aegean islands (NAe) (0.43%), and Ionian Islands (IoI) (0.74%) (Table 3).

Table 3. Woodland and forest area (km^2) and cover (%) in each floristic region of Greece.

Floristic Regions of Greece (Code)	Woodland and Forest Area Per Floristic Region (km^2)	Woodland and Forest Cover Per Region (%)
East Aegean islands (EAe)	1398.72	3.43%
East Central Greece (EC)	1119.01	2.75%
Ionian Islands (IoI)	301.12	0.74%
Kriti and Karpathos (KK)	680.63	1.67%
Kiklades (KiK)	20.08	0.05%
North Aegean islands (NAe)	175.34	0.43%
North Pindos (NPi)	4927.72	12.10%
North Central Greece (NC)	5961.83	14.64%
North-East Greece (NE)	10,518.52	25.82%
Peloponnisos (Pe)	4610.17	11.32%
South Pindos (SPi)	4911.51	12.06%
Sterea Ellas (StE)	4539.95	11.15%
West Aegean islands (WAe)	1570.03	3.85%
Total	**40,734.63**	**100.00%**

Data source: CORINE Land Cover dataset, 2018 [38].

The distribution of woodland and forest ecosystem type is thematically presented in Figure 3a, under the EEA 10 × 10 km reference grid, and depicts the actual area cover of the ecosystem type at each grid cell under the "very low" to "very high" rating scale. Darker cells highlight areas where woodland and forest ecosystem type is abundant; mountain tops in the mainland host the majority of cells with "high" or "very high" designation, while lowlands and island regions follow. More precisely, the overwhelming majority of cells characterized as "Very high" (for woodland and forest) for area cover are located northern of Peloponnisos, and only two of them are present in Peloponnisos, at Mt. Menalo and Mt. Taygetos. We should pinpoint the presence of two cells characterized as "Very high" in the region of West Aegean Islands, and in particular, on the mountain ranges of northern Evia (WAe). In general, island regions are found to have a significantly low cover of woodland and forest area compared to the mainland regions; however at Evia (WAe), at three major EAe islands (i.e., Lesvos, Samos, and Rhodes) and in southwest Crete (at the surrounding area of Samaria river gorge) (KK) there are cells with a high area cover of woodland and forests.

In Figure 3b, each cell depicts the relative woodland and forest area with respect to a given floristic region total area cover for woodland and forest. Since all values are normalized to a 0 to 1 scale, the results of each cell can be directly compared with any other cell of the grid. This thematic representation highlights areas (cells) within each floristic region that are important for woodland and forest assessments in the region; and by this, each cell's rating is considered as a score for MAES studies prioritization. The most characteristic example lies in the floristic region of Kiklades (KiK), where areas in the central and northern part (i.e., at Naxos, Tinos, and Andros islands) are scored as important ("medium" to "very high" scores) for their woodland and forest area cover in the region.

Detailed data information for each grid cell is provided in the Supplementary Materials (Table S1).

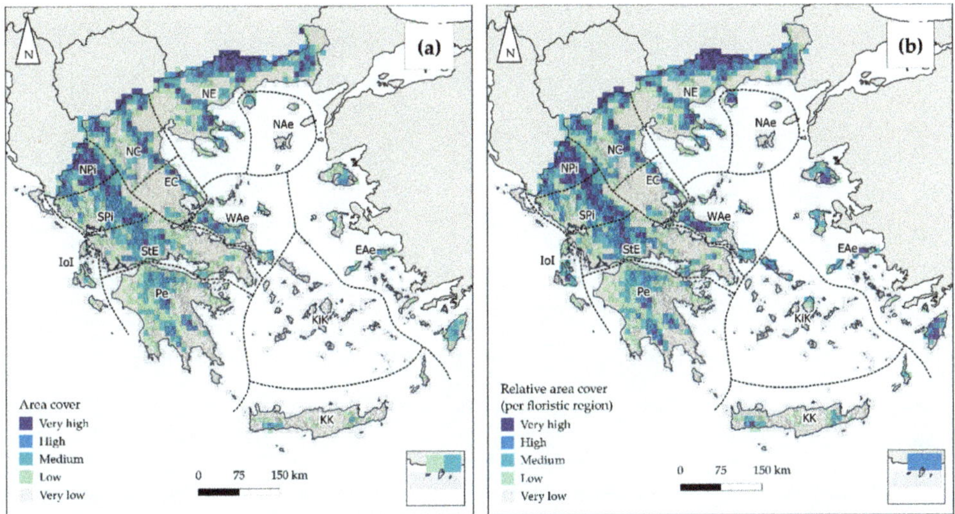

Figure 3. Thematic representation for 'woodland and forest' ecosystem type extent at the 10 × 10 EEA reference grid level: (**a**) Ecosystem extent expressed as the total area of woodland and forest per cell; (**b**) ecosystem extent expressed as the relative area of woodland and forest per cell, i.e., woodland and forest area per cell divided by the total area of woodland and forest in each floristic region (normalized). Floristic regions of Greece [40,42] are also depicted: East Aegean islands (EAe), East Central Greece (EC), Ionian Islands (IoI), Kriti and Karpathos (KK), Kiklades (KiK), North Aegean islands (Nae), North Pindos (Npi), North Central Greece (NC), North-East Greece (NE), Peloponnisos (Pe), South Pindos (Spi), Sterea Ellas (StE), West Aegean islands (Wae).

3.3. Ecosystem Condition and Plant Diversity

Plant diversity is considered as a proxy for ecosystem condition, and the results of the analyses are summarized, as follows.

Figure 4 depicts the thematic representation for total plant diversity, in terms of species richness within the woodland and forest ecosystem type. Figure 4a presents a gradient map for the total number of species present in each grid cell, classified under the "very low" to "very high" rating scale. Cells rated as "very high" or "high" consist of 3.2% (226 out of 7112 cells) of the total number, distributed scattered throughout all floristic regions, and highlight mountainous areas of various altitudes. When we applied a similar analysis using the relative total species richness, as described in the methodology, the pattern changes (Figure 4b) and 4.7% (337 out of 7112 cells) of the cells are rated as "very high" or "high". More precisely, different cells are now considered important with respect to their relative species richness; characteristic examples are found in the regions of Kiklades (KiK), Peloponissos (Pe) and Kriti and Karpathos (KK), where cells with "low" or "very low" species richness (Figure 4a), are now pinpointed as of significant importance for the region (Figure 4b) (e.g., cells in Kiklades, southern Peloponnisos and northwestern Kriti). Detailed data information for each grid cell is provided in the Supplementary Materials.

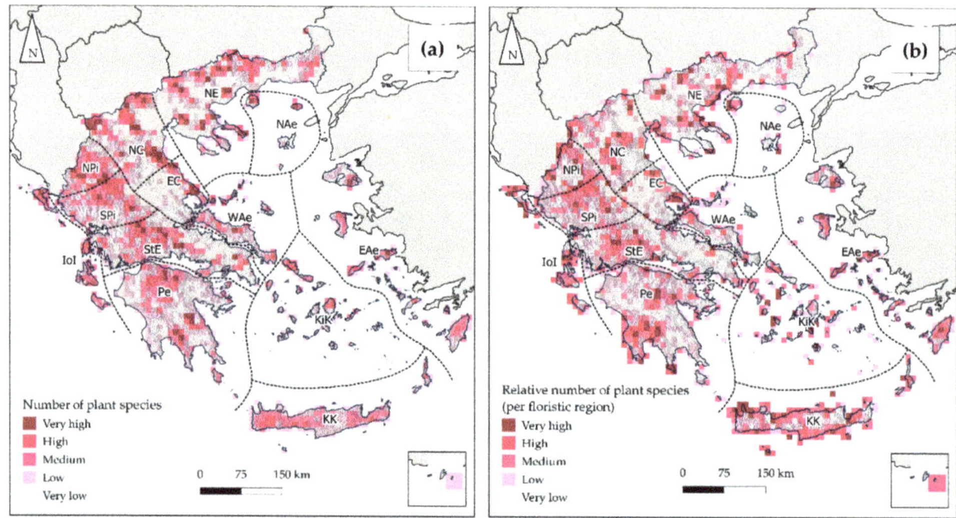

Figure 4. Thematic representation for 'woodland and forest' ecosystem type species richness at the 10×10 EEA reference grid level: (**a**) Number of species present in woodland and forest per cell; (**b**) relative number of species present in woodland and forest per cell, i.e., number of plant species present in woodland and forest per cell divided by the number of plant species present in woodland and forest in each floristic region (normalized). Floristic regions of Greece [40,42] are also depicted: East Aegean islands (EAe), East Central Greece (EC), Ionian Islands (IoI), Kriti and Karpathos (KK), Kiklades (KiK), North Aegean islands (NAe), North Pindos (NPi), North Central Greece (NC), North-East Greece (NE), Peloponnisos (Pe), South Pindos (SPi), Sterea Ellas (StE), West Aegean islands, (WAe).

Figure 5 presents the thematical representation of the assessment results for three additional biodiversity (plant diversity) categories selected for the present study. More precisely:

(a) Relative number of plant species exclusively present in woodland and forest (Figure 5a): This analysis highlights 250 cells (3.5% of the total) as of "high" (165 cells) or "very high" (85 cells) importance. The distribution is scattered throughout the floristic regions, with a significant concentration in Pindos mountain range (NPi, SPi), North Central Greece (NC), Sterea Ellada (StE), southwestern Peloponisson (Pe) and in western Kriti (K). Thasos (NAe), Samos (EAe) and Kerkira (IoI) islands are considered as local hotspots.

(b) Relative number of endemic plant species present in woodland and forest (Figure 5b): This analysis highlights 160 cells (2.2% of the total) as of "high" (126 cells) or "very high" (34 cells) importance considering Greek endemic species. The distribution pattern suggests as hotspots the central part of the Pindos mountain range (NPi, SPi), central and southern Peloponissos (Pe), Kriti (KK), Kiklades (KiK) and East Aegean Islands (EAe) and in particular the island of Rhodes in the southeastern part of the region.

(c) Relative number of endemic plant species exclusively present in woodland and forest (Figure 5c): This analysis highlights 171 cells (2.4% of the total) as of "high" (105 cells) or "very high" (66 cells) importance considering Greek endemics exclusively present in woodland and forest. Cells of Ionian islands (IoI) include 50% (33 cells) of the cells rated as "very high" in the Greek territory, and it is notable that all cells in the IoI are rated as "very high". The general pattern follows the one described for the endemic plant species (Figure 5b), suggesting almost identical hotspots among the floristic regions.

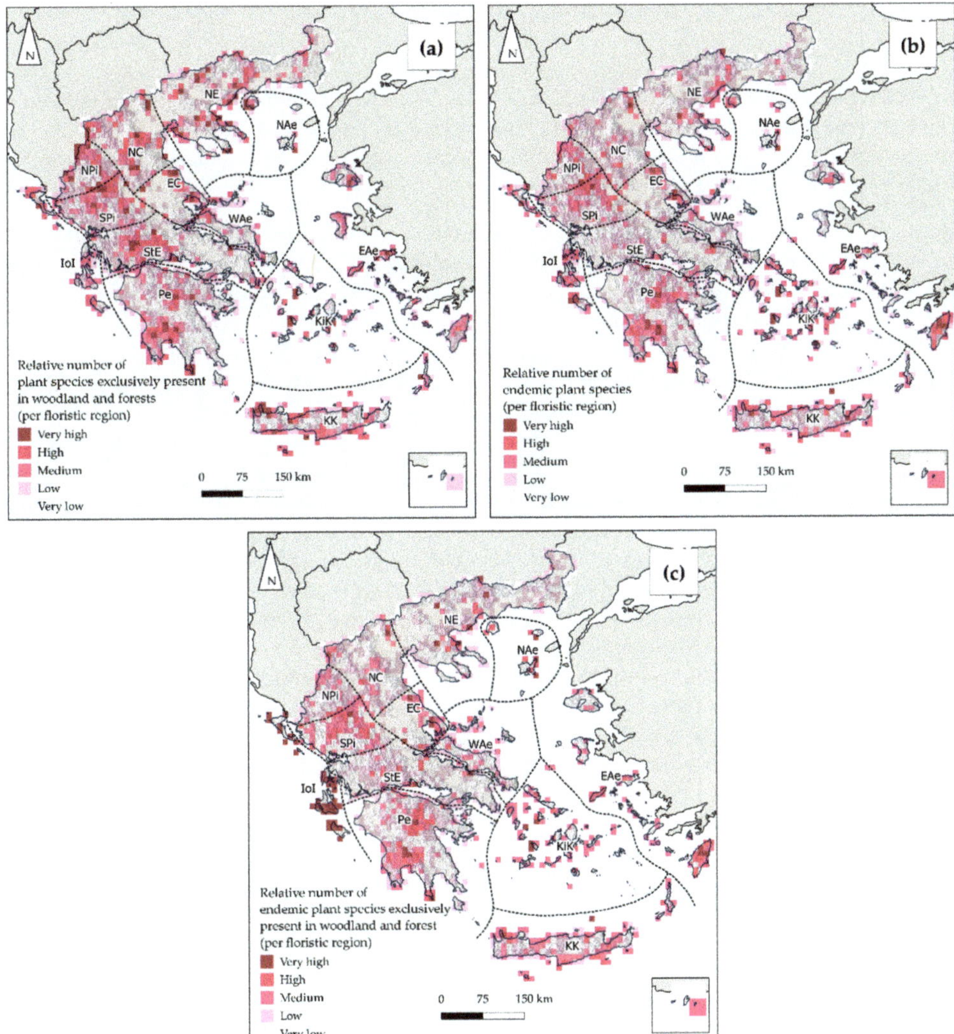

Figure 5. Thematic representation for 'woodland and forest' ecosystem type species richness categories at the 10 × 10 EEA reference grid level: (**a**) Relative number of species exclusively present in woodland and forest per cell, i.e., number of plant species exclusively present in woodland and forest per cell divided by the number of plant species exclusively present in woodland and forest in each floristic region (normalized); (**b**) relative number of endemic plant species present in woodland and forest per cell, i.e., number of plant species present in woodland and forest per cell divided by the number of endemic plant species present in woodland and forest in each floristic region (normalized); (**c**) relative number of endemic plant species exclusively present in woodland and forest per cell, i.e., number of endemic plant species exclusively present in woodland and forest per cell divided by the number of endemic plant species exclusively present in woodland and forest in each floristic region (normalized). Floristic regions of Greece [40,42] are also depicted: East Aegean islands (EAe), East Central Greece (EC), Ionian Islands (IoI), Kriti and Karpathos (KK), Kiklades (KiK), North Aegean islands (Nae), North Pindos (Npi), North Central Greece (NC), North-East Greece (NE), Peloponnisos (Pe), South Pindos (Spi), Sterea Ellas (StE), West Aegean islands, (Wae).

Detailed data information for each category per grid cell is provided in the Supplementary Materials.

3.4. Ecosystem Asset Proxy Indicators

The combination of the ecosystem extent (relative area cover) with the plant diversity categories resulted in the calculation of four relevant ecosystem asset proxy indicators and is thematically presented in Figure 6 and in detail is presented for each grid cell in the Supplementary Materials. For each proxy indicator the results are as follows:

(a)   Proxy indicator 1 (total plant species): The application of this indicator highlights 337 cells (4.7%) as of "high" (252 cells) or "very high" (85 cells) importance. Hotspots are scattered throughout the mountainous areas of Greece and especially in the Pindos mountain range (Npi, Spi), in the northeastern mountains of Peloponnisos (Pe), in southern Evia (Wae), on Mts Pelion, Olympus (EC), Athos (NE), in southwestern Kriti (KK) and in Rhodes island (Eae) (Figure 6a).

(b)   Proxy indicator 2 (total plant species exclusively present in woodland and forest): The application of this indicator provides similar results with proxy indicator 1, i.e., 334 cells (4.75% of the total) are rated as of "high" (247 cells) or "very high" (87 cells) importance, following almost identical spatial patterns (Figure 6b).

(c)   Proxy indicator 3 (endemic species): The application of this indicator highlights 339 cells (4.8% of the total) as of "high" (234 cells) or "very high" (105 cells) importance. Similar spatial distribution patterns occur, and Pindos mountain range (NPi, SPi) continues to be the main hotspot; however secondary, but equally important hotspots are now more clearly highlighted and represented by cells rated as of "very high" importance, e.g., mountain tops of northeastern Peloponnisos (Pe) and Rhodes island (EAe) (Figure 6c).

(d)   Proxy indicator 4 (endemic species exclusively present in woodland and forest): The application of this indicator highlights 325 cells (4.5%) as of "high" (261 cells) or "very high" (64 cells) importance. The general spatial pattern of hotspots is similar to the results of the proxy indicators 2 and 3; particular importance of specific areas is highlighted, e.g., Kafalonia (IoI) and Samothraki islands (NAe).

4. Discussion

This is the first approach of a national-scale assessment that combines spatial, plant diversity data with area cover, and acts as a pilot, baseline assessment, and a starting point for future studies on the MAES implementation and natural capital accounting in Greece. It is highlighted how an extensive and detailed biodiversity dataset (i.e., vascular plants of Greece dataset [39]) can be incorporated into MAES procedure towards scientific documentation, environmental consulting, and decision-making. The development of proxy indicators adds value to the adopted National Set of MAES Indicators in Greece [26] and identifies advantages, limitations, and shortcomings, through the 'woodland and forest' ecosystem type pilot. Given the availability of the extensive floristic database in Greece, the selection of plant diversity for our study is also supported by the conclusions of Quijas et al. [44] who highlight the paramount role of plant diversity in the provision of ecosystem services and to conservation planning and management. Moreover, Balvanera et al. [45] pinpoint the role of plant diversity on ecosystem function and ecosystem services, while at large spatial scales, Costanza et al. [46] used plant species richness to show that over half of the spatial variation in net productivity in North America could be explained by biodiversity patterns (if the effects of temperature and precipitation were taken into account).

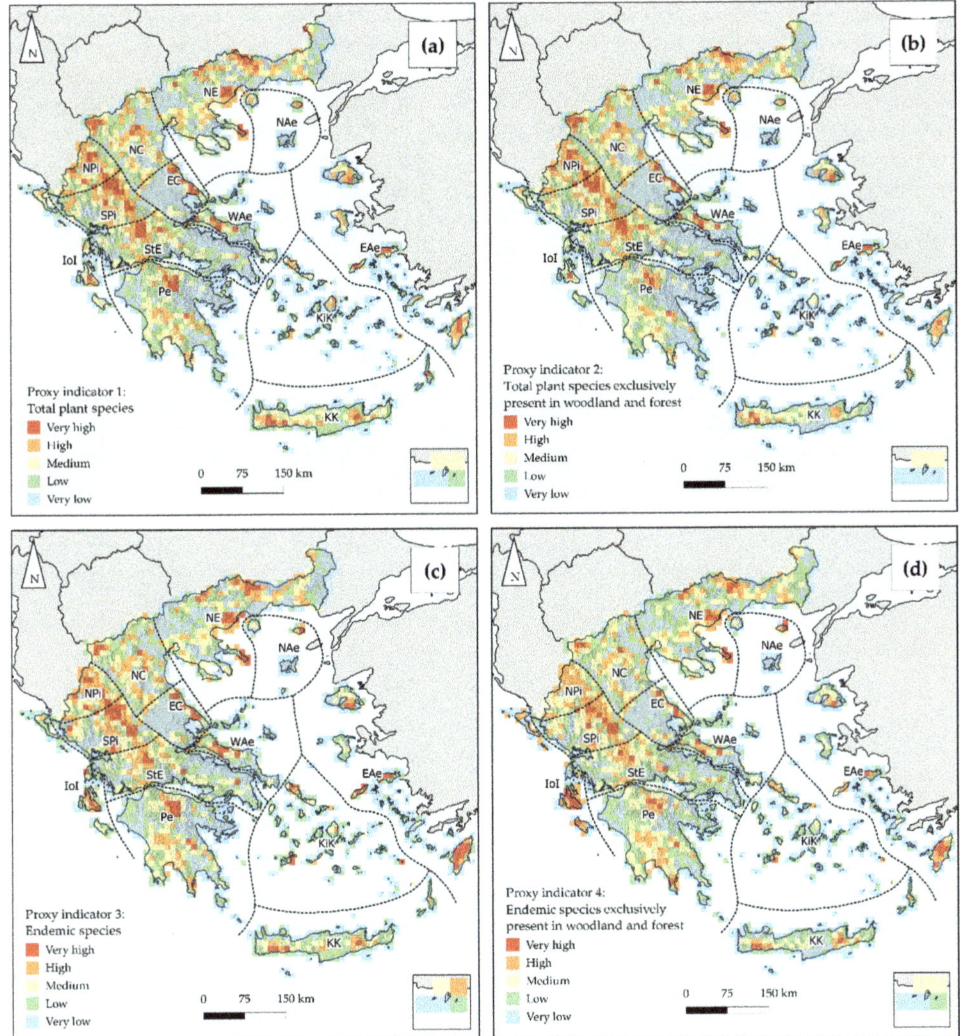

Figure 6. Thematic representation of the four ecosystem asset proxy indicators: (**a**) Total plant species index; (**b**) total plant species exclusively present in woodland and forest index; (**c**) endemic species index; (**d**) endemic species exclusively present in woodland and forest index. Floristic regions of Greece [31,34] are also depicted: East Aegean islands (EAe), East Central Greece (EC), Ionian Islands (IoI), Kriti and Karpathos (KK), Kiklades (KiK), North Aegean islands (NAe), North Pindos (NPi), North Central Greece (NC), North-East Greece (NE), Peloponnisos (Pe), South Pindos (SPi), Sterea Ellas (StE), West Aegean islands, (WAe).

4.1. Ecosystem Extent and Condition

The baseline dataset for implementing any MAES related study is the ecosystem area extent and its condition, of which any kind of ES is supplied or potentially supplied. A key feature of the SEEA-EEA accounting model is the delineation of these spatial areas and their ecosystem assets within these areas [35]. In this study, we presented the actual area extent of woodland and forest at the cell level, which is a straightforward way to express the extent of the ecosystem. Moreover, the inclusion in the methodology of the relative ecosystem area with respect to total ecosystem area in each given floristic region is considered as of high importance, since it provides information for conservation and decision-making integrating each floristic regions' specific characteristics. For example, cells including small woodland and forest areas in forest-poor regions are designated as of equal importance with cells including more extensive areas in forest-rich regions. This is also the case for the ecosystem condition based on plant diversity assessment, which highlights the importance for conservation and management even of areas (grid-cells) with minimum woodland and forest area cover, but of very high plant diversity (e.g., endemic species exclusively present in woodland and forest). Moreover, the integration in the analysis of total ecosystem area extent, as well as species richness based on each floristic region's data, encapsulates a comparison to an ideal best-case situation, which can be considered as the reference value.

4.2. Ecosystem Services

The provision (or potential provision) of ecosystem services is directly related to ecosystem condition, which indicates the state of the ecosystem and its capacity to generate ES flows [47], and thus, is strongly linked to human wellbeing [7,23]. The results of the study suggest areas (rated as "high or "very high") where fine-scale MAES studies should be implemented, including the potential to supply provisioning, regulating and maintenance and cultural services [48] and proceed to valuation methods for the prevailing and/or most important ones. Ecosystem services management and climatic scenarios should also be developed at the local and regional scale following the methods proposed by References [49–51].

4.3. Limitations of the Study

The ecosystem extent is calculated using the CORINE Land Cover dataset, which also provides time-series data since 1990, and thus, is useful for accounting purposes. However, this dataset can only be used for national and regional MAES studies, due to its scale. To overcome this limitation, Greece prepares the ecosystem type map of Greece, via the LIFE-IP 4 NATURA project [31], using a typology for ecosystem types, corresponding to 30 MAES level-3 ecosystem types [26]. Moreover, the use of various categories of plant species diversity may be a commonly used measure for biodiversity, however, more study is needed on the other dimensions of biodiversity, i.e., functional, structural, and taxonomic diversity [52], which should be integrated into the ecosystem condition studies.

4.4. Future Steps and Management Implications

Based on the results of the study, scientists and conservation practitioners should begin incorporating at the 10 × 10 km EEA reference grid-cell level, all available information for biodiversity in Greece. A characteristic example with ready-for-use, compatible data, is the recent work by Cheminal et al. [53] which provides the first review of existing knowledge on the Lamiaceae species in Greece and presents the results under the 10 × 10 km EEA reference grid-cell; it provides information for Lamiaceae diversity and its potential to provide services, based on each species components and characteristics. The results from endemic species categories should be further studied, due to their importance of hosting genetic, medicinal, functional, and morphological characteristics, most of them unexplored or underexplored; these results highlight areas where relevant studies should focus. Moreover, and as underlined by Kallimanis et al. [54], the adoption of higher-taxon surrogacy can be applied in cases when detailed biodiversity data are not available, or full biodiversity survey is not feasible. For instance, for other living organisms, such as the invertebrates, where extensive and detailed datasets are missing, diversity richness at the genus or family level can be used as an indicator and also contribute to the total biodiversity index development as proposed in the National Set of MAES Indicators in Greece [26]. Additionally, research on species abundance and relative abundance is also needed, as a proposed indicator for MAES assessments [23]; however, this information is missing at the scale of our study and is mainly available at a local level (e.g., from case-study assessments in conservation studies) and in most cases only for selected species (e.g., Annex II species of the Habitats' Directive or for other endangered, e.g., Red Data Book species). One the other hand, it is also important to incorporate data that correspond to potential ecosystem disservices [55], such as ruderal and alien species information. For example, the work on ruderal plant species of Greece, also deployed under 10 × 10 km EEA reference grid-cell, highlights the positive correlation among the various ruderal species categories with the different ecosystem types (including forest) [56]. Simultaneously, alien tree species are found to have invasive behavior, threatening native plant communities; e.g., *Eucalyptus camaldulensis* Dehnh. poses a threat on alluvial forests [57], thus, its spatial distribution is needed for future ecosystem condition assessments. Subsequently, non-native plants are important to be assessed at all ecosystem types in terms of invasiveness, regarding ecosystem condition as well in terms of functioning (for non-invasive species) alongside native plants and other organisms. More efforts are also needed to identify the intra- and inter-ecosystem flows and in combination with ecosystem characteristics identification, e.g., extent, structure, and condition will provide the adequate information to delineate ecosystem asset, the baseline of the general ecosystem accounting model of SEEA-EEA [28] (Figure 7), which finally leads to individual and societal wellbeing.

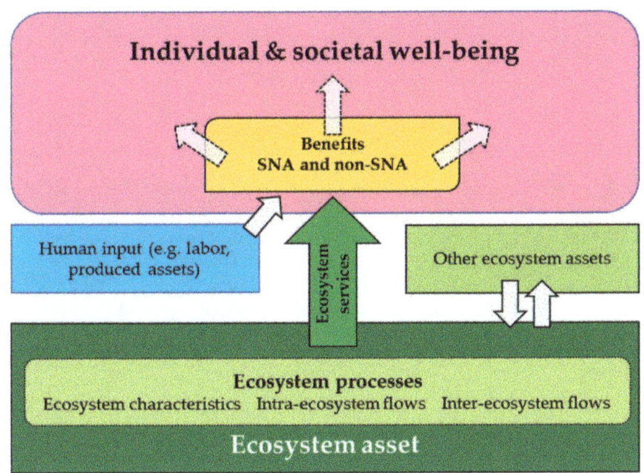

Figure 7. The general ecosystem accounting model, as proposed by SEEA-EEA [28] (redesigned).

This study also contributes to the thematic target for biodiversity set by the European Green Deal [12], as well as to the national forest policy, which sets forest biodiversity conservation among its priorities [58].

5. Conclusions

This study presents a methodological approach for integrating plant diversity data into MAES implementation, using woodland and forest ecosystem type as a pilot case-study. It is based on the national set of MAES indicators in Greece and provides the first test of its guidelines. The results revealed that indicators using well-developed biodiversity datasets in combination with ecosystem extent data could provide the baseline for ecosystem condition assessment, ecosystem asset delineation, and support operational MAES studies. The relation among biodiversity, ecosystem condition, and ecosystem services is not a linear equation and detailed, fine-scale assessments are needed to identify and interpret all aspects of biodiversity. The results pinpoint areas of importance throughout Greece and provide guidance for case-study selection, conservation strategy, and decision-making under the perspective of national and EU environmental policies.

Supplementary Materials: The following are available online at http://www.mdpi.com/1999-4907/11/9/956/s1, Table S1: Detailed data information for each grid cell used for the analyses.

Author Contributions: Conceptualization, K.K., I.P.K. and P.D.; methodology, K.K., I.P.K. and P.D.; validation, K.K., I.P.K. and P.D.; formal analysis, K.K. and I.P.K.; investigation, K.K. and I.P.K.; resources, A.S. and P.D.; data curation, K.K., I.P.K., A.S. and P.D.; writing—original draft preparation, K.K. and I.P.K.; writing—review and editing, K.K., I.P.K., A.S. and P.D.; visualization, K.K. and I.P.K.; supervision, P.D.; project administration, P.D.; funding acquisition, P.D. All authors have read and agreed to the published version of the manuscript.

Funding: This research was funded by the European Commission LIFE Integrated Project, LIFE-IP 4 NATURA "Integrated Actions for the Conservation and Management of Natura 2000 sites, species, habitats and ecosystems in Greece", Grant Number: LIFE 16 IPE/GR/000002.

Conflicts of Interest: The authors declare no conflict of interest.

References

1. Westman, W.E. How much are nature's services worth? *Science* **1977**, *197*, 960–964. [CrossRef] [PubMed]
2. Ehrlich, P.; Ehrlich, A. *Extinction: The Causes and Consequences of the Disappearance of Species*; Random House: New York, NY, USA, 1981.
3. De Groot, R.S. Environmental functions as a unifying concept for ecology and economics. *Environmentalist* **1987**, *7*, 105–109. [CrossRef]
4. Fisher, B.; Turner, R.K.; Morling, P. Defining and classifying ecosystem services for decision making. *Ecol. Econ.* **2009**, *68*, 643–653. [CrossRef]
5. Costanza, R.; de Groot, R.; Sutton, P.; van der Ploeg, S.; Anderson, S.J.; Kubiszewski, I.; Farber, S.; Turner, R.K. Changes in the global value of ecosystem services. *Glob. Environ. Chang.* **2014**, *26*, 152–158. [CrossRef]
6. Costanza, R.; de Groot, R.; Braat, L.; Kubiszewski, I.; Fioramonti, L.; Sutton, P.; Farber, S.; Grasso, M. Twenty years of ecosystem services: How far have we come and how far do we still need to go? *Ecosyst. Serv.* **2017**, *28*, 1–16. [CrossRef]
7. Maes, J.; Teller, A.; Erhard, M.; Liquete, C.; Braat, L.; Berry, P.; Egoh, B.; Puydarrieus, P.; Fiorina, C.; Santos, F.; et al. *Mapping and Assessment of Ecosystem and Their Services. An Analytical Framework for Ecosystem Assessments under Action 5 of the EU Biodiversity Strategy to 2020*; Publications office of the European Union: Luxemburg, 2013; ISBN 9789279293696.
8. Millennium Ecosystem Assessment. *Ecosystems and Human Well-Being: Synthesis*; Island Press: Washington, DC, USA, 2005; ISBN 1-59726-040-1.
9. Czúcz, B.; Arany, I.; Potschin-Young, M.; Bereczki, K.; Kertész, M.; Kiss, M.; Aszalós, R.; Haines-Young, R. Where concepts meet the real world: A systematic review of ecosystem service indicators and their classification using CICES. *Ecosyst. Serv.* **2018**, *29*, 145–157. [CrossRef]
10. IPEBS. *Global Assessment Report on Biodiversity and Ecosystem Services of the Intergovernmental Science-Policy Platform on Biodiversity and Ecosystem Services*; Brondizio, E.S., Settele, J., Díaz, S., Ngo, H., Eds.; IPBES Secretariat: Bonn, Germany, 2019.
11. *European Commission EU Biodiversity Strategy for 2030: Bringing Nature Back into Our Lives*; Communication from the Commission to the European Parliament, the Council, the European Economic and Social Committee and the Committee of the Regions: Brussels, Belgium, 2020; pp. 1–22, COM/2020/380 final.
12. The European Green Deal. *Communication from the Commission to the European Parliament, the European Council, the Council, the European Economic and Social Committee and the Committee of the Regions*; The European Green Deal: Brussels, Belgium, 2019.
13. Willemen, L.; Burkhard, B.; Crossman, N.; Drakou, E.G.; Palomo, I. Editorial: Best practices for mapping ecosystem services. *Ecosyst. Serv.* **2015**, *13*, 1–5. [CrossRef]
14. Mace, G.M.; Norris, K.; Fitter, A.H. Biodiversity and ecosystem services: A multilayered relationship. *Trends Ecol. Evol.* **2012**, *27*, 19–26. [CrossRef]
15. Braat, L.; ten Brink, P. *The Cost of Policy Inaction: The Case not Meeting the 2010 Biodiversity Target*; Alterra: Wageningen, The Netherlands, 2008.
16. Harrison, P.A.; Berry, P.M.; Simpson, G.; Haslett, J.R.; Blicharska, M.; Bucur, M.; Dunford, R.; Egoh, B.; Garcia-Llorente, M.; Geamănă, N.; et al. Linkages between biodiversity attributes and ecosystem services: A systematic review. *Ecosyst. Serv.* **2014**, *9*, 191–203. [CrossRef]

17. Grunewald, K.; Syrbe, R.U.; Walz, U.; Richter, B.; Meinel, G.; Herold, H.; Marzelli, S. Germany's Ecosystem Services–State of the Indicator Development for a Nationwide Assessment and Monitoring. *One Ecosyst.* **2017**, *2*, e14021. [CrossRef]

18. Steur, G.; Verburg, R.W.; Wassen, M.J.; Verweij, P.A. Shedding light on relationships between plant diversity and tropical forest ecosystem services across spatial scales and plot sizes. *Ecosyst. Serv.* **2020**, *43*, 101107. [CrossRef]

19. Alkemade, R.; Burkhard, B.; Crossman, N.D.; Nedkov, S.; Petz, K. Quantifying ecosystem services and indicators for science, policy and practice. *Ecol. Indic.* **2014**, *37*, 161–162. [CrossRef]

20. Crossman, N.D.; Burkhard, B.; Nedkov, S.; Willemen, L.; Petz, K.; Palomo, I.; Drakou, E.G.; Martín-Lopez, B.; McPhearson, T.; Boyanova, K.; et al. A blueprint for mapping and modelling ecosystem services. *Ecosyst. Serv.* **2013**, *4*, 4–14. [CrossRef]

21. Burkhard, B.; Crossman, N.; Nedkov, S.; Petz, K.; Alkemade, R. Mapping and modelling ecosystem services for science, policy and practice. *Ecosyst. Serv.* **2013**, *4*, 1–3. [CrossRef]

22. Feld, C.K.; Sousa, J.P.; da Silva, P.M.; Dawson, T.P. Indicators for biodiversity and ecosystem services: Towards an improved framework for ecosystems assessment. *Biodivers. Conserv.* **2010**, *19*, 2895–2919. [CrossRef]

23. Maes, J.; Teller, A.; Erhard, M.; Grizzetti, B.; Barredo, J.I.; Paracchini, M.L.; Condé, S.; Somma, F.; Orgiazzi, A.; Jones, A.; et al. *Mapping and Assessment of Ecosystems and their Services: An Analytical Framework for Mapping and Assessment of Ecosystem Condition in EU*; Publications office of the European Union: Luxemburg, 2018; ISBN 978-92-79-74288-0.

24. Van Reeth, W. *Ecosystem Service Indicators in Flanders: Are We Measuring What We Want to Manage? Rapporten van het Instituut voor Natuur-en Bosonderzoek*; Instituut voor Natuuren Bosonderzoek: Belgium, Brussel, 2014.

25. Bratanova-Doncheva, S.; Chipev, N.; Gocheva, K.; Stoyan, V.; Fikova, R. Methodological Framework for Assessment and Mapping of Ecosystem Condition and Ecosystem Services in Bulgaria. Conceptual Bases and Principles of Application. 2017. Available online: https://www.researchgate.net/publication/329773739_METHODOLOGICAL_FRAMEWORK_FOR_ ASSESSMENT_AND_MAPPING_OF_ECOSYSTEM_CONDITION_AND_ECOSYSTEM_SERVICES_IN_ BULGARIA_Part_A_Conceptual_basis_and_principles_of_application (accessed on 30 July 2020).

26. Kokkoris, I.P.; Mallinis, G.; Bekri, E.S.; Vlami, V.; Zogaris, S.; Chrysafis, I.; Mitsopoulos, I.; Dimopoulos, P. National set of MAES indicators in Greece: Ecosystem services and management implications. *Forests* **2020**, *11*, 595. [CrossRef]

27. Hein, L.; Obst, C.; Edens, B.; Remme, R.P. Progress and challenges in the development of ecosystem accounting as a tool to analyse ecosystem capital. *Curr. Opin. Environ. Sustain.* **2015**, *14*, 86–92. [CrossRef]

28. United Nations; European Commission; Food and Agricultural Organization of the United Nations; Organization for Economic Co-operation and Development; World Bank. *System of Environmental-Economic Accounting 2012: Experimental Ecosystem Accounting*; The World Bank: Washington, DC, USA, 2014; ISBN 9789210559263.

29. United Nations. *Technical Recommendations in Support of the System of Environmental-Economic Accounting 2012-Experimental Ecosystem Accounting*; United Nations: New York, NY, USA, 2019.

30. Remme, R.P.; Hein, L.; Van Swaay, C.A.M. Exploring spatial indicators for biodiversity accounting. *Ecol. Indic.* **2016**, *70*, 232–248. [CrossRef]

31. LIFE-IP 4 Natura-Integrated Actions for the Conservation and Management of Natura 2000 Sites, Species, Habitats and Ecosystems in Greece (LIFE16 IPE/GR/000002). Available online: https://ec.europa.eu/ environment/life/project/Projects/index.cfm?fuseaction=search.dspPage&n_proj_id=6520 (accessed on 30 July 2020).

32. European Environment Agency Forest Information System for Europe: Greece. Available online: https: //forest.eea.europa.eu/countries/greece (accessed on 30 July 2020).

33. Kokkoris, I.P.; Dimopoulos, P.; Xystrakis, F.; Tsiripidis, I. National scale ecosystem condition assessment with emphasis on forest types in Greece. *One Ecosyst.* **2018**, *3*, e25434. [CrossRef]

34. Kokkoris, I.P.; Drakou, E.G.; Maes, J.; Dimopoulos, P. Ecosystem services supply in protected mountains of Greece: Setting the baseline for conservation management. *Int. J. Biodivers. Sci. Ecosyst. Serv. Manag.* **2018**, *14*, 45–59. [CrossRef]

35. Obst, C.; Brooks, T.; Maes, J.; Czucz, B.; Nicholson, E.; Alfieri, A.; Javorsek, M. Options for Incorporating biodiversity in the SEEA. In Proceedings of the 2019 Forum of Experts in SEEA Experimental Ecosystem Accounting, Glen Cove, NY, USA, 26–27 June 2019; United Nations, UN Environment, The World Bank, European Union: Geln Cove, NY, USA, 2019; pp. 1–12.

36. European Comission. *Natural Capital Accounting: Overview and Progress in the European Union (6th Report–Final)*; Publications office of the European Union: Luxemburg, 2019; ISBN 978-92-79-89744-3.

37. EEA Reference Grid. Available online: https://www.eea.europa.eu/data-and-maps/data/eea-reference-grids-2 (accessed on 12 April 2020).

38. CORINE Land Cover. Available online: https://land.copernicus.eu/pan-european/corine-land-cover (accessed on 12 April 2020).

39. Dimopoulos, P.; Raus, T.; Strid, A. Flora of Greece Web: Vascular Plants of Greece an Annotated Checklist. Available online: http://portal.cybertaxonomy.org/flora-greece/intro (accessed on 12 May 2020).

40. Dimopoulos, P.; Raus, T.; Bergmeier, E.; Constantinidis, T.; Iatrou, G.; Kokkini, S.; Strid, A.; Tzanoudakis, D. Vascular plants of Greece: An annotated checklist. *Englera* **2013**, *31*, 370.

41. Dimopoulos, P.; Raus, T.; Bergmeier, E.; Constantinidis, T.; Iatrou, G.; Kokkini, S.; Strid, A.; Tzanoudakis, D. Vascular plants of Greece: An annotated checklist. Supplement. *Willdenowia* **2016**, *46*, 301–347. [CrossRef]

42. Arne, S.; Tan, K. *Flora Hellenica*; Koeltz: Königstein, Germany, 1997.

43. European Union CORINE Land Cover Nomeclature. Available online: https://land.copernicus.eu/user-corner/technical-library/corine-land-cover-nomenclature-guidelines/html (accessed on 30 July 2020).

44. Quijas, S.; Schmid, B.; Balvanera, P. Plant diversity enhances provision of ecosystem services: A new synthesis. *Basic Appl. Ecol.* **2010**, *11*, 582–593. [CrossRef]

45. Balvanera, P.; Pfisterer, A.B.; Buchmann, N.; He, J.S.; Nakashizuka, T.; Raffaelli, D.; Schmid, B. Quantifying the evidence for biodiversity effects on ecosystem functioning and services. *Ecol. Lett.* **2006**, *9*, 1146–1156. [CrossRef] [PubMed]

46. Costanza, R.; Fisher, B.; Mulder, K.; Liu, S.; Christopher, T. Biodiversity and ecosystem services: A multi-scale empirical study of the relationship between species richness and net primary production. *Ecol. Econ.* **2007**, *61*, 478–491. [CrossRef]

47. Hein, L.; Bagstad, K.; Edens, B.; Obst, C.; De Jong, R.; Lesschen, J.P. Defining ecosystem assets for natural capital accounting. *PLoS ONE* **2016**, *11*, e0164460. [CrossRef]

48. Haines-Young, R.; Potschin, M. Common International Classification of Ecosystem Services (CICES): 2011 Update. In *Report to the European Environmental Agency*; The University of Nottingham: Nottingham, UK, 2011.

49. Kokkoris, I.P.; Bekri, E.S.; Skuras, D.; Vlami, V.; Zogaris, S.; Maroulis, G.; Dimopoulos, D.; Dimopoulos, P. Integrating MAES implementation into protected area management under climate change: A fine-scale application in Greece. *Sci. Total Environ.* **2019**, *695*, 133530. [CrossRef]

50. Kougioumoutzis, K.; Kokkoris, I.P.; Panitsa, M.; Trigas, P.; Strid, A.; Dimopoulos, P. Plant Diversity Patterns and Conservation Implications under Climate-Change Scenarios in the Mediterranean: The Case of Crete (Aegean, Greece). *Diversity* **2020**, *12*, 270. [CrossRef]

51. Kougioumoutzis, K.; Kokkoris, I.P.; Panitsa, M.; Trigas, P.; Strid, A.; Dimopoulos, P. Spatial Phylogenetics, Biogeographical Patterns and Conservation Implications of the Endemic Flora of Crete (Aegean, Greece) under Climate Change Scenarios. *Biology* **2020**, *9*, 199. [CrossRef]

52. Lyashevska, O.; Farnsworth, K.D. How many dimensions of biodiversity do we need? *Ecol. Indic.* **2012**, *18*, 485–492. [CrossRef]

53. Cheminal, A.; Kokkoris, I.P.; Strid, A.; Dimopoulos, P. Medicinal and aromatic lamiaceae plants in greece: Linking diversity and distribution patterns with ecosystem services. *Forests* **2020**, *11*, 661. [CrossRef]

54. Kallimanis, A.S.; Mazaris, A.D.; Tsakanikas, D.; Dimopoulos, P.; Pantis, J.D.; Sgardelis, S.P. Efficient biodiversity monitoring: Which taxonomic level to study? *Ecol. Indic.* **2012**, *15*, 100–104. [CrossRef]

55. Lyytimäki, J.; Sipilä, M. Hopping on one leg-The challenge of ecosystem disservices for urban green management. *Urban For. Urban Green.* **2009**, *8*, 309–315. [CrossRef]

56. Panitsa, M.; Iliadou, E.; Kokkoris, I.; Kallimanis, A.; Patelodimou, C.; Strid, A.; Raus, T.; Bergmeier, E.; Dimopoulos, P. Distribution patterns of ruderal plant diversity in Greece. *Biodivers. Conserv.* **2020**, *29*, 869–891. [CrossRef]

57. Badalamenti, E.; Cusimano, D.; La Mantia, T.; Pasta, S.; Romano, S.; Troia, A.; Ilardi, V. The ongoing naturalisation of Eucalyptus spp. in the Mediterranean Basin: New threats to native species and habitats. *Aust. For.* **2018**, *81*, 239–249. [CrossRef]
58. Spanos, K.; Gaitanis, D.; Skouteri, A.; Petrakis, P.; Meliadis, I. Implementation of Forest Policy in Greece in Relation to Biodiversity and Climate Change. *Open J. Ecol.* **2018**, *8*, 174–191. [CrossRef]

MDPI

St. Alban-Anlage 66

4052 Basel

Switzerland

Tel. +41 61 683 77 34

Fax +41 61 302 89 18

www.mdpi.com

Forests Editorial Office

E-mail: forests@mdpi.com

www.mdpi.com/journal/forests